Handbook of Environmental Science

Handbook of
Environmental Science

Edited by Paige Tucker

SYRAWOOD
PUBLISHING HOUSE

New York

Published by Syrawood Publishing House,
750 Third Avenue, 9th Floor,
New York, NY 10017, USA
www.syrawoodpublishinghouse.com

Handbook of Environmental Science
Edited by Paige Tucker

International Standard Book Number: 978-1-64740-140-5 (Hardback)

Cataloging-in-Publication Data

Handbook of environmental science / edited by Paige Tucker.
        p. cm.
Includes bibliographical references and index.
ISBN 978-1-64740-140-5
1. Environmental sciences. 2. Ecology. I. Tucker, Paige.
GE105 .H36   2022
363.7--dc23

TABLE OF CONTENTS

PREFACE

Environmental science is an interdisciplinary field which studies the environment. It seeks to find solutions to environmental problems by integrating physical, biological and information sciences. The components of environmental science are geosciences, atmospheric sciences, ecology and environmental chemistry. Geosciences are further subdivided into environmental geology, environmental soil science, volcanic phenomena and evolution of the Earth's crust. There are other areas of study which are closely related to environmental science such as environmental engineering. The environmental issues studied under this field include pollution control, climate change and its effects as well as alternative energy. Some of the diverse topics covered in this book address the varied branches that fall under the category of environmental sciences. It will also provide stimulating topics for research which interested readers can take up. The extensive content of this book provides the readers with a thorough understanding of the subject.

The information contained in this book is the result of intensive hard work done by researchers in this field. All due efforts have been made to make this book serve as a complete guiding source for students and researchers. The topics in this book have been comprehensively explained to help readers understand the growing trends in the field.

I would like to thank the entire group of writers who made sincere efforts in this book and my family who supported me in my efforts of working on this book. I take this opportunity to thank all those who have been a guiding force throughout my life.

Editor

Preparation and characterization of bioadsorbent beads for chromium and zinc ions adsorption

Suhaib S. Salih[1,2]* and Tushar K. Ghosh[3]

*Corresponding author: Suhaib S. Salih, Chemical Engineering, University of Missouri-Columbia, 510 high street, Columbia, MI 65201, USA; Department of Chemical Engineering, University of Tikrit, Iraq
E-mail: sss43b@mail.missouri.edu
Reviewing editor: Arno Rein, Technical University of Munich, Germany

Abstract: Low-cost chitosan beads were prepared by dropping chitosan solution into an alkaline bath and then were used for Cr(VI) and Zn(II) ions adsorption from aqueous solution. Prepared chitosan beads were characterized by Fourier transform infrared spectroscopy (FTIR), thermal gravimetric analysis (TGA), Brunauer–Emmett–Teller (BET) measurements, scanning electron microscopy (SEM), and energy dispersive X-ray spectroscopy (EDS). The effect of solution-pH, contact time, initial ion concentration, and temperature on both metal ions adsorption was investigated. The kinetics of adsorption suggested a pseudo-second-order model fits better than pseudo-first-order model for both metals. The equilibrium adsorption isotherm of both metals matches well with the Langmuir isotherm model. The maximum adsorption capacity of chitosan beads was 79.56 mg Cr(VI)/g and 109.18 mg Zn(II)/g at initial ion concentration 1,000 mg/L, and temperature 10 °C. Thermodynamic parameters showed that the adsorption of Cr(VI) and Zn(II) ions onto chitosan beads was feasible, spontaneous, and exothermic under the studied conditions. While the chitosan beads enabled a good adsorption application, further lab work and field studies are necessary before using in a practical adsorption process.

ABOUT THE AUTHOR

Suhaib S. Salih is currently a research assistant at the University of Missouri-Columbia majoring in chemical engineering. He received his BE degree in 2006 and MSc in 2012, in chemical engineering from the University of Tikrit, Tikrit, Iraq. In 2015, he joined the department of chemical engineering, University of Missouri-Columbia, as a PhD student. His current research interests are in adsorption processes. He has expertise in lab management, operation of atomic absorption spectrophotometer, intra red spectrophotometer, ultra violet spectrophotometer, and HPLC spectrophotometer.

PUBLIC INTEREST STATEMENT

Cr(VI) and Zn(II) ions in water and wastewater streams are not biodegradable in nature and they have a significant effect on the ecosystem and human health. So, it is important to remove them before supplying water to the public. The objective of this study is to examine the performance of new composite adsorbents that were prepared from sustainable material, chitosan, to remove Cr(VI) and Zn(II) ions from aqueous solution instead of using traditional adsorbents such as activated carbon which is more expensive. In addition, we investigated the optimum adsorption conditions to enhance the adsorption capacity of Zn(II) and Cr(VI) ions from aqueous solution. Overall results from this research suggest that the prepared chitosan beads could be employed as a cost-effective and excellent alternative adsorbent for Cr(VI) and Zn(II) ions removal from water and wastewater.

Subjects: Environmental Studies; Environmental Management; Environmental Issues; Environment & Health; Research Methods in Environmental Studies; Environment & the City; Industrial Engineering & Manufacturing; Chemical Engineering; Civil, Environmental and Geotechnical Engineering

Keywords: Chitosan; Cr(VI) ions; Zn(II) ions; adsorption kinetics; sustainable adsorbent

1. Introduction

The most important and essential compound on the ground is water for all living creatures. However, rapidly growing population, climate change, and environmental deterioration affect the quality of water supplies. Water contamination caused by heavy metals has been identified as serious environmental issues worldwide, since heavy metals are highly toxic, non-biodegradable, non-metabolizable, and can cause many biological abnormalities and tend to accumulate in the food chains. Heavy metals mainly exist in the wastewaters of many industries such as metal plating, mining operations, electric device manufacturing, and battery production (Min et al., 2015). There are many heavy metal ions including chromium and zinc that appear in the US Environmental Protection Agency's priority list of pollutants due to their high toxicity, prevalence, existence, and persistence in the environment (Satya, 2015). The accumulation of Cr(VI) in human body causes a stomach erosion, hemorrhaging, and death is likely. The main symptoms of Zn(II) poisoning are an electrolyte imbalance, stomachache, dehydration, nausea, dizziness, and incoordination in muscles (Jain, Singhal, & Sharma, 2004). Physical, chemical, and biological processes for heavy metals removal from wastewater have been extensively researched and used such as adsorption, reverse osmosis, ion exchange, evaporation, solvent extraction, chemical precipitation, filtration, flotation, membrane, coagulation and flocculation, and electrochemical methods (Hua et al., 2012). Adsorption has been considered as the most cost-effective method for heavy metals removal from aqueous solution because the process is simple in design and chemical consumption or/and waste generation are not a significant issue compared with other methods. Traditional adsorbents such as activated carbon are not efficient in a low concentration of pollutants, and therefore it needs additives with a higher surface area to enhance the adsorption capacity. In recent decades, bioadsorbent materials for biological origin have emerged as an attractive material for removing heavy metals from wastewater streams, largely because these materials are sustainable, abundant in nature, cheap, effective in low and medium metal level, and easy to regenerate for reuse (Babel & Kurniawan, 2003). Many biological materials have shown a good adsorption capacities for heavy metals from aqueous solution such as alginate (Pandey, Bera, Shukla, & Roy, 2007), seaweed (Ghimire, Inoue, Ohto, & Hayashida, 2008), husk (Ricordel, Taha, Cisse, & Dorange, 2001), sugar beet pulp (Ozer & Tumen, 2005), and chitosan (Zhao, Repo, Yin, & Sillanpää, 2013).

Chitosan is a natural organic material, commercially produced by partial deacetylation of chitin, which is the second plentiful organic component in nature next to cellulose polymer. Due to the presence of amino (–NH2) and hydroxyl (–OH) groups in its structure, chitosan is capable to adsorb heavy metals from aqueous solution by creating an electronic bond between these active sites and metal ions (Jagtap et al., 2009; Ngah, Ab Ghani, & Kamari, 2005). In addition, chitosan has been investigated by many studies as an excellent material for heavy metal adsorption such as cadmium and copper (Shyam & Arun, 2014), arsenic (Min et al., 2015), and chromium and zinc (Annaduzzaman, 2015; Qudsieh, Mirghani, Kabbashi, & Muyibi, 2014; Sofiane & Sofia, 2015; Toledo et al., 2014).

In this study, chitosan was prepared in spherical shape instead of using its original form powder, which has been applied in previous studies, to enhance its total surface area, reduce its agglomeration in acidic media, and to increase the metal ions spreading onto active sites, and then it was used to remove Cr(VI) and Zn(II) ions from aqueous solution. In addition, the performance of prepared chitosan beads was characterized by Fourier transform infrared spectroscopy (FTIR), thermal gravimetric analysis (TGA), Brunauer–Emmett–Teller (BET) measurements, scanning electron microscopy (SEM), and energy dispersive X-ray spectroscopy (EDS). Moreover, the adsorption process was examined as a function of solution-pH, contact time, initial metal ion concentration, and temperature.

Eventually, adsorption isotherms, thermodynamic parameters, and the mechanism of adsorption were investigated.

2. Materials and methods

2.1. Materials

Medium molecular weight chitosan biopolymer (deacetylation degree is 87%, the molecular weight is ~190,000–310,000 g/mol) was procured from Aldrich Chemical Corporation and used to synthesize chitosan beads. Zinc sulfate heptahydrate ($H_{14}O_{11}SZn$) and potassium dichromate ($K_2Cr_2O_7$) were used to prepare a stock solution containing 1,000 mg/L of Cr(VI) or Zn(II) ions. Sodium hydroxide, oxalic acid, and deionized water were used to adjust the solution-pH and to dissolve chitosan. All chemicals that were used in this study were analytical grade and used as received without further purification.

2.2. Preparation of chitosan beads

Chitosan gel solution was prepared by dissolving 20 g of chitosan into 1 L of 0.2 mol/L oxalic acid solution under continuous stirring. The chitosan solution was heated to 65–70 °C to facilitate acylation. Spherical chitosan beads were prepared using the dropwise method—dropwise addition of the chitosan mixture gel solution into a precipitation bath containing 0.8 mol/L NaOH solution. The objective of dropping chitosan gel solution into a basic solution is to rapidly neutralize the drops and get a spherical shape of chitosan beads. The chitosan hydrogel beads remained in the NaOH solution with slowly stirred mixing (60 rpm) for 6 h for hardening. The hardened spherical chitosan beads were washed by deionized water for neutralizing and removing sodium ions that might be attached to the chitosan beads. Eventually, the chitosan beads were dried by vacuum furnace for 24 h at 70 °C. The final dry chitosan beads had an average diameter of 1 mm and they were available for adsorption experimental work as shown in Figure 1.

2.3. Adsorption experiments

Batch adsorption experiments of Cr(VI) and Zn(II) ions adsorption onto chitosan beads from aqueous solution were studied at initial metal concentrations of 50, 100, 250, 500, and 1,000 mg/L. In a series of 125 mL Erlenmeyer flasks, 50 mL of Cr(VI) or Zn(II) liquid solution at different concentrations and 0.5 g of chitosan beads were agitated in an orbital shaker at 200 rpm for 8 h. The volume of the liquid solution without chitosan beads was 42.5 mL, measured by burette after 8 h of stirring, and that volume was used to determine the chitosan beads adsorption capacity. After a certain time, samples were taken from the flasks and filtrated to determine the final metal ion concentrations in the solution samples. Inductively Coupled Plasma-Mass Spectrometer (ICP-MS) was used to determine the metal ion concentration in the samples. When the steady state of adsorption attained, the adsorption capacity of chitosan beads was determined using Equation (1) that generated from the mass balance of metal ions (Zhou, Li, Jin, Lian, & Han, 2017),

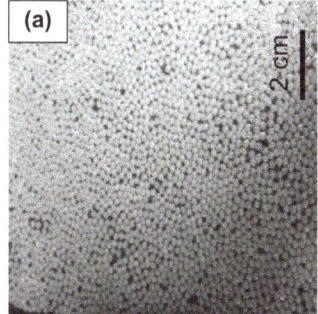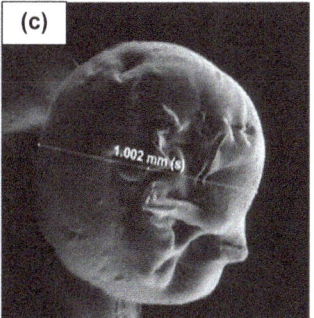

Figure 1. (a) wet chitosan beads, (b) dry chitosan beads, (c) single chitosan bead (scanning electron microscopy image, 500 μm and 1,000 kV image magnification).

$$q_e = \frac{(C_0 - C_e)V}{m} \tag{1}$$

where q_e is the equilibrium adsorption capacity (mmol/g), C_0 is the initial metal concentration in the solution (mmol/L), and C_e is the equilibrium metal concentration in the solution (mmol/L), V is the volume of solution (L), and m is the mass of chitosan beads (g). Batch adsorption runs were studied at different temperatures, different initial metal ion concentrations, different solutions-pH, and different contact times.

3. Results and discussion

3.1. Characterization of chitosan beads

Chitosan beads were characterized to estimate some of their physical and chemical properties. They provide a better interpretation of Zn(II) and Cr(VI) adsorption mechanism associated with the adsorption process. Some physical and chemical properties of dried chitosan beads are summarized in Table 1. We can see that the BET surface area of chitosan beads was found to be 1.9 m^2/g, which is not too large compared to other adsorbents like activated carbon and silica because chitosan material in its natural form is soft. The solution-pH of chitosan beads at the point of zero charges (pHZPC) was 5, that means the surface of chitosan beads has a positive charge at pH less than 5, which is presumably created by the protonation of amine groups in the chitosan. Therefore, the negative species are favorite to adsorb at low pH and species that have positive charges (heavy metals) are easily adsorbed and interacted on chitosan beads at pH greater than 5.

Thermal gravimetric analysis (TGA) (Figure S1, supplementary material) was done to investigate the thermal stability of chitosan beads. Figure S1 shows that there are three distinctive steps of weight loss. First, about 5% weight loss occurred at 144 °C due to the loss of water content (dehydration). Second, 24% weight loss occurred at 370 °C which may correspond to the degradation behavior of the chitosan. Third, about 38% weight loss happened at 686 °C which may correspond to the decomposition (thermal and oxidative) of chitosan. More details can be found in the Supplementary Material

Results of Fourier transform infrared spectroscopy (FTIR) analysis are shown in Figure S2. It can be seen that the chitosan beads exhibit a large band of the amine group (N-H) stretching between 3,200 and 3,510 cm^{-1}, and a peak between 1,400 and 1,655 cm^{-1} corresponds to C=N stretching bond. These observations indicate that the amine groups are present in chitosan beads and successfully bonded with heavy metals, indicating that the metal ions are adsorbed onto chitosan beads (Qi,

Table 1. Physical and chemical properties of the chitosan beads

Parameters	Value
Physical nature	Porous
Average particle diameter (mm)	1
Surface area (m^2/g)*	1.9
Pore size (mm)*	3.73×10^{-6}
Pore volume (m^3/g)*	8.2×10^{-9}
pH at the point of zero charge (pHZPC)	5
Compositions**:	
Carbon content (wt%)	57.11
Oxygen content (wt%)	18.26
Nitrogen content (wt%)	17.60
Sodium content (wt%)	6.25

*Determined by Brunauer–Emmett–Teller (BET) measurement.

**Determined by Energy dispersive X-ray spectroscopy (EDS).

Xu, Jiang, Hu, & Zou, 2004). It can be seen from scanning electron microscopy (SEM) images (Figure S3) that the surface of fresh chitosan beads is wrinkle and porous, whereas the surface of loaded chitosan beads with heavy metals is less porous due to the adsorption of heavy metals.

In order to determine the elemental composition of the chitosan beads, energy dispersive X-ray spectroscopy (EDS) analysis was attained. EDS analysis clearly shows the presence of C, N, and O, which can be seen in Figure S4. The higher percentage of O along with N enhances the interaction between chitosan beads and heavy metals.

3.2. Effect of initial pH on adsorption

Figure 2 shows the effect of solution-pH on the adsorption of Zn(II) and Cr(VI) ions onto chitosan beads. These results show that the Cr(VI) ion adsorption was raised up from 1.2 mmol/g at pH 2 to 1.53 mmol/g at pH 5 and then was raised down to 1.35 mmol/g at pH 8. The maximum rate of Cr(VI) removal from aqueous solution onto chitosan beads was about 1.53 mmol/g (79.56 mg/g) at pH 5. In acidic solution, the Cr(VI) ions are mostly in form of H_2CrO_4 which has a neutral charge and does not adsorb onto chitosan beads. It is assumed that the competition happened between the remaining Cr(VI) ions as $CrO_4^{(2-)}$ and the positive hydrogen ions (H[+]) at low pH, between 2 and 4, which presumably reduced the adsorption capacity of Cr(VI). These results are in a good agreement with Toledo et al. (2014); Arvand and Pakseresht (2013); Hu et al. (2011). In their experiments, in basic solution (pH > 7), the protonation of amine groups was reduced and the concentration of hydroxyl groups (OH[-]) increased which creates a negative surface charge on chitosan beads. That leads to produce a repulsive interaction between $CrO_4^{(2-)}$ ions and chitosan beads, which decreases the adsorption capacity of Cr(VI) in basic solution (Malkoc, Nuhoglu, & Dundar, 2006; Toledo et al., 2014).

The increase in Zn(II) ions removal with increasing solution-pH from 2 to 6 could be explained by reducing the hydrogen proton (H[+]) and positive Zn(II) ions competition at the same active sites on chitosan beads. In acidic media, chitosan beads surface is positively charged due to the protonation of amino groups, which decreases the interaction between active sites in chitosan beads and Zn(II) ions (Karthikeyan, Anbalagan, & Andal, 2004). After solution-pH exceeded 6, it was observed that the zinc hydroxide was formed and precipitated in the solution. The results showed that the Zn(II) adsorption onto chitosan beads from aqueous solution was affected strongly by solution-pH and the maximum adsorption capacity for Zn(II) ion was 1.67 mmol/g (109.18 mg/g) at pH 6. In addition, it was observed that at low solution-pH (pH = 3), swelling and agglomeration of chitosan beads occurred and tended to form a gel.

3.3. Effect of initial ion concentration on adsorption

The effect of initial metal concentration on adsorption capacity of Cr(VI) and Zn(II) ions onto chitosan beads was investigated by considering different ranges of initial Cr(VI) concentrations (0.96, 1.92, 4.8, 9.6, and 19.23 mmol/L) at pH 6 and temperature 10 °C. And for Zn(II) ions, initial concentrations were 0.76, 1.53, 3.8, 7.65, and 15.3 mmol/L at pH 5, temperature 10 °C, and 0.25–8 h contact

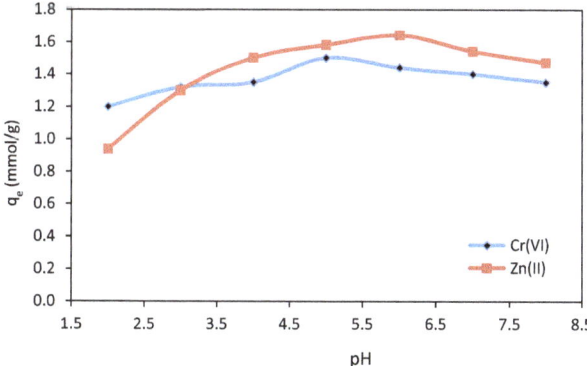

Figure 2. Effect of initial solution-pH on Zn(II) and Cr(VI) adsorption onto chitosan beads (initial concentration (C_o) for Zn(II) = 15.3 mmol/L; C_o for Cr(VI) = 19.23 mmol/L; temperature = 10 °C; mass of chitosan beads = 0.5 g).

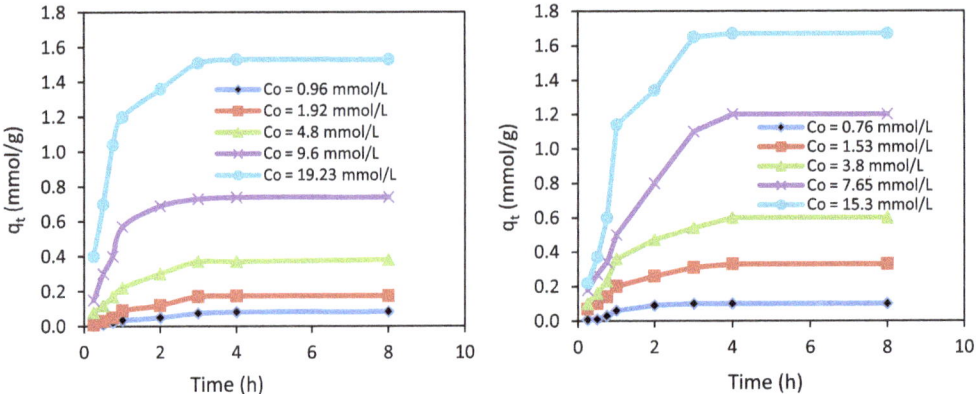

Figure 3. Adsorption capacity (q_t) as a function of time for different initial metal concentrations in solution (C_o) for (a) Cr(VI) ions and (b) Zn(II) ions (pH = 5; mass of chitosan beads = 0.5 g; temperature = 10 °C).

time. Varying the initial metal concentration can affect the driving forces of adsorption and consequently affect the adsorption behavior of metal ions onto chitosan beads. The results are illustrated in Figure 3. Cr(VI) ions removal was significantly increased from 0.084 to 1.53 mmol/g when the initial Cr(VI) concentration increased from 0.96 to 19.23 mmol/L. However, increasing the initial Zn(II) concentration from 0.76 to 15.3 mmol/L caused an increase in Zn(II) removal from 0.1 to 1.67 mmol/g onto chitosan beads. That is because, at high initial metal concentration, the concentration gradient between bulk solution and chitosan beads surface overcomes the mass transfer resistance of Zn(II) and Cr(VI) ions. The adsorption of both metals was rapid in the first 1.5 h and then slowed down until it reached the equilibrium in around 3 h in most of the adsorption runs. It is understood that all the adsorption sites are available at the initial stage (fast adsorption), while the adsorption sites are gradually occupied with the adsorption process until all of them are not be free. These results show that the adsorption process of Cr(VI) and Zn(II) ions onto chitosan beads from aqueous solution is fast and affected significantly by initial metal ion concentration.

3.4. Effect of temperature on adsorption

The effect of temperature on the adsorption of Cr(VI) and Zn(II) ions onto chitosan beads was investigated at 10, 20, 30, and 40 °C. As shown in Figure 4, the results showed that the low temperatures were favorite for Cr(VI) and Zn(II) ions adsorption onto chitosan beads. It can be seen that due to a temperature increase from 10 to 40 °C, the Zn(II) adsorption was reduced to 1.34 mmol/g from 1.67 mmol/g, whereas, the Cr(VI) adsorption was decreased from 1.53 at 10 °C to 1.12 mmol/g at 40 °C. The reason for this phenomenon is when the temperature increases, the solubility of metal ions species increases in the solution. Consequently, the van der Waals interaction forces between the metal ions and solution are stronger than those between metal ions and adsorbent (Saha & Chowdhury, 2011). As a result, the metal ions are more difficult to adsorb at high temperature.

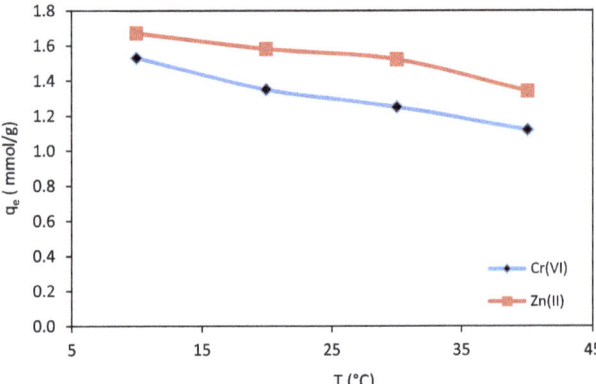

Figure 4. Equilibrium adsorption capacity (q_e) as a function of temperature (T) for Cr(VI) and Zn(II) (C_o for Zn(II) = 15.3 mmol/L; C_o for Cr(VI) = 19.23 mmol/L; pH = 5–6; mass of chitosan beads = 0.5 g).

Thermodynamic parameters including Gibb's free energy ($\Delta G°$), enthalpy ($\Delta H°$), and entropy ($\Delta S°$) were determined for Cr(VI) and Zn(II) adsorption onto chitosan beads using Equation (2) (Tellinghuisen, 2006):

$$\Delta G^\smile = \Delta H^\smile - T\Delta S^\smile \qquad (2)$$

where ΔGo can be expressed as,

$$\Delta G° = -RT \ln K_d \qquad (3)$$

Substituting Equation (2) into (3) provides the van't Hoff equation (Salih & Ghosh, 2018; Tellinghuisen, 2006),

$$\ln K_d = \frac{\Delta S°}{R} - \frac{\Delta H°}{RT} \qquad (4)$$

$$K_d = \frac{q_e}{C_e} \qquad (5)$$

where R is the universal gas constant (8.314 J/mol.K), T is the absolute temperature (K), $\Delta G°$ is Gibb's free energy (J/mol), and K_d is the equilibrium partition coefficient between solid and liquid solution (L/g), C_e is the equilibrium concentration of metal ions in the solution (mmol/L), q_e is the equilibrium adsorption capacity onto chitosan beads (mmol/g).

By plotting ($\ln K_d$) against $1/T$, the values of $\Delta H°$ (J/mol) and $\Delta S°$ (J/mol) can be determined (Figure 5). The determined $\Delta H°$ may be considered as the heat of adsorption when the adsorption capacity of chitosan beads reaches its saturated capacity of Cr(VI) or Zn(II) ions. Obtained values of $\ln K_d$, $\Delta H°$, $\Delta S°$, and $\Delta G°$ are listed in Table 2. The results of metal ions adsorption onto chitosan beads show that the adsorption process was exothermic and the negative values of $\Delta G°$ at the low temperature specified that the adsorption was spontaneous. The thermodynamic parameter values that were obtained in this study are in a good agreement with Aydin and Aksoy (2009) and Hawari, Rawajfih, and Nsour (2009).

Isosteric heat of adsorption (ΔH_x) is a useful pertinent thermodynamic factor for characterizing the temperature effect on the adsorption process. It is defined as the heat of adsorption occurring at constant adsorption capacities of metal ions onto adsorbent using the well-known Clausius–Clapeyron Equation (Saha & Chowdhury, 2011),

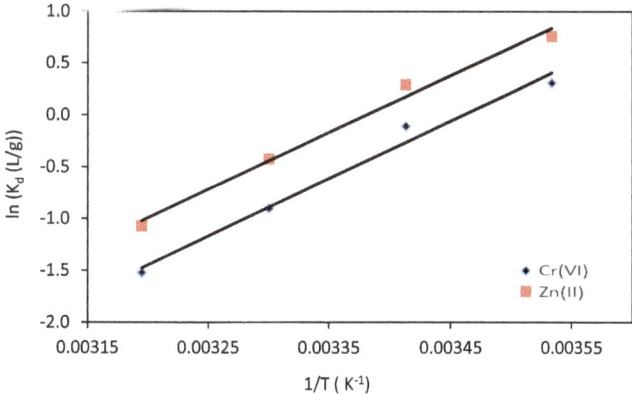

Figure 5. ln K_d as a function of $1/T$ for Cr(VI) and Zn(II). Lines indicate linear fits (C_0 = 1,000 mg/L for both metals).

$$\frac{d(\ln C_e)}{dT} = -\frac{\Delta H_x}{RT^2} \tag{6}$$

Integrating Equation (6), assuming that the ΔH_x does not depend on temperature, gives Equation (7) (Saha & Chowdhury, 2011),

$$\ln C_e = -\left(\frac{\Delta H_x}{R}\right)\frac{1}{T} + K_q \tag{7}$$

where Kq is the integral constant (mmol/L), C_e is the equilibrium concentration of metal ions in the solution (mmol/L), R is the universal gas constant (8.314 J/mol.K), T is the absolute temperature (K). From the slope of plotting (ln C_e) against $1/T$ (Figures 6 and 7), the value of ΔH_x can be determined. Values of ΔH_x were obtained at different temperatures (10, 20, 30, and 40 °C) and at different adsorption capacities of Cr(VI) or Zn(II) ions onto chitosan beads. From Table 3, it could be seen that the variation of ΔH_x values with adsorption capacities of Cr(VI) or Zn(II) is indicative that the chitosan beads have energetically homogeneous surfaces. In addition, the variation of ΔH_x values with the adsorption capacities can be attributed to the possibility of the competition between the adsorbed heavy metal ions and other ions in solution such as Na, SO_4, originating from metal salts onto the same active groups (–NH2, –OH). The results show that the ΔH_x was strongly depended on the adsorption capacities, which decreased with increasing the adsorption capacities. This can be seen as a hint on a rather homogeneous surface of chitosan beads. Otherwise, in case the surface of chitosan beads would be heterogeneous, the ΔH_x of metal ions adsorption onto chitosan beads would be constant even at varying adsorption capacities of metal ions (Saha & Chowdhury, 2011; Doke & Khan, 2013). It can be assumed that the high value of ΔH_x at low adsorption capacities (q_e) refers to high initial interactions between adsorbed metal ions and chitosan beads due to the presence of highly active amine and hydroxyl groups within the chitosan (Saha & Chowdhury, 2011).

Table 2. Thermodynamic parameters and fitted ln K_d of Cr(VI) and Zn(II) adsorption onto chitosan beads

Cr(VI)					Zn(II)			
T(K)	ln K_d (L/g)	$\Delta G°$ (kJ/mol)	$\Delta H°$ (kJ/mol)	$\Delta S°$ (kJ/mol K)	ln K_d (L/g)	$\Delta G°$ (k l/mol)	$\Delta H°$ (kJ/mol)	$\Delta S°$ (kJ/mol K)
283	0.31	−1.208	−48.045	−0.165	0.75	−1.974	−45.65	−0.154
293	0.16	0.446			0.29	−0.431		
303	−0.9	2.101			−0.4	1.112		
313	−1.5	3.756			−1.0	2.655		

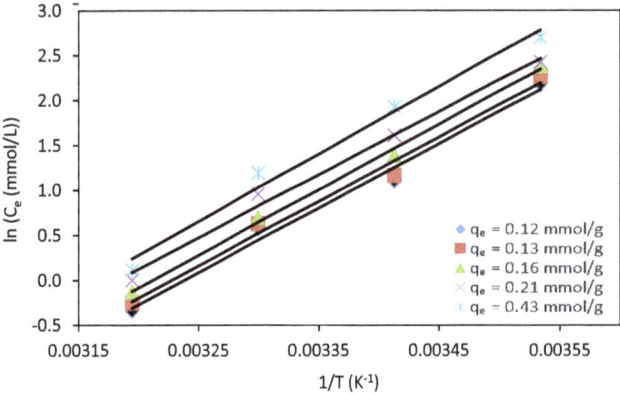

Figure 6. ln C_e as a function of $1/T$ for Cr(VI) adsorption onto chitosan beads at different adsorption capacities (q_e). Lines indicate linear fits.

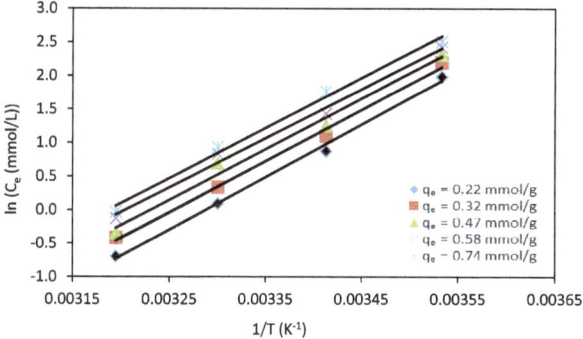

Figure 7. ln C_e as a function of $1/T$ for Zn(II) adsorption onto chitosan beads at different adsorption capacities (q_e). Lines indicate linear fits.

Table 3. Isosteric heat of Cr(VI) and Zn(II) adsorption onto chitosan beads at constant adsorption capacities (q_e) with Kq-constant values, and coefficients of determination

Heavy metals	Adsorption capacity (q_e) (mmol/g)	Isosteric heat of adsorption (ΔH_x) (kJ/mol)	Kq- constant (mmol/L)	Coefficient of determination (R^2)
Cr(VI)	0.12	−49.786	20.307	0.9898
	0.13	−53.117	21.551	0.9919
	0.16	−53.623	21.969	0.9961
	0.21	−55.782	23.038	0.9903
	0.43	−61.477	25.532	0.9867
Zn(II)	0.22	−57.733	23.215	0.9998
	0.32	−58.763	23.679	0.9996
	0.47	−59.056	24.117	0.9908
	0.58	−59.874	24.383	0.9951
	0.74	−64.187	26.406	0.9971

3.5. Adsorption kinetics

The adsorption kinetics of Cr(VI) and Zn(II) adsorption onto chitosan beads were determined by carrying out measurements after different contact times (0.25, 0.5, 0.75, 1, 2, 3, 4, and 8 h), a temperature of 10 °C, a solution-pH of 6, and an initial metal concentration of 1,000 mg/L. To explore adsorption behavior, both pseudo-first-order and pseudo-second-order models were used to fit adsorption data, as shown in Figure 8. Generally, the linear pseudo-first-order and pseudo-second-order models are, respectively, expressed by Equations (8) and (9) (Ho & McKay, 1998; HO & McKay, 1999),

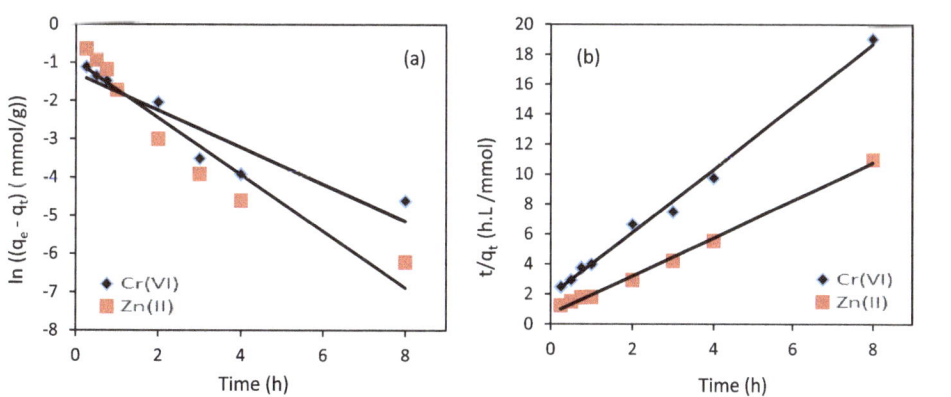

Figure 8. (a) Pseudo first-order kinetics (solid lines), **(b)** pseudo second-order kinetics (solid lines) fitted for Cr(VI) and Zn(II) adsorption experiments (pH = 5–6; mass of chitosan beads = 0.5 g; C_0 = 1,000 mg/L for both metals; temperature = 10 °C).

$$\ln(q_e - q_t) = \ln(q_e) - \frac{k_1}{2.303} \tag{8}$$

$$\frac{t}{q_t} = \frac{1}{k_2 q_e^2} + \frac{t}{q_e} \tag{9}$$

where q_e and q_t (mmol/g) are the adsorption capacities of metal ions onto chitosan beads at equilibrium and at time t (h), respectively, k_1 is the rate constant of the first-order adsorption process (h−1), k_2 is the rate constant of the second-order adsorption process [g/(mmol.h)].

It can be clearly observed from Figure 8 that the adsorption data of both metals were fitted better by the pseudo-second-order model (Figure 8(b)) than the pseudo-first-order model (Figure 8(a)), suggesting that the whole adsorption process is mainly controlled by chemisorption that is involved in valence forces for sharing or exchanging electrons through coordination or chelation between amine groups in chitosan beads and heavy metal ions (Zhang et al., 2017).

3.6. Adsorption isotherms

Adsorption equilibrium was studied at different initial metal concentrations, for Cr(VI) ions at 0.96, 1.92, 4.8, 9.6, and 19.23 mmol/L, and for Zn(II) ions at 0.76, 1.53, 3.8, 7.65, and 15.3 mmol/L. The adsorption experiments were carried out at 10 °C for 8 h. The Langmuir and Freundlich models are the most commonly used isotherm models. The Langmuir isotherm model is based on monolayer adsorption onto an adsorbent surface containing a limited number of adsorptive sites with uniform energies. The Freundlich isotherm model, can be used for the adsorption onto heterogeneous surfaces and multilayer adsorption with different energies (Caner, Ahmet, & Mustafa, 2015).

Langmuir and Freundlich isotherm models were fitted to the equilibrium adsorption data to determine the relationship between the adsorption isotherm parameters and to shed light on how the heavy metals interact with the chitosan beads surfaces. In linearized form, Langmuir isotherm is expressed as follows (Chou, Wang, Chang, & Chang, 2010),

$$\frac{C_e}{q_e} = \frac{1}{K_L q_m} + \frac{C_e}{q_m} \tag{10}$$

where C_e is the equilibrium concentration of metal ions in the solution (mmol/L), q_e and q_m are the equilibrium and theoretical maximum adsorption capacity (mmol/g), respectively. K_L is the Langmuir constant (L/g) related to binding energy.

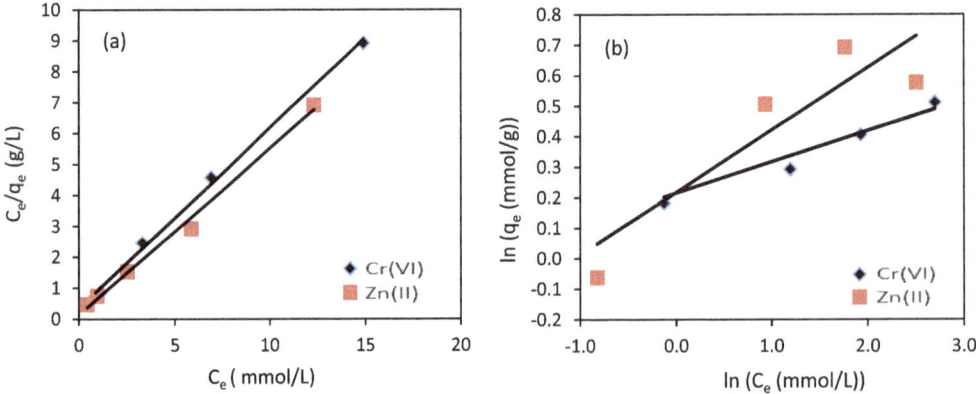

Figure 9. (a) Langmuir isotherms (solid lines), (b) Freundlich isotherms (solid lines) fitted for Cr(VI) and Zn(II) adsorption experiments (pH = 5–6; mass of chitosan beads = 0.5 g; temperature = 10 °C).

Table 4. Langmuir and Freundlich isotherm parameters for Cr(VI) and Zn(II) adsorption onto chitosan beads

Heavy metals	Langmuir parameters			Freundlich parameters		
	q_m (mmol/g)	KL(mmol/L)	R_L^2	$1/n$	KF (mmol/g)	R_F^2
Cr(VI)	1.713	1.6715	0.997	0.102	1.24	0.915
Zn(II)	1.86	3.88	0.915	0.204	1.24	0.829

The Freundlich isotherm is given as follows (Chou et al., 2010),

$$\ln q_e = \frac{1}{n}\ln C_e + \ln K_F \tag{11}$$

where K_F is the Freundlich constant (mmol/g) and n is the Freundlich exponent (-). Langmuir and Freundlich isotherms for Cr(VI) and Zn(II) adsorption onto chitosan beads from aqueous solution are presented in Figure 9, and fitted kinetic parameters are shown in Table 4. Based on the contrasting results of the correlation coefficients R^2 of Langmuir isotherm (R_L^2) and Freundlich model (R_F^2), the correlation coefficients of Langmuir isotherm (R_L^2) were all above 0.91 for both metal ions, indicating that the data were fitted better by Langmuir isotherm than by Freundlich isotherm, suggesting that Langmuir isotherm model could well interpret the adsorption procedure, which indicates that the adsorption process of both metals on the chitosan beads is driven by the formation of a heavy metal monolayer on the adsorbent surfaces. In this case, such surfaces show a homogeneous morphology of chitosan beads and a finite number of identical active sites (Chou et al., 2010; Vaz, Pereira, Fajardo, Azevedo, & Rodrigues, 2017).

4. Conclusions

This study indicates that the removal rate of Cr(VI) and Zn(II) ions from aqueous solution onto chitosan beads was excellent and up to 1.67 mmol/g (109.18 mg/g) for Zn(II), and 1.53 mmol/g (79.56 mg/g) for Cr(VI). The unique properties of chitosan make it an exciting and promising agent for purification of industrial wastewater purposes. The capacities of Cr(VI) and Zn(II) adsorption onto chitosan were strongly dependent on solution-pH, contact time, and temperature. The highest adsorption capacities were observed at pH 6 for Zn(II) and at pH 5 for Cr(VI), and at temperature 10 °C. Adsorption kinetic data were well-fitting with the pseudo-second-order kinetic model for both metals. Cr(VI) and Zn(II) ions adsorption onto chitosan beads could be described well by the Langmuir model with a R^2 equal to 0.997 and 0.915, respectively. Thermodynamic quantities such as Gibb's free energy ($\Delta G°$), enthalpy ($\Delta H°$), entropy ($\Delta S°$), and isoteric heat of adsorption (ΔH_x) indicated that both metal ions adsorption onto chitosan beads was exothermic, spontaneous, and the surface of chitosan beads was energetically homogeneous. It was observed that at low pH, swelling of chitosan beads occurred and tended to form a gel. Therefore, it is necessary to modify chitosan physically or chemically before applicable use in adsorption.

Acknowledgments
The authors thank the Chemical Engineering, and Nuclear Science and Engineering Institute at the University of Missouri-Columbia for financial support of the experimental studies and in general help with other analyses.

Funding
The authors received no direct funding for this research.

Competing interests
The authors declare no competing interest

Author details
Suhaib S. Salih[1,2]
E-mail: sss43b@mail.missouri.edu
ORCID ID: http://orcid.org/0000-0002-5276-3566
Tushar K. Ghosh[3]
E-mail: GhoshT@missouri.edu
[1] Chemical Engineering, University of Missouri-Columbia, 510 high street, Columbia, MI, 65201, USA.
[2] Department of Chemical Engineering, University of Tikrit, Iraq.
[3] Nuclear Science and Engineering Institute, University of Missouri-Columbia, 416 S. Sixth Street, E2434 Lafferre Hall, Columbia, MI, 65211, USA.

References

Annaduzzaman, M. (2015). *Chitosan biopolymer as an adsorbent for drinking water treatment: Investigation on arsenic and uranium* ((Doctoral dissertation). KTH Royal Institute of Technology, Stockholm.

Arvand, M., & Pakseresht, M. A. (2013). Cadmium adsorption on modified chitosan-coated bentonite: Batch experimental studies. *Journal of Chemical Technology and Biotechnology, 88*(4), 572–578. https://doi.org/10.1002/jctb.2013.88.issue-4

Aydın, Y. A., & Aksoy, N. D. (2009). Adsorption of chromium on chitosan: Optimization, kinetics, and thermodynamics. *Chemical Engineering Journal, 151*(1–3), 188–194. https://doi.org/10.1016/j.cej.2009.02.010

Babel, S., & Kurniawan, T. A. (2003). Low-cost adsorbents for heavy metals uptake from contaminated water: A review. *Journal of Hazardous Materials, 97*(1–3), 219–243. https://doi.org/10.1016/S0304-3894(02)00263-7

Caner, N., Ahmet, S., & Mustafa, T. (2015). Adsorption characteristics of mercury (II) ions from aqueous solution onto chitosan-coated diatomite. *Industrial & Engineering Chemistry Research, 54*(30), 7524–7533. https://doi.org/10.1021/acs.iecr.5b01293

Chou, W. L., Wang, C. T., Chang, W. C., & Chang, S. Y. (2010). Adsorption treatment of oxide chemical mechanical polishing wastewater from a semiconductor manufacturing plant by electrocoagulation. *Journal of Hazardous Materials, 180*(1–3), 217–224. https://doi.org/10.1016/j.jhazmat.2010.04.017

Doke, K. M., & Khan, E. M. (2013). Adsorption thermodynamics to clean up wastewater; critical review. *Reviews in Environmental Science and Bio/Technology, 12*(1), 25–44. https://doi.org/10.1007/s11157-012-9273-z

Ghimire, K. N., Inoue, K., Ohto, K., & Hayashida, T. (2008). Adsorption study of metal ions onto crosslinked seaweed *Laminaria japonica. Bioresource Technology, 99*(1), 32–37. https://doi.org/10.1016/j.biortech.2006.11.057

Hawari, A., Rawajfih, Z., & Nsour, N. (2009). Equilibrium and thermodynamic analysis of zinc ions adsorption by olive oil mill solid residues. *Journal of Hazardous Materials, 168*(2–3), 1284–1289. https://doi.org/10.1016/j.jhazmat.2009.03.014

Ho, Y. S., & McKay, G. (1998). Sorption of dye from aqueous solution by peat. *Chemical Engineering Journal, 70*(2), 115–124. https://doi.org/10.1016/S0923-0467(98)00076-1

Ho, Y. S., & McKay, G. (1999). Pseudo-second order model for sorption processes. *Process Biochemistry, 34*(5), 451–465. https://doi.org/10.1042/bj3440451

Hu, X. J., Wang, J. S., Liu, Y. G., Li, X., Zeng, G. M., Bao, Z. L., & Long, F. (2011). Adsorption of chromium (VI) by ethylenediamine-modified cross-linked magnetic chitosan resin: Isotherms, kinetics and thermodynamics. *Journal of Hazardous Materials, 185*(1), 306–314. https://doi.org/10.1016/j.jhazmat.2010.09.034

Hua, M., Zhang, S., Pan, B., Zhang, W., Lv, L., & Zhang, Q. (2012). Heavy metal removal from water/wastewater by nanosized metal oxides: A review. *Journal of Hazardous Materials, 211*, 317–331. https://doi.org/10.1016/j.jhazmat.2011.10.016

Jagtap, S., Thakre, D., Wanjari, S., Kamble, S., Labhsetwar, N., & Rayalu, S. (2009). New modified chitosan-based adsorbent for defluoridation of water. *Journal of Colloid and Interface Science, 332*(2), 280–290. https://doi.org/10.1016/j.jcis.2008.11.080

Jain, C. K., Singhal, D. C., & Sharma, M. K. (2004). Adsorption of zinc on bed sediment of River Hindon: Adsorption models and kinetics. *Journal of Hazardous Materials, 114*(1), 231–239. https://doi.org/10.1016/j.jhazmat.2004.09.001

Karthikeyan, G., Anbalagan, K., & Andal, N. M. (2004). Adsorption dynamics and equilibrium studies of Zn(II) onto chitosan. *Journal of Chemical Sciences, 116*(2), 119–127. https://doi.org/10.1007/BF02708205

Malkoc, E., Nuhoglu, Y., & Dundar, M. (2006). Adsorption of chromium (VI) on panacea olive oil industry waste: Batch and column studies. *Journal of Hazardous Materials, 138*(1), 142–151. https://doi.org/10.1016/j.jhazmat.2006.05.051

Min, L. L., Yuan, Z. H., Zhong, L. B., Liu, Q., Wu, R. X., & Zheng, Y. M. (2015). Preparation of chitosan-based electrospun nanofiber membrane and its adsorptive removal of arsenate from aqueous solution. *Chemical Engineering Journal, 267*, 132–141. https://doi.org/10.1016/j.cej.2014.12.024

Ngah, W. W., Ab Ghani, S., & Kamari, A. (2005). Adsorption behavior of Fe(II) and Fe(III) ions in aqueous solution on chitosan and cross-linked chitosan beads. *Bioresource Technology, 96*(4), 443–450. https://doi.org/10.1016/j.biortech.2004.05.022

Ozer, A., & Tumen, F. (2005). Cu(II) adsorption from aqueous solutions on sugar beet pulp carbon. *The European Journal of Mineral Processing and Environmental Protection, 5*(1), 26–34.

Pandey, A., Bera, D., Shukla, A., & Ray, L. (2007). Studies on Cr(VI), Pb(II) and Cu(II) adsorption-desorption using calcium alginate as biopolymer. *Chemical Speciation & Bioavailability, 19*(1), 17–24. https://doi.org/10.3184/095422907X198031

Qi, L., Xu, Z., Jiang, X., Hu, C., & Zou, X. (2004). Preparation and antibacterial activity of chitosan nanoparticles. *Carbohydrate Research, 339*(16), 2693–2700. https://doi.org/10.1016/j.carres.2004.09.007

Qudsieh, I. Y., Mirghani, E. S., Kabbashi, N. A., & Muyibi, S. A. (2014). Central composite design of zinc removal from model water using Chitosan biopolymer. *Advances in Environmental Biology*, 658–667.

Ricordel, S., Taha, S., Cisse, I., & Dorange, G. (2001). Heavy metals removal by adsorption onto peanut husks carbon: Characterization, kinetic study, and modeling. *Separation and Purification Technology, 24*(3), 389–401. https://doi.org/10.1016/S1383-5866(01)00139-3

Saha, P., & Chowdhury, S. (2011). *Insight into adsorption thermodynamics.* INTECH. https://doi.org/10.5772/558

Salih, S. S., & Ghosh, T. K. (2018). Adsorption of Zn (II) ions by chitosan coated diatomaceous earth. *International Journal of Biological Macromolecules, 106C*, 602–610.

Satya, C. W. S. (2015). *Development of magnetic separation using modified magnetic chitosan for removal of pollutants in solution* (PhD theses). Sapporo: Hokkaido University.

Shyam, R. K., & Arun, R. R. (2014). Adsorption studies for the removal heavy metal by chitosan-g-poly (acrylic acid-co-acrylamide) composite. *Science Journal of Analytical Chemistry, 2*(6), 67–70.

Sofiane, B., & Sofia, K. S. (2015). Biosorption of heavy metals by chitin and the chitosan. *Dev Pharma Chemica, 7*, 54–63.

Tellinghuisen, J. (2006). Van't Hoff analysis of K°(T): How good … or bad? *Biophysical Chemistry, 120*(2), 114–120. https://doi.org/10.1016/j.bpc.2005.10.012

Toledo, T. V., Bellato, C. R., Souza, C. H., Domingues, J. T., Silva, D. D. C., Reis, C., & Fontes, M. P. F. (2014). Preparation and evaluation of magnetic chitosan particles modified with ethylenediamine and Fe(III) for the removal of Cr(VI) from aqueous solutions. *Química Nova, 37*(10), 1610–1617.

Vaz, M. G., Pereira, A. G., Fajardo, A. R., Azevedo, A. C., & Rodrigues, F. H. (2017). Methylene blue adsorption on chitosan-g-poly (acrylic acid)/rice husk ash superabsorbent composite: Kinetics, equilibrium, and thermodynamics. *Water, Air, & Soil Pollution, 228*(1), 14. https://doi.org/10.1007/s11270-016-3185-4

Zhang, H., Dang, Q., Liu, C., Cha, D., Yu, Z., Zhu, W., & Fan, B. (2017). Uptake of Pb(II) and Cd(II) on chitosan microsphere surface successively grafted by methyl acrylate and diethylenetriamine. *ACS Applied Materials & Interfaces, 9*(12), 11144–11155. https://doi.org/10.1021/acsami.7b00480

Zhao, F., Repo, E., Yin, D., & Sillanpää, M. E. (2013). Adsorption of Cd(II) and Pb(II) by a novel EGTA-modified chitosan material: Kinetics and isotherms. *Journal of Colloid and Interface Science, 409*, 174–182. https://doi.org/10.1016/j.jcis.2013.07.062

Zhou, T., Li, C., Jin, H., Lian, Y., & Han, W. (2017). Effective Adsorption/Reduction of Cr(VI) Oxyanion by Halloysite@ Polyaniline Hybrid Nanotubes. *ACS Applied Materials & Interfaces, 9*(7), 6030–6043. https://doi.org/10.1021/acsami.6b14079

2

Oil development in the grasslands: Saskatchewan's Bakken formation and species at risk protection

Andrea Olive[1]*

*Corresponding author: Andrea Olive, Political Science and Geography, University of Toronto Mississauga, 3359 Mississauga Rd, Mississauga, ON L5L 1C6, Canada
E-mail: andrea.olive@utoronto.ca
Reviewing editor: Fei Li, Zhongnan University of Economics and Law, China

Abstract: This paper considers the possible impacts of oil development on wildlife in the grasslands ecosystem, particularly in the province of Saskatchewan. The Bakken formation, a major North American shale play, overlaps with one of the largest areas for grassland birds in Canada the US. Access to the oil is made possible through fracking and horizontal drilling, which are controversial techniques that have been regulated and banned in other parts of North America and the world. Drawing on analysis of recovery documents for listed species at risk, this paper illustrates that oil development is impacting species through habitat destruction, oil and noise pollution, invasive species, and road infrastructure. Current wildlife policy in Saskatchewan is inadequate to protect species at risk in the Bakken formation.

Subjects: Conservation - Environment Studies; Energy Policy; Environmental Politics; Federalism

Keywords: species at risk; oil development; Bakken formation; grassland conservation

1. Introduction

The Saskatchewan prairies lie at the northern edge of the North American Great Plains, which is an eco-region that stretches from Canada through the United States and down into Mexico. Saskatchewan's prairie includes mixed grasslands and aspen parklands as well as a plethora of flora and fauna (see Savage, 2011). From a global perspective, Saskatchewan prairie is important habitat for grassland and migratory birds (Henwood, 2010; Savage, 2011). For example, there are two Western Hemisphere Shorebird Reserve Network sites in the prairies, Chaplin Lake and Quill Lakes, and the oldest bird sanctuary in North America, known as Lake Mountain Bird Sanctuary (Savage,

ABOUT THE AUTHOR

Andrea Olive is a professor of Political Science and Geography at the University of Toronto. She is the author of Land, Stewardship, and Legitimacy: Endangered Species Policy in Canada and the US and the Canadian Environment in Political Context. She is also the co-editor of a forthcoming book entitled Transboundary Environmental Governance across the World's Longest Border. Her main area of research is wildlife conservation on private lands in North America. The research presented here is part of a larger project examining the political ecology of the Bakken formation in Saskatchewan and North Dakota. Of specific interest is the environmental externalities of hydraulic fracturing in the grasslands ecosystem.

PUBLIC INTEREST STATEMENT

The Canadian prairie province of Saskatchewan is home to the grasslands ecosystem and a large oil reserve known as the Bakken formation. In the past decade, Saskatchewan has become a significant producer of oil due to new technological advances in recovering oil from shale rock. However, oil development in this region is impacting species at risk through habitat destruction, oil and noise pollution, invasive species, and road infrastructure. Current wildlife policy in Saskatchewan is inadequate to protect species at risk in the Bakken formation.

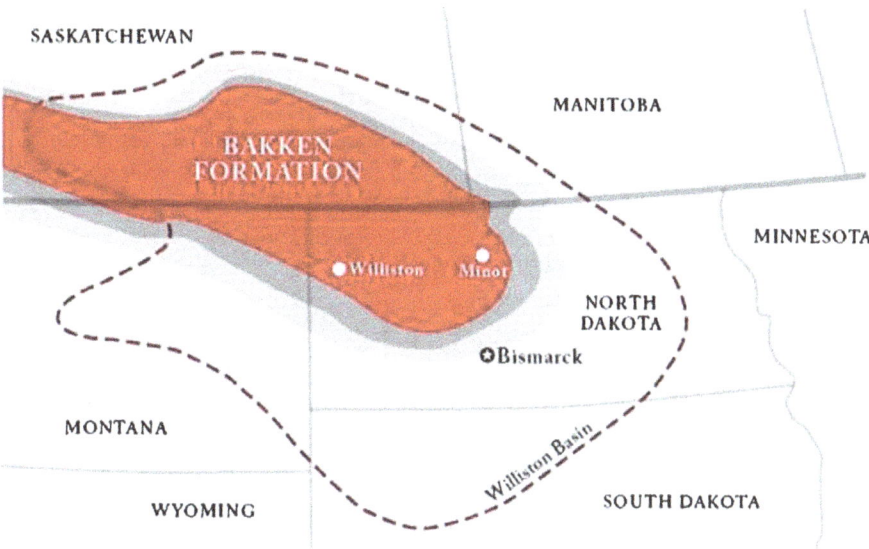

Figure 1. Map of the Bakken formation in Canada and the US.

Source: United States Geological Survey (n.d.).

Figure 2. Stages in shale hydraulic fracturing.

2011). Today the grasslands are in jeopardy as they "are one of the most significantly modified landscapes in Canada" (Nasen, Noble, & Johnstone, 2011, p. 195) and "constitute one of the most endangered ecosystem types on Earth" (Roch & Jaeger, 2014, p. 2506).

Saskatchewan is also home to the Bakken formation, which is an area of shale rock with vast oil potential (see Carter, Fraser, & Zalik, 2017; Zink & Eaton, 2016). This shale play extends through North Dakota and Montana into the southeast corner of Saskatchewan and the southwest corner of Manitoba. See Figure 1. For the past 15 years, the region has been producing immense quantities of oil—enough to make North Dakota the second largest producer of oil in the US, after Texas, and Saskatchewan the second largest producer of oil in Canada, after Alberta (Carter et al., 2017). The United States Geological Survey suggests potential for up to 503 million barrels of oil in the Bakken formation (Langdon, 2008).

There has been limited research on the direct impacts posed by oil development in the Bakken region to flora and fauna (as exceptions, see Braun, Oedekoven, & Aldridge, 2002; Hill & Olson, 2013; Thompson, Johnson, Niemuth, & Ribic, 2015). More generally, however, research shows a link between oil and gas development and the decline of grasslands species in North America (see, for example, Bernath-Plaisted & Koper, 2016; Hill & Olson, 2013; Nasen et al., 2011; Northrup & Wittemeyer 2012). Nasen et al. (2011) conducted a cumulative impact assessment of oil and gas development in the mixed grasslands of southwestern Saskatchewan, and found significant long-term, and spatially long-range, impacts. Notably, there were changes in soil structure around well sites, which can have ripple effects across the ecosystem. Unfortunately, no similar research has been conducted in the Bakken region.

This paper examines the potential impact of oil production on prairie grassland biodiversity. Specifically, the main focus is species at risk with range in Saskatchewan's Bakken region. The paper begins with a brief overview of the oil development process as well as the regulatory frameworks for oil and wildlife in Saskatchewan. The next section presents the research questions and methodology. The data, taken from government documents for listed species, is given in detail before a discussion regarding the main impacts of oil in the grasslands.

2. Oil and fracking process and policy

Saskatchewan has been producing oil since the 1950s, but more recent technologies in hydraulic fracturing (fracking) and horizontal drilling have caused a boom in oil production in the Bakken (Carter et al., 2017; Zink & Eaton, 2016). Thanks to the technique, which involves injecting high-pressure water and chemicals deep underground to unlock stores of trapped oil, Saskatchewan produced 527,000 barrels of oil per day on average in 2015 (National Energy Board, 2016). Unlike other shale plays in North American, such as the Marcellus on the East coast or the Barnett in Texas, which are rich in natural gas, the Bakken is predominately oil rich. Between 2006 and 2012, over 2000 new wells were drilled in Saskatchewan's Bakken area (Zink & Eaton, 2016, p. 19).

Since most of the oil is trapped in shale rock, fracking is the predominate method of oil extraction. This is a process that can occur in six stages as illustrated in Figure 2.

After an oil firm determines a suitable location and obtains the mineral rights (stage 1), a well is drilled. Most oil sites contain as many as five well heads and infrastructure like solution tanks and disposal facilities for waste water (Nasen et al., 2011). These sites also require access roads for trucks and equipment. In stage 2, the oil firm installs a permanent well-head and the necessary infrastructure to beginning hydraulic fracturing. Hydraulic fracturing takes place in stage 3 where water and a chemical mixture are injected at high pressure into the well. This pressure causes the shale to fracture. In stage 4, the well is completed and a permanent wellhead is installed so that oil can be pulled to the surface. During the production stage, oil, gas, and water (including the chemicals and sand) flow from the surface wellhead through the flow lines to tanks that separate the oil from the rest of the mixture. Oil is often stored in large containers until it is ready for pick-up and delivery to a pipeline or refinery. The water used in fracking is either recycled for use at another well location or it is stored and disposed of according to government regulations. After all the economically viable oil has been recovered from the well, the final stage is well abandonment. Often wells are filled with cement and surface equipment is removed from the landscape. For a detailed description of the process see Lade and Rudik, 2017, Shrestha, Chilkoor, Wilder, Gadhamshetty, & Stone, 2017 or the Petroleum Services Association of Canada, 2018).

Environmental degradation occurs unevenly across the stages of fracking. Most scientific research has focused on water quality and quantity impacts across the different stages (see Shrestha et al., 2017 for an example of water impacts in the Bakken formation) or greenhouse gas emissions across the stages (see Jiang et al., 2011; for example). Other research has focused more broadly on the general environmental impacts, which range from altered land use (Fore, Overmore, & Hill, 2015; Moran, Cox, Wells, Benichou, & McClung, 2015), chemical and air pollution (Prenni et al., 2016; Rabe, 2014), human health risks (Webb et al., 2014), and stressed transportation infrastructure (Rahm, Fields, & Farmer, 2015). In the scientific literature, these risks are not examined systematically across the different stages of the fracking process.

2.1. Governing oil in Saskatchewan

Saskatchewan has taken a hands-off approach to regulating oil development, largely because it depends on the industry for revenue and job creation (Carter & Eaton, 2016). According to Carter et al. (2017), "over the 2009–2014 period, Saskatchewan's oil sector directly contributed 13 to 24 percent of the province's total revenue" (p. 64). The oil and gas sector is regulated by the Saskatchewan Ministry of the Economy, whereas the Ministry of Environment has little to no role in decisions concern oil development in the province (Carter et al., 2017). Saskatchewan has chosen not to regulate fracking as a "distinct and specifically risky practice," and instead relies on general oil and gas development rules (Carter & Eaton, 2016). There is no regulatory distinction between conventional and unconventional oil development. Most wells are approved in Saskatchewan without undergoing an environmental impact assessment (Carter & Eaton, 2016). This is because each application for individual lease sites are "rarely considered significant enough to trigger a formal EA" (Nasen et al., 2011, p. 196). And all oil and gas exploration activities are exempt from environmental assessment (Bowden, 2010; Carter & Eaton, 2016). These exemptions are part of the government's "results-based regulatory regime," which the government describes as one that "establishes clear performance expectations while eliminating ineffectual scrutiny and attention to process, especially for routine, well understood, and low-risk activities" (Government of Saskatchewan, 2014). This hands-off approach creates a regulatory atmosphere that is favorable for oil and gas industry.

3. Wildlife policy in Saskatchewan

In Canada, wildlife is co-managed by the provinces and federal government, with the former holding the bulk of responsibility (Olive, 2014a). Since the federal government, under the Constitution of Canada, has some jurisdiction over navigable waters and federal lands as well as transboundary issues, that government is responsible for some aquatic species, wildlife on federal lands (such as national parks or First Nation reserves), and migratory birds that are part of the Canada–US Migratory Birds Convention Act 1917. Everything else is predominately provincial responsibility as all provinces have jurisdiction over their own natural resources (such as oil), private lands, and provincial crown lands.

Species at risk are subsequently managed at two levels of government. The federal government has a longer and more global relationship to biodiversity loss and species at risk. In 1976, Canada established a non-government scientific body, the Committee on the Status of Endangered Wildlife in Canada (COSEWIC), to assess the status of all flora and fauna across the country.[1] In 1992 the federal government signed the UN Convention on Biodiversity and set about creating a national biodiversity strategy for the country. The federal government passed the Species at Risk Act (SARA) in 2002, which makes it illegal to kill or harm listed species and alter critical habitat. However, the main legal provisions of the law extend only to listed species on federal lands, some aquatic species, and some migratory birds (see Olive, 2014a).

COSEWIC assesses a species and grants it a scientific status: endangered, threatened, species concern, not at risk, or data deficient. The federal government has nine months to accept the status and add the species to SARA or can ask COSEWIC for more data. If nine months elapse without a decision, the species is automatically listed on SARA with the status granted by COSEWIC (see Mooers et al., 2010; Olive, 2014b). For endangered and threatened species, the government has 2 and 3 years, respectively, to write a recovery plan, which outlines the biological description and needs of the species, it geographical range, and the major threats or reasons for decline. After a recovery plan is finalized and made public, the government is required to write an action plan for the species, which is a document intended to the recovery process into policy guided actions with economic considerations. For species listed as special concern on SARA, the government must prepare a management plan, which is similar to a recovery plan but not supported by the force of law (see Mooers et al., 2010; Olive, 2014b). Special concern species are excluded from the main legal provisions of SARA. Overall, SARA protects COSEWIC designated endangered and threatened species and makes it illegal to harm said species, but the federal government can only implement policy and work toward the protection and recovery of these species on federal lands.

Saskatchewan is one of only four provinces in Canada to lack stand-alone species-at-risk legislation (Olive, 2014a). Instead, the province uses its 1998 Wildlife Act to regulate all flora and fauna. Under the Act, native wild species can be designated as vulnerable, threatened, endangered, or extirpated from the province.[2] While similar to COSEWIC's categories, these designations in Saskatchewan are at the discretion of the provincial Minister of Environment. The province uses a conservation status ranking system through its own Saskatchewan Conservation Data Centre. This center is responsible for maintaining a centralized database for all scientific information pertaining to the native species (see Saskatchewan Conservation Data Centre, 2017).

Table 1. List of species at risk with habitat range in the Bakken formation

	Latin name	Taxon	COSEWIC status	SARA status	SK wildlife act status
Alkaline Wing-nerved Moss	*Pterygoneurum kozlovii*	Mosses	Threatened	Threatened	No status
American Badger taxus subspecies	*Taxidea taxus taxus*	Mammal	Special Concern	No status	No status
Baird's Sparrow	*Ammodramus bairdii*	Bird	Special Concern	Special Concern	No status
Bank Swallow	*Riparia riparia*	Bird	Threatened	No status	No status
Barn Swallow	*Hirundo rustica*	Bird	Threatened	No status	No status
Black-tailed Prairie Dog	*Cynomys ludovicianus*	Mammal	Threatened	Special Concern	No status
Bobolink	*Dolichonyx oryzivorus*	Bird	Threatened	No status	No status
Buff-breasted Sandpiper	*Tryngites subruficollis*	Bird	Special Concern	Special Concern	No status
Buffalograss	*Bouteloua dactyloides*	Vascular Plant	Special Concern	Special Concern	No status
Burrowing Owl	*Athene cunicularia*	Bird	Endangered	Endangered	Endangered
Chestnut-collared Longspur	*Calcarius ornatus*	Bird	Threatened	Threatened	No status
Chimney Swift	*Chaetura pelagica*	Bird	Threatened	Threatened	No status
Common Nighthawk	*Chorediles minor*	Bird	Threatened	Threatened	No status
Dakota Skipper	*Hesperia dacotae*	Arthropod	Endangered	Threatened	No status
Eastern Whip-poor-will	*Antrostomus vociferous*	Bird	Threatened	Threatened	No status
Eastern Yellow-bellied Racer	*Coluber constrictor flaviventris*	Reptile	Threatened	Threatened	No status
Ferruginous Hawk	*Buteo regalis*	Bird	Threatened	Threatened	No status
Great Plains Toad	*Anaxrus cognatus*	Amphibian	Special Concern	Special Concern	No status
Greater Sage Grouse urophasianus subspecies	*Centrocercus urophasianus urophasianus*	Bird	Endangered	Endangered	Endangered
Greater Short-horned lizard	*Phrynosoma hernandesi*	Reptile	Endangered	Endangered	No status
Greenish-white Grasshopper	*Hypochlora alba*	Arthropod	Special Concern	No Status	No status
Hairy Prairie-clover	*Dalea villosa*	Vascular Plant	Special Concern	Special Concern	No status
Horned Grebe Western Population	*Podiceps auritus*	Bird	Special Concern	Special Concern	No status
Loggerhead Shrike Prairie subspecies	*Lanius ludovicianus excubitorides*	Bird	Threatened	Threatened	No status
Long-billed Curlew	*Numenius americanus*	Bird	Special Concern	Special Concern	No status
Monarch	*Danaus plexippus*	Arthropod	Endangered	Special Concern	No status
Mormon Metalmark Prairie Population	*Apodemia mormo*	Arthropod	Special Concern	Threatened	No status
Nine-spotted Lady Beetle	*Coccinella novemnotata*	Arthropod	Endangered	No status	No status
Pale Yellow Dune Moth	*Copablepharon grandis*	Anthropods	Special Concern	Special Concern	No Status
Piping Plover circumcinctus subspecies	*Charadrius melodus circumcinctus*	Bird	Endangered	Endangered	Endangered
Plains Bison	*Bison bison bison*	Mammal	Threatened	No Status	No status
Prairie Rattlesnake	*Crotalus viridis*	Reptile	Special Concern	No status	No status
Red Knot rufa subspecies	*Calidris canutus rufa*	Bird	Endangered	Endangered	No status
Red-headed Woodpecker	*Melanerpes erythrocephalus*	Bird	Threatened	Threatened	No status
Red-neck Phalarope	*Phalaropus labatus*	Bird	Special Concern	No status	No status
Short-eared Owl	*Asio flammenus*	Bird	Special Concern	Special Concern	No status
Sprague's Pipit	*Anthus spraguelli*	Bird	Threatened	Threatened	No status
Western Tiger Salamander Prairie population	*Ambystoma mavortium*	Amphibian	Special Concern	No Status	No status

Under the Wildlife Act, all flora and fauna is protected from "being disturbed, collected, harvested, captured, killed, sold or exported without a permit." For species deemed at risk, "the den, house, nest, dam, or usual place of habitation" is also protected from disturbance and destruction. As of 2018, the Wildlife Act designates five species as extirpated, nine species as endangered, and one as threatened. The list of species has not changed since the original list was created in 1999. However, according to COSEWIC, in 2018, there are 39 special concern species, 33 threatened, 21 endangered, and 2 extirpated species with range in Saskatchewan. The federal government, using COSEWIC data, is adding species to SARA and creating regulatory obligations for that level of government.

4. Research question & methodology

We know that oil development has been significantly increasing in Saskatchewan's grassland prairie region for at least the past 10 years (Carter et al., 2017; Zink & Eaton, 2016). We know the province has very few environmental regulations in place and its overall approach to oil is "hands off" so as to not impact the economic side of the industry (Carter & Eaton, 2016). What we know less about is the impact oil development has on wildlife in the province. This paper examines the extent to which oil development is thought to be impacting species at risk in Saskatchewan's Bakken region. Already at-risk species are important because they are the most vulnerable and because the federal (and provincial) government is already legally responsible for their protection. The main research question is: In what ways is oil development impacting species at risk in the Bakken formation?

4.1. Methodology

In order to establish a list of known species at risk we used the SARA Public Registry to identify every species COSEWIC categorized as endangered, threatened, or special concern with range in Saskatchewan (see Government of Canada, n.d.). This search resulted in 93 species. We examined the critical habitat and range listing for each of these species and eliminated those without range in the Bakken formation. This narrowed the list to 38 species (listed in Table 1 below).

For each species with range in the Bakken, we checked to see if the federal government published an official recovery document. Under SARA, the government is required to prepare a recovery strategy for endangered and threatened species and a management plan for special concern species (Olive, 2014b). If a published document was located, we examined it for mentions of oil development. This was done using a key word search for "fracking" or "oil" or "petroleum" or "energy." Each mention was carefully read in context for meaning. All recovery strategies and management plans are legally required to list a species' threats, according to peer-reviewed science and Aboriginal traditional knowledge. This section of each document was read and carefully analyzed to determined overall threats to each species. See Figure 2 for a depiction of our method. The goal was to determine if the federal government considers oil development to be a threat.

5. Results and discussion

Table 1 provides a list of the 38 species, as well as their taxon, and their status according to COSEWIC, as well as their standing on SARA and the Saskatchewan Wildlife Act.

Overall, 35 of these have no status on the Wildlife Act and, therefore, are not protected by the provincial government. However, 28 of the species are listed on SARA. In the case of the other 10 species, such as the American badger, the Barn swallow, or the Western Tiger salamander, these species are waiting to be granted designation on SARA. In the meantime, SARA does not protect them.

In terms of the recovery strategies, the federal government has prepared strategies for 13 of the 18 endangered or threatened species. Species like the chimney swift and the red-headed woodpecker should have a strategy in place but do not. It is not uncommon in Canada for species to be listed and the government lags in preparing a recovery strategy (Bird & Hodges, 2017; Olive, 2014a). In terms of management plans, the government has prepared 7 plans (including 2 delayed plans awaiting approval) out of 11 special concern species. See Table 2 for an overview of these findings.

Table 2. Link between species-at-risk in the Bakken and oil & gas, as established through recovery plans and peer review literature

Endangered or threatened species	Federal recovery strategy or management plan	Strategy or plan mentions oil
Alkaline Wing-nerved Moss	No	–
Burrowing Owl	Yes	Yes
Chestnet-collared Longspur	Yes	Yes
Chimney Swift	No	–
Common Nighthawk	Yes	No
Dakota Skipper	Yes	No
Eastern Whip-poor-will	Yes	Yes
Eastern Yellow-bellied Racer	Yes	Yes
Ferruginous Hawk	No	–
Greater Sage Grouse urophasianus subspecies	Yes	Yes
Greater Short-horned lizard	Yes	Yes
Loggerhead Shrike Prairie subspecies	Yes	No
Mormon Metalmark Prairie Population	Yes	No
Piping Plover circumcinctus subspecies	Yes	Yes
Red Knot rufa subspecies	Yes	No
Red-headed Woodpecker	No	–
Red-neck Phalarope	No	–
Sprague's Pipit	Yes	Yes
Special concern species		
Barid's Sparrow	No	–
Black-tailed Prairie Dog	Yes	No
Buff-breasted Sandpiper	No	–
Buffalograss	No	–
Great Plains Toad	Yes	Yes
Hairy Prairie-clover	Delayed	Yes
Horned Grebe Western Population	No	–
Long Billed Curlew	Yes	Yes
Monarch	Yes	No
Pale Yellow Dune Moth	Yes	Yes
Short-eared Owl	Delayed	Yes

In total we examined 20 documents: 13 recovery strategies and 7 management plans. As Table 2 illustrates, the federal government considers oil development to negatively impact at least 13 of these species. It is worth briefly reviewing what the government considers to be the primary threats to these species. The species are reviewed here in alphabetical order, beginning with the endangered and threatened species, followed by the management plans for the special concern species. After reviewing all the documents, threats from oil were categorized into five main threat types as illustrated in Table 3.

5.1. Reviewing recovery strategies

The recovery strategy for the Burrowing owl says "no single factor" is causing population reduction, but a cumulative impact of several factors is "likely" responsible (Environment Canada, 2012b, p. 13).

These factors include food shortages, habitat loss, loss of burrows, vehicle collisions, and environmental contaminants. These factors are linked—directly or indirectly—to oil development in the province. For example, it is noted that natural grasslands are lost through "expanding agriculture, petroleum exploration and extraction, and urban sprawl" (2012b, p. 14), but it is unclear how grassland loss directly affects the owl population. For instance, owl populations have outpaced grassland loss and been similar across Manitoba, Saskatchewan, and Alberta despite different amounts of land and oil development in those three provinces (2012b, p. 14). As an another example, the strategy also explains that vehicle collisions are a problem on roads, such as those created for oil exploration, and nests crushed by heavy machinery, including from "cultivation, highway repair, oil and gas activities, or lawn maintenance" are problematic (2012b, p. 16). Thus, while no single issue—such as oil or fracking—is causing owl populations to decrease, the government does recognize that oil development is causing habitat loss, vehicle collisions, and accidental deaths of the owls.

For the chestnut-collared longspur oil development is named alongside farming, ranching, and invasive species as a key threat since all these factors contribute to native prairie habitat loss (ECCC, 2016c, p. iii). The strategy discusses oil extraction activities explicitly and concludes that while most research has been on natural gas development, longspurs are less abundant at sites containing active oil wells than at sites without wells (ECCC, 2016c, p. 11). Moreover, the impacts of roads are noted as longspurs are less common along roads than can be explained by the loss of suitable habitat (ECCC, 2016c, p. 12). The development of roads has been attributed to the booming oil and gas industry.

The primary threats to the Common nighthawk are unknown but "possible threats" include loss of prey, habitat loss, climate change, pollution, and invasive species (EC, 2016c, p. v). This bird is migratory and spends about 10% of its time in Canada where it's habitat ranges across the country. The species is known to breed in the prairie grasslands, but why a 49% decline in Canada has occurred over the past three generations is still unknown (EC, 2016c, p. 1). Oil development is also not mentioned in the recovery plan for the Dakota skipper. Instead, the primary threats are thought to be habitat loss due to cultivated native prairie, habitat degradation from burning, overgrazing, and haying, invasive species, pollution from pesticides and herbicides, climate change, and the collection of the species for natural history specimens (EC, 2007, p. iii).

The eastern whip-poor-will is a migratory bird that breeds in Southern Saskatchewan and other Canadian provinces. The bird's population has been steadily declining at a rate of 2.77–5.53% between 2002 and 2012 (EC, 2015c, p. iii). Energy exploration and transportation, including oil, is considered a cause of habitat loss for the eastern whip-poor-will, which is the primary threat to the species (EC, 2015c, p. 12). The recovery strategy also points out that activities associated with energy development can cause nest destruction (EC, 2015c, p. 12).

The yellow-bellied racer recovery strategy lists primary threats as habitat loss due to human activities, small population size, road mortality, and human disturbance of hibernacula (Parks Canada Agency, 2010, p. v). However, information gaps about the species are major, including the snake's overall population numbers and distribution. At present, only seven hibernacula are known. Industrial activities are minimal within those areas, but their existence in nearby areas "may" reduce snake populations (Parks Canada Agency, 2010, p. 6).

The recovery strategy for the Greater sage grouse lists the main threats as "sensory disturbance from vertical structures and noise, habitat loss and degradation, increased predator pressure, drought and extreme weather conditions, West Nile virus, alteration of natural hydrology" (EC, 2013c, p. iii). The recovery strategy discusses oil development explicitly. For example, Sage grouse avoid anthropogenic sites like oil well pads. Experiments have shown that noise alone (typical of oil/ gas drilling) discourages breeding, and the species does not habituate itself to increased noise levels (EC, 2013c, p. 9). Related, tall vertical structures (power lines, oil and gas structures, wind turbines, buildings) discourage nesting nearby (EC, 2013c, p. 11). Moreover, the oil and gas industry is also heavily implicated in the destruction of Sage grouse habitat. The recovery strategy points out that habitat conversion to cropland occurred mainly before 1981, meaning it cannot explain more recent population declines (EC, 2013c, p. 11). Habitat conversion to petroleum industry development has been closely linked temporally to Sage grouse population declines (EC, 2013c, pp. 11–12). Conversion of habitat to roads also introduces harmful noise, invasive species, and reduced natural vegetation. Sage grouse tend to avoid anthropogenic "edges" such as roads and trails (EC, 2013c, p. 12). Essentially, this means that a combination of threats is introduced by oil industry development in Saskatchewan.

The recovery strategy for the Greater short horned lizard suggests that oil development may be a problem. The listed primary threats include conversion of natural habitat to industrial infrastructure, conversion of habitat to roads, dams and irrigation, conversion of habitat to cropland, invasion of exotic plants, inclement or extreme weather, traffic and pet mortality "due to urban expansion," mortality from oil spills, and collection (EC, 2015a, p. iii). Specially, the strategy points out that oil and gas development affects habitat by clearing vegetation, stripping and mixing soil, compacting soil, "possible localized soil contamination," clearing pathways into formerly inaccessible areas, and disturbances affecting insect (prey) populations (EC, 2015a, p. 8). Above-ground infrastructure associated with wells can attract mammalian predators, snakes, and provide perches for bird predators (EC, 2015a, p. 9). Road and trail construction, associated with the oil and gas industry, "can lead to an increase in occurrence of weeds and invasive plant species"—affecting lizard movements, prey availability, micro-thermal conditions (EC, 2015a, p. 9). Importantly, the document points out that agriculture is not the threat to habitat as conversion of badlands/natural habitat to cropland has "never been a serious threat" because lizard habitat terrain/soil is not suitable to crop and forage production (EC, 2015a, p. 9).

The loggerhead shrike's recovery strategy does not focus on oil at all. Instead, the primary threats are considered land use changes on wintering grounds, cultivation of natural grasslands on breeding grounds, predation, severe weather, disease and parasitic infections. Other possible threats include pesticides, environmental contaminants and vehicle collisions (EC, 2015b). While it could be the case that oil development is contributing to these threats, such as habitat loss through roads and vehicle collisions, the recovery plan does not make that explicit connection.

Similar to other grassland birds, the primary threats facing the Mormon Metalmark are habitat loss, invasive species, pollution, and climate change. Mormon Metalmark habitat is found within the Grasslands National Park where no oil development occurs. However, the recovery strategy does point out that "oil and gas exploration and development could cause destruction and degradation of metalmark habitat" so it would become an issue if habitat is found outside the park (Pruss, Henderson, Fargey, & Tuckwell, 2008, p. 9).

The recovery strategy for the piping plover mentions oil, but the top threat has been identified as predation (Environment Canada, 2006, p. 9). Other threats include habitat loss and degradation, which can occur through "oil and gas development". However, the plan does not go on to consider this link or otherwise establish an connection between oil and piping plover populations.

Oil and gas is not acknowledged as a threat to the Red knot. The recovery strategy is for three subspecies of the bird. The main threats are residential and commercial development, agriculture and aquaculture, energy production and mining, human intrusions and disturbance, natural system modifications, invasive and problematic species and genes, pollution, and climate change (ECCC, 2016a, p. iv). These threats apply to all three subspecies, whereas threats from oil drilling and mining/quarrying are described only for the Arctic subspecies (ECCC, 2016a, p. 16).

Lastly, the recovery strategy for the Sprague's pipit reveals key threats as the loss and degradation of breeding habitat, including through the encroachment of invasive species or shrubs/woody vegetation (EC, 2012a). The species requires large areas of open grassland, which is associated with low-moderate grazing, or periodic haying or burning (EC, 2012a, p. iv). Pipits breed in higher numbers in native grasslands compared to non-native grasslands (EC, 2012a, p. 6). As of 2001, 75% of native grasslands on the Canadian prairies had been lost, "primarily to cultivation, succession, road construction, gravel extraction, petroleum exploration and extraction, and settlement" (EC, 2012a, p. 6).

5.2. Reviewing management plans

The black tailed prairie dog's management plan indicates the primary threat is sylvatic plague. Other listed threats in the management plan include natural diseases (tularemia), habitat loss and degradation, predation, pest control, drought, floods, and severe winters (p. v). The plan does not discuss oil production explicitly.

The primary threat to the Great Plains toad is habitat loss in the prairies from the drainage of wetlands and the cultivation of native grasslands. Other threats listed include pesticides and fertilizers, overgrazing, oil and gas development and operations, traffic, disease, and climate change (EC, 2013b, p. ii). Oil and gas development is considered a threat because of habitat loss, chemical pollution, and traffic collisions (EC, 2013b).

The management plan for the hairy prairie-clover is the only document in this study to specifically mention the word "fracking." The document points out that "sand and gravel extracted from sand dunes is used for road construction, oil and gas activities (e.g. fracking), agriculture (e.g. potato farming), and personal use" (Environment & Climate Change Canada, 2017, p. 12). The plant is found mainly in Manitoba and its range likely extends into southeastern Saskatchewan. More of a concern is that other locations of known hairy prairie-clover in the sands dunes of Saskatchewan are under potential threat because sand is being extracted to support fracking in the Bakken formation and in Alberta (Environment & Climate Change Canada, 2017). Fracking requires large amounts of sand to blast into the shale rock in order to extract oil.

Oil development is cited as a major threat to long-billed curlew. The management plan lists major current threats as conversion of grasslands to agriculture, fire suppression resulting in forest and shrub encroachment, urbanization (for curlews in British Columbia), energy development, human developments in wintering habitat (US/Mexico), predation, and invasive plant species. The plan suggests that energy development in Alberta and Saskatchewan reduces habitat, and "may enhance other threats"—such as introducing invasive plant species, increasing traffic, and vehicle collisions (EC, 2013a, p. 10).

Monarch butterflies are under threat from "degradation and loss of overwintering habitat in Mexico and along the Californian coast, the widespread use of pesticides and herbicides throughout their breeding grounds, climate change, severe weather events, succession and conversion of breeding and nectaring habitat, and for the Eastern population, the impacts of Bark Beetles on overwintering habitat" (ECCC, 2016b, p. iii). Oil development is not considered at all in the management plan of the Monarch. The only form of energy mention is wind energy can be a threat if wind mills are placed in the migratory pattern of the Monarch.

Similar to the hairy prairie clover, the pale yellow dune moth is also threatened by sand extraction. The management plan points to a low-level concern over the "construction of roads and energy infrastructure and sand extraction in sand dunes" (EC, 2016b, p. 11). It is thought that "suitable habitat will likely decrease and/or degrade" over the long term given these threats (EC, 2016b, p. 11).

Lastly, energy and mining exploration are considered a threat to short-eared owls. Specifically, the plan suggests that oil, gas, coal, and hydroelectricity contribute to habitat loss but that the "direct impacts of these threats on the Short-eared owl populations have not yet been demonstrated" (EC, 2016a, p. 13). These owls breed in all Canadian provinces, but are found most commonly in the

Table 3. Summary of the impact of conventional oil and fracking on SARA listed species in the Bakken formation according to federal recovery documents

Habitat loss & destruction	Oil & noise pollution	Invasive species (from oil industry)	Road infrastructure/ vehicle collisions
Burrowing Owl	Greater short horned	Greater Sage Grouse	Burrowing Owl
Chestnut-collared	lizard	Greater short horned lizard	Chestnut-collared longspurs
longspur	Greater Sage Grouse	Loggerhead Shrike	Yellow-bellied racers
Eastern Whip-poor-will	Sprauge's Pipit		Greater Sage Grouse
Greater Sage Grouse			Greater short horned lizard
Greater short horned lizard			Great Plains toad
Piping Plover			Loggerhead Shrike
Sprague's Pipit			
Great Plains Toad			
Hairy Prairie Clover			
Loggerhead Shrike			
Pale Yellow Dune			
Moth			
Short-eared Owls			

prairies. Unfortunately, the species has experienced a 23% decline in the last decade with lost of habitat considered to be the primary threat to the owl (EC, 2016a).

5.3. Summary of threats

We reviewed 20 federal recovery documents for species listed on the Species at Risk Act with range in Saskatchewan's Bakken Formation. The documents suggest that oil development impacts 13 species in 4 main ways: habitat loss, oil and noise pollution, invasive species, and road infrastructure. Table 3 categorizes these threats and lists the species that are impacted by each threat based on our document analysis.

5.4. Habitat destruction

Habitat destruction caused by oil development is mentioned in a recovery document for twelve species in this study. Essentially, the federal government—using peer-reviewed science—is acknowledging that oil development is destroying important habitat for species in the Bakken formation. In some cases, like the Greater sage grouse, the connection between habitat and oil/fracking is very clear. In fact, for the Greater sage grouse this link was so clear that in 2013 the federal government invoked the "safety net" clause in SARA to protect the bird inside the provinces of Alberta and Saskatchewan, where it was otherwise not being adequately protected (SAR Public Registry, n.d.). The federal government has used the safety net only twice since the law was enacted in 2003, thus demonstrating how severe the situation is for the Greater sage grouse in the prairies. While the Saskatchewan Wildlife Act does list the bird as endangered, the province has done very little in the form of regulation to protect or recover the Grouse. Instead, the 2014 provincial conservation plan discusses various voluntary management practices that could be implemented and suggests that the best habitat for the sage grouse is in the federal Grasslands National Park (Weiss & Prieto, 2014, p. 14), thereby passing the responsibility back to the federal government.

The relationship between oil development and habitat destruction and fragmentation is less clear in other cases, such as the eastern yellow bellied racer or the long-billed curlew. It is important to point out that habitat destruction (in general) is considered a threat for all species with a recovery strategy in this study. In some cases, oil or energy development is not specifically mentioned, but it could be because no peer review science yet exists for the specific species in Saskatchewan. For example, recent studies in the Bakken area have confirmed that "avoidance of unconventional oil wells and roads" is making habitat loss more severe for many species, such as the Sprague's Pipit, which was found to be more sensitive to oil and gas activity, avoiding them more than some other grasslands birds (Thompson et al., 2015). The case of the short-eared owl or the Mormon Metalark are other examples of species threatened by habitat loss and fragmentation, but where oil has not been directly named as a cause of that habitat loss.

Related, there are data that discuss changes in the landscape that alter species behavior. Very recent scientific studies have illustrated that for grassland songbirds oil and gas structures are an ecological trap, providing structures for the birds but also for their predators (Bernath-Plaisted & Koper, 2016). One study observed more loggerhead shrikes in areas with high levels of oil field development, then in no or low development areas (Fiehler, Cypher, & Saslaw, 2017). Shrikes were observed using power lines as perches, as well as other anthropogenic structures such as fences. The combination of structures with prey availability (insects, lizards) "may facilitate shrike persistence in highly disturbed oil production landscapes" (Fiehler et al., 2017, p. 139).

5.5. Pollution

For oil and noise pollution, the federal government suggests an impact on at least three species. In the case of the greater short-horned lizard mortality from oil spills is explicitly mentioned as well as localized soil contamination. For the Greater Sage Grouse, noise pollution is problematic because it discourages breeding and because species do not want to be near increased noise levels.

However, none of the recovery documents mention water pollution from oil or fracking. This is somewhat surprising given all attention to water pollution in fracking literature (see Lauer, Harkness, & Vengosh, 2016; Olive & Delshad, 2017; Rozell & Reaven, 2011). Peer review literature does suggest that water pollution from oil development poses a threat to grassland birds and other species (Daigh & Klaustermeier, 2016). Trail (2006) addresses the dangers of "oil pits," or open tanks and areas with accessible waste fluids, that can resemble water and trap birds. Some birds may be killed in the pits, or otherwise harmed by "oiling" or exposure to toxic chemicals if they escape (Trail, 2006). In a more recent study, oiling was found to impede flight and cause more frequent stops, which could result in landing in areas of poor food quality—relevant for long-distance migrating birds such as red knot (Perez, Moye, Cacela, Dean, & Pritsos, 2017). While the recovery strategies for species in the Bakken do not mention water pollution, this is an area where more data are needed.

5.6. Invasive species

Invasive species, introduced through oil development and activity, is considered a threat to at least three species in the Bakken, according to recovery strategies. In these cases, trucks and other equipment introduce new species into the area and/or wellpad and road construction creates ideal conditions for invasive species. This points back to the cumulative impact assessment done in southwestern Saskatchewan where Nasen et al. (2011) found that changes to soil structure around well sites can "favor the establishment of fast-growing, weedy, and often invasive plants" (Nasen et al., 2011, p. 201). For sage grouse, short-horned lizards, and loggerhead shrikes changes in habitat are linked directly to invasive species and overall harm.

5.7. Roads and infrastructure

Lastly, roads created by the oil industry are a threat to at least seven listed species in the Bakken. As mentioned, the construction of roads destroys habitat and opens up new areas for invasive species to thrive. But the existence of roads also introduces traffic that can cause collisions with wildlife, such is the case for burrowing owls. Roads can also put prey at risk as it creates unprotected spaces were some species become vulnerable to predators, such as loggerhead shrikes who expose themselves to prey in order to gain access to insects or lizards on roads.

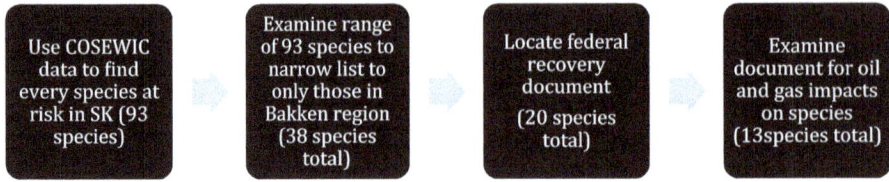

Figure 3. Process used to determine impact of oil development on Saskatchewan species at risk.

6. Limitations and future research

This paper has not considered the link between oil development and climate change, which is another possible driver of biodiversity loss and species at risk in the grasslands. Indeed, oil is double-linked to climate change in the grasslands because when "native grasslands are converted to other land use types, carbon is released contributing significantly to greenhouse gas emissions" (Roch & Jaeger, 2014, p. 2506). Thus, oil production involves both the destruction of a carbon sink as well as the simultaneous production of more greenhouse emissions—through oil production and use as well as through methane production in the flaring of excess natural gas produced through the fracking process (see Carter & Eaton, 2016). Future work could examine recovery documents for links between climate change and grassland species.

Similarly, it would be helpful to break-down the impacts of oil development by stage of the fracking process. Hydraulic fracture is a multi-stage process (see Figure 2 above) and impacts on species would vary across the different stages of the production cycle. For example, the relative contributions of oil production at the site identification stage would be different than the production stage. Namely, landscape changes, road construction, noise, and traffic could be potentially harmful in stage 1, whereas water pollution, oil spills, soil contamination, and traffic could be problematic in stage 5. A systematic analysis of the fracking process and its potential impacts on listed species should be conducted (Figure 3).

Next steps for expanding this research could also include examining peer-reviewed literature and Aboriginal Traditional Knowledge on all 38 listed species at risk. This paper was limited to examining only the 20 species with documents prepared by the federal government. Opening the search up to all existing knowledge on each species would create a more precise understanding of the relationship between oil development and grassland species. Moreover, future work could examine species in the Bakken formation that are not listed as a species at risk by COSEWIC. All species in the grassland ecosystem are important and oil development could be having cumulative impacts on the ecosystem that cannot be discovered by a species-by-species analysis (see Nasen et al., 2011).

Lastly, research must examine why the oil industry in Saskatchewan is not being regulated so as to protect these known species at risk under threat from oil development. This research suggests the federal government knows oil and fracking is a threat to at least 13 species in the Bakken formation. We already know that part of the answer is federalism and the federal government cannot protect species in the Bakken because SARA only extends to federal lands. In southern Saskatchewan there is only one federal park, the Grasslands National Park, which lies at the extreme western end of the Bakken formation. The federal government has created a *Multi-Species Action Plan for Grassland National Park of Canada*, but there is no oil development in the park and, thus, no oil industry that the government could regulate.

From the provincial government's standpoint, it is only illegal to kill or harm the burrowing owl, greater sage grouse, and piping plover on provincial lands or private lands. Those are the only species the Wildlife Act regulates. All the other species discussed in this paper are not legally protected on private lands or provincial crown lands, which comprise most of the land in Saskatchewan. It appears that Saskatchewan's "hands off" regulatory approach to fracking and oil development is also a "hands off" approach to species at risk. More research is needed into the rationale for Saskatchewan's lack of policy as well as the long-term impacts for the grasslands ecosystem. Interviews with government officials, non-government organizations, and scientists should be conducted to gain better insight into the regulatory regime in the province.

Related, regulations and technologies to minimize environmental and ecosystem impact risks should be examined in the context of Saskatchewan's regulatory system. For example, Small et al. (2014) argues that environmental risks can be "mitigated significantly" through technologies and procedures like wastewater recycling, the addition of tracers to injection fluids, methane migration, contamination monitoring programs, and other best practices for the industry (pp. 8290–8291). The extent to which the Saskatchewan government is encouraging these practices is unknown. There use is not discussed in federal recovery documents. However, if oil is impacting species in the Bakken then these solutions, as well as others, are worth serious government consideration and more analysis.

7. Conclusion

The grasslands are one of the most threatened ecosystems on earth and "one of the least protected biomes, with only 4% under protection" (Roch & Jaeger, 2014, p. 2506). Oil development, such as that occurring in the Bakken formation, poses significant potential challenges to native grassland species. The national scientific committee that assesses Canadian wildlife, COSEWIC, designates 38 species at risk—as endangered, threatened or special concern—in the Bakken formation. This determination was made using the best available science, Aboriginal Traditional Knowledge, and local knowledge. The federal government accepted COSEWIC's designation for 28 of these species and placed them on the federal Species at Risk Act. The province of Saskatchewan, using its own Conservation Data Centre, has listed only 3 of the species on its Wildlife Act. Thus, 28 species are being protected by the federal government, 3 species are being protected by both the federal and provincial government, and there are 10 species without protection by either level of government. Moreover, there are potentially more species at risk in southeastern Saskatchewan, but they have yet to be assessed by COSEWIC or the Conservation Data Centre.

While our scientific understanding of the cause–effect relationships between oil and the grasslands is still in its early stages, we do have a plethora of data to suggest that oil development is potentially negatively impacting species at risk. As this paper has illustrated, data collected by the federal government links oil to at least four main threats: habitat destruction, oil and noise pollution, invasive species, and road infrastructure. For 12 of the 20 examined species, one or more of these threats is having a negative impact on the long-term health and viability of the species.

The governments of Canada are failing to adequately protect species from expanding oil development in the Bakken. The oil industry in Saskatchewan is one of the least regulated in North America (Carter & Eaton, 2016). The provincial government offers little in the way of environmental protections (Carter et al., 2017) and has shown limited interest in regulating and protecting wildlife from the oil industry. Despite the federal governments' acknowledgment of the link between oil and species at risk, it can do very little to regulate the oil industry in Saskatchewan. The federal government does not have the jurisdiction to do so. Instead, it can prevent oil development in the Grasslands National Park and other federal wildlife sanctuaries. The other option is for the federal government to invoke the "safety net clause" of SARA for Bakken Species, as it did for the Greater sage grouse in 2013. This would likely be an unpopular move by the federal government, as it does not regularly infringe on areas under provincial jurisdiction. Instead, Saskatchewan needs to step up and do the dirty work of balancing the environment with the oil economy.

Funding
This work was supported by Social Sciences and Humanities Research Council of Canada.

Author details
Andrea Olive[1]
E-mail: andrea.olive@utoronto.ca

[1] Political Science and Geography, University of Toronto Mississauga, 3359 Mississauga Rd, Mississauga, ON L5L 1C6, Canada.

Notes
1. Once assessed, using science, Aboriginal Traditional Knowledge, and local knowledge, a species is determined to be data deficient, not at risk, special concern,

threatened with extinction, endangered with extinction, extirpated from Canada, or extinct in the wild.
2. Vulnerable means a species has low and/or declining numbers but is not threatened or endangered. Threatened means a species is likely to become endangered is the factors causing its decline are not reversed. Endangered means a species is threatened with imminent extirpation or extinction. And extirpated means a species no longer exists in Saskatchewan, but still exists in the wild.

References

Bernath-Plaisted, J., & Koper, N. (2016). Physical footprint of oil and gas infrastructure, not anthropogenic noise, reduces nesting success of some grassland songbirds. *Biological Conservation, 204*, 434–441. https://doi.org/10.1016/j.biocon.2016.11.002

Bird, S. C., & Hodges, K. E. (2017). Critical habitat designation for Canadian listed species: Slow, biased, and incomplete. *Environmental Science & Policy, 71*, 1–8. https://doi.org/10.1016/j.envsci.2017.01.007

Bowden, M. A. (2010). Environmental assessment reform in Saskatchewan: Taking care of business. *Journal of Environmental Law and Practice, 21*, 261–277.

Braun, C. E., Oedekoven, O. O., & Aldridge, C. L. (2002, April). Oil and gas development in western North America: Effects on sagebrush steppe avifauna with particular emphasis on sage-grouse. In *Transactions of the North American Wildlife and Natural Resources Conference* (Vol. 67, pp. 337–349).

Carter, A. V., & Eaton, E. (2016). *Saskatchewan's 'Wild West' approach to fracking.* Canadian Centre for Policy Alternatives. Retrived from https://www.policyalternatives.ca/publications/monitor/saskatchewan's-"wild-west"-approach-fracking

Carter, A., Fraser, G. S., & Zalik, A. (2017). Environmental policy convergence in canada's fossil fuel provinces? Regulatory streamlining, impediments, and drift. *Canadian Public Policy, 43*(1), 61–76. https://doi.org/10.3138/cpp.2016-041

Daigh, A., & Klaustermeier, A. W. (2016). Approaching brine spill remediation from the surface: A new *in situ* method. *Agricultural and Environmental Letters.* doi:10.2134/ael2015.12.0013.

Environment Canada. (2006). *Recovery strategy for the Piping Plover (Charadrius melodus circumcinctus) in Canada. Species at risk act recovery strategy series.* Ottawa: Author.

Environment Canada. (2007). *Recovery strategy for the Dakota Skipper (Hesperia dacotae) in Canada. Species at risk act recovery strategy series* (p. vi + 25). Ottawa: Environment Canada.

Environment Canada. (2012a). *Amended recovery strategy for the Sprague's Pipit (Anthus spragueii) in Canada. Species at risk act recovery strategy series.* Ottawa: Author.

Environment Canada. (2012b). *Recovery strategy for the burrowing owl (Athene cunicularia) in Canada. Species at risk act recovery strategy series.* Ottawa: Author.

Environment Canada. (2013a). *Management plan for the long-billed Curlew (Numenius americanus) in Canada. Species at risk act management plan series.* Ottawa: Author.

Environment Canada. (2013b). *Management plan for the Great Plains Toad (Anaxyrus cognatus) in Canada. Species at risk act management plan series* (pp. Iii–16). Ottawa: Author.

Environment Canada. (2013c). *Amended recovery strategy for the Greater Sage-Grouse (Centrocercus urophasianus urophasianus) in Canada [Proposed]. Species at risk act recovery strategy series* (p. vi + 49). Ottawa: Author.

Environment Canada. (2015a). *Recovery strategy for the Greater Short-horned Lizard (Phrynosoma hernandesi) in Canada. Species at risk act recovery strategy series.* Ottawa: Author.

Environment Canada. (2015b). *Recovery strategy for the Loggerhead Shrike Prairie subspecies (Lanius ludovicianus excubitorides) in Canada. Species at Risk Act Recovery Strategy Series.* Ottawa: Author.

Environment Canada. (2015c). *Recovery strategy for the Eastern Whip-poor-will (Antrostomus vociferus) in Canada [Proposed]. Species at risk act recovery strategy series* (pp. 5–59). Ottawa: Author.

Environment Canada. (2016a). *Management plan for the Short-eared Owl (Asio flammeus) in Canada [Proposed]. Species at risk act management plan series* (p. v + 35). Ottawa: Environment Canada.

Environment Canada. (2016b). *Management plan for the Pale Yellow Dune Moth (Copablepharon grandis) in Canada. Species at risk act management plan series* (p. iii + 21). Ottawa: Environment Canada.

Environment Canada. (2016c). *Recovery strategy for the Common Nighthawk (Chordeiles minor) in Canada. Species at risk act recovery strategy series* (p. vii + 49). Ottawa: Author.

Environment and Climate Change Canada. (2016a). *Recovery strategy and management plan for the Red Knot (Calidris canutus) in Canada [Proposed]. Species at risk act recovery strategy series.* Ottawa: Author.

Environment and Climate Change Canada. (2016b). *Management plan for the Monarch (Danaus plexippus) in Canada. Species at risk act management plan series* (p. iv + 45). Ottawa: Author.

Environment and Climate Change Canada. (2016c). *Recovery strategy for the Chestnut-collared Longspur (Calcarius ornatus) in Canada [Proposed]. Species at risk act recovery strategy series* (p. v + 31). Ottawa: Author.

Environment and Climate Change Canada. (2017). *Management plan for the Hairy Prairie-clover (Dalea villosa) in Canada [Proposed]. Species at risk act management plan series* (p. iv + 40). Ottawa: Author.

Fiehler, C. M., Cypher, B. L., & Saslaw, L. R. (2017). Effects of oil and gas development on vertebrate community composition in the southern San Joaquin Valley, California. *Global Ecology and Conservation, 9*, 131–141. https://doi.org/10.1016/j.gecco.2017.01.001

Fore, S., Overmore, K., & Hill, M. J. (2015). Grassland conservation in North Dakota and Saskatchewan: Contrasts and similarities in protected areas and their management. *Journal of Land Use Science, 10*(3), 298–322.

Government of Canada. (n.d). *Species at risk public registry: A to Z species list.* Retrieved September 6, 2017, from https://www.registrelep-sararegistry.gc.ca/sar/index/default_e.cfm

Government of Saskatchewan. (2014). *Changing how we do business: An introduction to results-based regulations and the Saskatchewan Environmental Code.* Regina: Ministry of Environment. Retrieved May 15, 2017, from http://publications.gov.sk.ca/documents/66/89766-Introduction%20to%20Results%20Based%20Regulations%20and%20the%20Code.pdf

Henwood, W. D. (2010). A strategy for the conservation and protection of the world's temperate grasslands. *Great Plains Research, 20*(1), 121–134.

Hill, M. J., & Olson, R. (2013). Possible future trade-offs between agriculture, energy production, and biodiversity conservation in North Dakota. *Regional Environmental Change, 13*, 311–328. https://doi.org/10.1007/s10113-012-0339-9

Jiang, M., Griffin, W. M., Hendrickson, C., Jaramillo, P., VanBriesen, J., & Venkatesh, A. (2011). Life cycle greenhouse gas emissions of Marcellus shale gas. *Environmental Research Letters, 6*, 3, 034014.

Lade, G. E., & Rudik I. (2017). *Costs of inefficient regulation: Evidence from the Bakken.* https://doi.org/10.3386/w24139

Langdon, J. (2008). Bakken formation: Will it fuel Canada's oil industry. *CBC Online.* Retrieved from http://www.cbc.ca/news/business/bakken-formation-will-it-fuel-canada-s-oil-industry-1.761789

Lauer, N. E., Harkness, J. S., & Vengosh, A. (2016). Brine spills associated with unconventional oil development in North Dakota. *Environmental Science & Technology, 50*(10), 5389–5397. https://doi.org/10.1021/acs.est.5b06349

Mooers, A. O., Doak, D. F., Findlay, C. S., Green, D. M., Grouious, C., Manne, L. L., ... Whitton, J. (2010). Science, policy, and species at risk in Canada. *BioScience, 60*(10), 843–849. https://doi.org/10.1525/bio.2010.60.10.11

Moran, M. D., Cox, A. B., Wells, R. L., Benichou, C. C., & McClung, M. R. (2015). Habitat loss and modification due to gas development in the Fayetteville Shale. *Environmental Management, 55*(6), 1276–1284. https://doi.org/10.1007/s00267-014-0440-6

Nasen, L. C., Noble, B. F., & Johnstone, J. F. (2011). Environmental effects of oil and gas lease sites in a grassland ecosystem. *Journal of Environmental Management, 92*, 195–204. https://doi.org/10.1016/j.jenvman.2010.09.004

National Energy Board. (2016). *Estimated production of canadian crude oil and equivalent.* Retrieved September 26, 2016, from https://www.neb-one.gc.ca/nrg/sttstc/crdlndptrlmprdct/stt/stmtdprdctn-eng.html

Northrup, J. M., & Wittemyer, G. (2012). Characterising the impacts of emerging energy development on wildlife, with an eye towards mitigation. *Ecology Letters, 16*[1], 112–125.

Olive, A. (2014a). *Land, stewardship, and legitimacy: Endangered species policy in the Canada and the United States.* Toronto, ON: University of Toronto Press.

Olive, A. (2014b). The road to recovery: Comparing Canada and US recovery strategies for shared endangered species. *The Canadian Geographer, 58*(3), 263–275. https://doi.org/10.1111/cag.v58.3

Olive, A., & Delshad, A. (2017). Fracking & framing: A comparative analysis of media coverage of yydraulic fracturing in Canadian and US newspapers. *Environmental Communication.* doi: https://doi.org/10.1080/17524032.2016.1275734

Parks Canada Agency. (2010). *Recovery strategy for Eastern Yellow-bellied Racer (Coluber constrictor flaviventris) in Canada. Species at risk act recovery strategy series.* Ottawa: Author.

Perez, C. R., Moye, J. K., Cacela, D., Dean, K. M., & Pritsos, C. A. (2017). Low level exposure to crude oil impacts avian flight performance: The Deepwater Horizon oil spill effect on migratory birds. *Ecotoxicology and Environmental Safety.* [In Press, Corrected Proof].

Petroleum Services Association of Canada. (2018). *Fracking explained.* Retrieved February 7, 2018 from https://oilandgasinfo.ca/all-about-fracking/fracking-explained/

Prenni, A. J., Day, D. E., Evanoski-Cole, A. R., Sive, B. C., Hecobian, A., Zhou, Y., ..., Schurman, M. I. (2016). Oil and gas impacts on air quality in federal lands in the Bakken region: An overview of the Bakken air quality study and first results. *Atmospheric Chemistry and Physics, 16*, 1401–1416. https://doi.org/10.5194/acp-16-1401-2016

Pruss, S. D., Henderson, A., Fargey, P., & Tuckwell, J. (2008). *Recovery strategy for the Mormon Metalmark (Apodemia mormo) Prairie population, in Canada. Species at risk act recovery strategy series* (p. vi + 23). Ottawa: Parks Canada Agency.

Rabe, B. G. (2014). Shale play politics: The intergovernmental Odyssey of American Shale Governance. *Environmental Science and Technology, 48*, 8369–8375. https://doi.org/10.1021/es4051132

Rahm, D., Fields, B., & Farmer, J. L. (2015). Transportation impacts of fracking in the Eagle Ford shale development in rural south Texas: Perceptions of local government officials. *Journal of Rural and Community Development. 10*(2).

Roch, L. & Jaeger, J. A. G. (2014). Monitoring an ecosystem at risk: What is the degree of grassland fragmentation in the Canadian prairie? *Environmental Monitoring and Assessment.* doi:10.1007/s10661-013-3557-9

Rozell, D. J., & Reaven, S. J. (2011). Water pollution risk associated with natural gas extraction from the Marcellus Shale. *Risk Analysis, 32*(8), 1382–1393.

SAR Public Registry. (n.d.). *Information on the Sage grouse recovery strategy and emergency protection order.* Retrieved September 9, 2017 from http://www.registrelep-sararegistry.gc.ca/document/default_e.cfm?documentID=1577

Saskatchewan Conservation Data Centre. (2017). *Species conservation rankings.* Retrieved from http://www.biodiversity.sk.ca/ranking.htm

Savage, C. (2011). *Prairie: A natural history.* Vancouver, BC: Greystone Books.

Shrestha, N., Chilkoor, G., Wilder, H., Gadhamshetty, V., & Stone, J. J. (2017). Potential water resource impacts of hydraulic fracturing from unconventional oil production in the Bakken shale. *Water Research, 108*, 1–24. https://doi.org/10.1016/j.watres.2016.11.006

Small, M. J., Stern, P. C., Bomberg, E., Christopherson, S. M., Goldstein, B. D., Israel, A. L., ... North, D. W. (2014). Risks and risk governance in unconventional Shale gas development. *Environmental Science and Technology, 48*, 8289–8297. https://doi.org/10.1021/es502111u

Thompson, S. J., Johnson, D. H., Niemuth, N. D., & Ribic, C. A. (2015). Avoidance of unconventional oil wells and roads exacerbates habitat loss for grassland birds in the North American great plains. *Biological Conservation, 192*, 82–90. https://doi.org/10.1016/j.biocon.2015.08.040

Trail, P. W. (2006). Avian mortality at oil pits in the United States: A review of the problem and efforts for its solution. *Environmental Management, 38*(4), 532–544. https://doi.org/10.1007/s00267-005-0201-7

United States Geological Survey. (n.d.). *Map of the Bakken formation.* Retrieved from https://www.esrl.noaa.gov/csd/groups/csd7/measurements/2014topdown/

Webb, E., Bushkin-Bedient, S., Cheng, A., Kassotis, C.D., Balise, V., & Nagel, S.C. (2014). Developmental and reproductive effects of chemicals associated with unconventional oil and natural gas operations. *Reviews on Environmental Health, 29* (4), 307–318.

Weiss, M., & Prieto, B. (2014). *A conservation plan for Greater Sage-Grouse in Saskatchewan.* Regina

Wildlife Act. (1998). *c.W-13.12, s.2; 2015, c.27, s.3.*

Zink, V., & Eaton, E. (2016). *Fault lines: Life and landscape in Saskatchewan's oil economy.* Winnipeg: University of Manitoba Press.

3

The role of solar energy for carbon neutrality in Helsinki Metropolitan area

Karna Dahal[1]*, Jari Niemelä[1] and Sirkku Juhola[1,2]

*Corresponding author: Karna Dahal, Department of Environmental Sciences, University of Helsinki, P.O. Box 65, Viikinkaari 2, Helsinki FI-00014, Finland
E-mail: karna.dahal@helsinki.fi

Reviewing editor: Conor Buggy, University College Dublin, Ireland

Abstract: Helsinki Metropolitan area possesses significant solar potential, which can be utilized by installing solar panels and collectors in the cities' public and private premises to fulfill the emission reduction targets. However, current development of solar energy production in the region is in its infancy. This paper outlines how current state of solar energy utilization can be improved in public and private buildings and utility companies in the Helsinki Metropolitan area in terms of costs and financial mechanisms. We applied document analysis and semi-structured interview methods to study the role of solar energy for carbon neutrality in this area. The analyses showed that the Metropolitan cities do not have clear electricity production targets from solar energy yet. Furthermore, their subsidy schemes and financial measures for solar energy productions are not attractive for the promotion of solar energy. Thus, we propose the Metropolitan cities to adopt a policy outlining that a certain percentage (e.g. 20%) of energy should be produced from solar energy to achieve the 20% renewable energy target by 2020. Financial incentives and subsidy schemes for solar power installations should be more tempting and accessible to private and public building owners.

ABOUT THE AUTHOR

Karna Dahal is a PhD candidate in the Department of Environmental Sciences at the University of Helsinki. His research work within the Urban Ecology Research Group (UERG) focuses on exploring new solution-oriented methods and policies for the development of carbon-neutral cities. His areas of expertise are energy technologies, climate policies, and green chemistry. UERG focuses on research and teaching in urban setting and sustainable urban development. This research paper helps to enhance the small-scale solar energy production in the building premises. It is also helpful for promoting solar energy in the dark climatic region and achieve the carbon neutral goal of the cities around the world.

PUBLIC INTEREST STATEMENT

Cities in many countries are identifying the means of cleaner production and sustainable consumption of local energy resources to tackle the climate change and to ensure energy security of their regions. Renewable energy promotion is one way of fighting with the climate change and ensuring energy security of the cities because renewable energy sources are clean and available locally. Thus, cities need to produce renewable energy from various available sources in their local area. This perspective article describes the situation of small-scale solar energy production in the cities of Helsinki Metropolitan area and various economic measures to promote solar energy productions in the buildings in this region. Based on the various document analyses and interviews, it was discovered that implementation of subsidies and attractive financial schemes and attitude changing activities are essential for cities to promote small-scale solar energy in their buildings in the Helsinki Metropolitan area.

Subjects: Environment & Economics; Environment & the City; Environmental Policy; Environmental Politics; Environmental Change & Pollution; Renewable Energy; Building Regulations; Building Techniques; Energy policy and economics; Solar energy

Keywords: carbon neutrality; public and private buildings; subsidy schemes; financial measures

1. Introduction

Cities are currently responsible for generating 60–70% of the global greenhouse gas (GHG) emissions (Kammen & Sunter, 2016). Due to national and international political agreements, cities around the world are obligated to reduce GHG emissions (Bridge, Bouzarovski, Bradshaw, & Eyre, 2013; Fuller, Portis, & Kammen, 2009). In research, focus has been placed on urban climate policies, energy transitions, urban infrastructure regimes, and socio-technical niches to understand how to reduce GHG emissions and establish low carbon or carbon neutral cities (Bulkeley, Castán Broto, & Maassen, 2014; Fuller et al., 2009; Rohracher & Späth, 2014).

The economic and environmental need to transition to a low-carbon economy is one major concern around the world (Fuller et al., 2009). In general, major electricity consumption occurs at buildings in cities and a critical transformation in buildings is needed to fulfill the GHG reduction targets. Approximately 75% of the total global heating demand, which is 40–50% of total energy demand, occurs within the building sector (Wang, Yang, Qiu, Zhang, & Zhao, 2015). Energy transition to low carbon or carbon neutrality requires a shift of energy production from fossil fuels to the cleaner energy sources (Kammen & Sunter, 2016). In addition, decentralization of energy generation can be a core strategy to the promulgation of the low carbon or carbon neutral logic (Bulkeley et al., 2014). Thus, energy production from solar technology becomes important for the cities to reduce significant GHG emissions and fulfill their emissions reduction targets. This also promotes energy security of a region.

Energy transition to renewable energy is an important factor for carbon neutrality. Energy transition is dependent on the geographic situations that cause the niche to evolve, be incorporated into the regimes, and landscape variations (Bridge et al., 2013). Solar energy production is becoming most popular among the renewable energy sources due to the falling prices of photovoltaics (PVs) and technological advancements (Campbell, 2017). Energy production from solar PVs and collectors is also worthwhile in the high-latitude Finnish residencies despite dark winter conditions (Zalpyte, Work Efficiency Institute (TTS), BEF-Latvia, & Hovi, 2013). Currently, Finland has targeted to produce 38% renewable energy mix by 2020 while the City of Helsinki has targeted to produce 20% of its energy from renewable sources by 2020 (Dahal & Niemelä, 2016; International Energy Agency, 2013).

Although several cities have legislated carbon neutral targets to address the climate change, much of the efforts have been focused on technology and policy solutions, with little attention given to how such aspiring goals can be realized through productive financial measures. To attract the city dwellers and energy utilities to install solar panels and thermal collectors at their residential houses and utility facilities, central government, city councils, and the energy authorities should establish several measures which include attractive economic and policy plans. Effectiveness of such measures depends on the roles of the government, city organizations, and utility companies in bringing the solar technology in realization. Thus, economic feasibility of the solar plants is the most important in the implementation process.

Very few studies have been conducted related to the feasibility of solar energy production in building premises in the Helsinki Metropolitan area. In addition, financial measures are missing in the available research and policy papers. This article focuses on two major research questions:

(1) How the current state of solar energy production can be improved in the public and private buildings in the Helsinki Metropolitan area in terms of costs and financial mechanisms?

(2) How solar energy production in the buildings can help cities in the Helsinki Metropolitan area to reach their 20% renewable energy production target by 2020 and carbon neutrality target by 2050.

The paper discusses various financial measures to install solar plants at buildings. To do so, we analyze documents from various sources and conducted interviews by solar energy experts, producers, and users in Helsinki Metropolitan area. Our findings show that Helsinki Metropolitan area possesses significant solar energy potential which can be utilized through various economic measures and motivating programs.

2. Background

Solar energy production in residential and commercial building sector is growing rapidly because of technological and economic advances (Fuller et al., 2009; Wirth, 2017). Real estate are becoming core of cities' energy policies to reduce GHG emissions (Fuller et al., 2009; International Energy Agency, 2014; Wirth, 2017). For instance, several cities in California (United States of America) have set targets for "net zero energy" for new buildings (Fuller et al., 2009). Installing solar plants in buildings also provides benefits such as new source of employment, reduced strain on the electric power system, and development of more comfortable and well-maintained buildings (Fuller et al., 2009). This will also help to establish zero emissions buildings and to create carbon neutral cities or regions.

A small-scale solar energy system has the ability to offer low to zero carbon emissions, offset capital-intensive investments for network upgrades, improve local energy independence and network security, and motivate social capital and cohesion (Kammen & Sunter, 2016). Both the small-scale and utility scale-solar energy productions have been growing slowly in Helsinki Metropolitan area in the recent 2–3 years, although the utility-scale solar energy production has more attractive supports from the Finnish Government than the supports for distributed solar energy production (MTV, 2016). One reason behind this can be recent changes in regulations for energy consumption in buildings, which are considered stricter than they were previously (Hakkarainen, Tsupari, Hakkarainen, & Ikäheimo, 2015).

The Helsinki Metropolitan area situated on the southern coast of Finland (at 60° N) receives about 870 kWh/m^2 insolation annually (European Commission, 2012). With this insolation, solar panels and thermal collectors effectively generate heat and electricity as most solar technologies function well in this solar irradiation level (Laleman, Albrecht, & Dewulf, 2011; Zalpyte et al., 2013). In Finland, long period of daylight in the summer compensates the winter darkness (Motiva, 2016a). Thus, energy production from solar PVs and collectors is also possible at Finnish residencies (Zalpyte et al., 2013). Despite this, only a limited number of private households and public buildings have installed solar panels and thermal collectors in the Helsinki metropolitan area.

Recent modeling Lidar data for solar insolation over roof parts at Helsinki Metropolitan area confirms that significant potential of solar energy exists in the area. It is estimated that a total of 1.93 TWh/year solar potential is available from 17.3569 km^2 area of suitable roofs in the Helsinki Metropolitan area (Kesäniemi & Räsänen, 2017). The City of Helsinki alone has 7.450 km^2 suitable roofs at 37 171 buildings which can yield 809.2 GWh/year of electricity, if all the roofs are covered by solar panels (Kesäniemi & Räsänen, 2017). The solar potential in building roofs in Helsinki area can fulfill 5.4% of total energy demand in the city and about 5–7% total energy demand in the Helsinki Metropolitan area. If the solar energy is produced from building walls, this amount will be largely increased.

A total of 1 984 MWh solar electricity was connected to grid in Helsinki Metropolitan area in 2015, which is a significant increase from 385 MWh in 2012. Two larger solar plants in Suvilahti (275 MWh annual production), and Kivikko (800 MWh annual production) alone produced total of 1 075 MWh electricity enough for 537 one-bedroom apartments (Galkin-Aalto, 2015; Uusitalo, 2015). In addition, energy extraction from buildings to cooling network is as considered solar energy and almost 32 GWh of district heat was produced during the summer 2012 from cooling network in Helsinki and about half of that was estimated to be solar energy (Hakkarainen et al., 2015).

3. Materials and methods

Document analysis and semi-structured interviews methods were used to study the role of solar energy for carbon neutrality in the Helsinki Metropolitan area. Semi-structured interviews consist of several key questions that help to define the areas to be explored, but also allows the interviewer or interviewee to diverge to pursue an idea or response in more detail (Gill, Stewart, Treasure, & Chadwick, 2008). In comparison to structured interviews, this method also allows flexibility on the expansion of information to participants that may not have been previously thought of by the interview team (Gill et al., 2008). A total of eight respondents in the Helsinki Metropolitan area were interviewed individually; one respondent from the energy industry (Helsinki energy company, Helen Limited), two respondents from Helsinki Region Environmental Services Authority (HSY), one respondent from the University of Helsinki, two respondents from the City of Helsinki, one respondent from solar technology providing company (Solnet Green Energy company), and one respondent was local citizen. All these organizations and companies have installed solar panels in their buildings. The local citizen is also a renewable energy expert and he has installed solar panels and thermal plants in his building. Due to the infancy state of solar energy development in the Helsinki Metropolitan area, a relatively small number of respondents were involved. However, the information obtained from these interviewees were beneficial to analyze the documents related to solar energy development scenarios in this region. The names of interviewees and some personal and institutional information related to data were kept confidential.

The main purpose of conducting interviews was to explore the views, experiences, expectations, and motivations of solar plant owners, energy experts, and solar energy and technology providers. The interview data were collected according to several themes used in the interview questions (see Appendix A). The categorized themes were:

- Solar plants owners' motivations and satisfaction with energy production
- Subsidy systems and financial measures
- Procurement processes of solar equipment
- Solar economy and technology situations
- Current problems and improvement measure for solar energy production

These categorized themes used in the interview transcripts were applied to the contents of documents.

Document analysis is a form of qualitative research in which documents are interpreted by the researcher to provide robust understanding around an assessment topic (Glenn, 2009). It also incorporates coding content into themes similar to how interview transcripts are analyzed (Glenn, 2009). Document analysis helped to supplement data from the semi-structured interviews. It provided (1) background and context of solar energy development, (2) information about the factors affecting the promotion of solar energy, (3) additional questions to be asked while conducting interviews, (4) supplementary data for the entire context of the research, and (5) verification of findings from interview data sources. We analyzed various types of documents, such as several solar energy data, periodic or final reports related to solar energy technologies, climate and carbon reduction strategies of the cities, and solar energy production data and assessment documents gained through the interviews. Most of these documents were obtained from the websites of the associated companies and organizations. Documents were also employed to verify and situation analyses of the interviews data.

Both the document analysis and interview methods proved complementary to each other for the collection and analysis of the data related to solar energy and carbon neutrality in the Helsinki Metropolitan area. There are very few policy and research documents related to solar energy in Helsinki Metropolitan area.

4. Results

4.1. Current state of solar energy in Helsinki Metropolitan area

Despite having such a big solar potential in this region, only very small amount (about 2 GWh/year) energy has been extracted from solar plants (Figure 1). Several utilities, companies, and organizations have started installing solar panels recently in their roofs and the grid connected solar energy production has started to increase (Figure 1) but private buildings and housing companies are still behind the progress (Hakkarainen et al., 2015). A comparable proportion of development to solar energy production happened in 4 years in three Metropolitan cities (Helsinki, Espoo, and Vantaa) but the City of Kauniainen does not have any solar energy production yet (Figure 1). The City of Helsinki is leading the progress as it produced 851 MWh in 2015 while the cities of Espoo and Vantaa produced 686 and 417 MWh, respectively, in 2015 (Figure 1). Yet, this amount is negligible in comparison to the available solar potential in the Helsinki Metropolitan area (theoretically over 800 GWh/year). The respondents commented that many people still have the misconception that solar energy production is not feasible in the dark winter climate of Helsinki. However, recent development shows that attitudes towards solar energy production in Helsinki Metropolitan area are changing. People are becoming more climate friendly and take interest in clean technologies (Jung, Moula, Fang, Hamdy, & Lahdelma, 2016). In the same time, the purchasing price for equipment used in solar plant has been lowered (International Renewable Energy Agency, 2016). However, the central government and Metropolitan cities do not have clear electricity production targets from solar energy which is not encouraging for the promotion of solar energy in this region and realization of the carbon neutrality target by 2050.

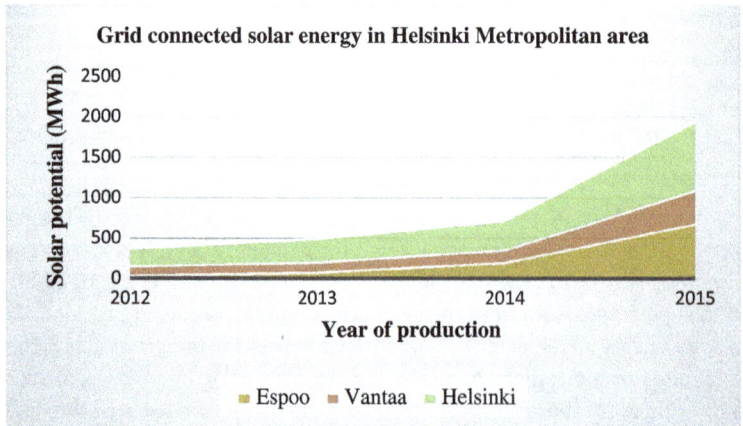

Figure 1. Grid connected solar energy in Helsinki Metropolitan area.

4.2. Solar energy prices and subsidies scenarios

Financial feasibility of solar energy production can be mapped through the investment cost, component prices, technological development, subsidies, taxes and distribution fees. General equation for the production cost, return on investment cost, and energy payback time of solar energy can be expressed as (Auvinen, 2016a):

$$\text{Production cost (€/MWh)} = \frac{\text{Life cycle cost}\left(\frac{€}{\text{MWh}}\right)/m^2, \text{pv or collector/year}}{\text{Total system output}\left(\frac{\text{MWh}}{m^2, \text{pv or collector/year}} \times \text{performance time (year)}\right)}$$

(1)

Life cycle cost includes system investment (€/m²) and operating and maintenance cost (€/m² collector or PV surface) (Auvinen, 2016a). Performance time for the solar energy system is usually 30 years (Auvinen, 2016a). The estimated production cost for a solar power plant of 3.5 MW capacity in general condition is about 16.3 c/kWh in Finland at this time.

$$\text{Annual return on investment (\%)} = \frac{\text{Net yearly energy saving cost (€)}}{\text{Cost of investment (€)}} \times 100\%$$

(2)

Net yearly energy saving cost can be obtained from adding the cost of solar energy production and cost of surplus energy sold in a year and subtracting all taxes and maintenance costs. Energy payback time can be calculated with Equation (3) (Bhandari, Collier, Ellingson, & Apul, 2015).

$$\text{Energy payback time (years)} = \frac{\text{Total investment cost (€)}}{\text{Yearly return on investment cost (€/years)}}$$

(3)

Currently, investment costs rebate on PVs with 25% subsidy is approximately 6.6% annual return for 13 years as per the average prices of the energy in municipalities in Finland (Auvinen, 2016b). This is still quite a long payback time with a small rebate margin of the investment to solar energy because most investors expect to achieve payback within few years of the investment. In addition, payback period for energy production from fossil fuels and some renewable sources are usually short. However, this payback time is not applied to private and public building owners.

Table 1 presents the solar and traditional sources of electricity in Helsinki Metropolitan area. The data excludes online purchase transfer fees, energy taxes, and 24% value added tax (VAT). Solar electricity purchase price with the 25% subsidies is still a bit higher in comparison to the purchase prices for traditional sources of energy.

As the household does not get investment subsidies for solar energy production, solar electricity price becomes much higher than the traditional sources of energy. Currently, small-scale solar electricity production price is about 11 c/kWh with payback time 20 years. The price for solar heat production is also higher in comparison to heat production from fossil fuels. This shows that without subsidies solar heat and power production at residential and public buildings is expensive.

However, one respondent commented that even without the subsidies, current investment on solar energy can be aligned with the average purchase price, levelized cost of energy (LCOE). But, it seems to be correct only if the subsidies are provided to produce solar energy. LCOE is the

Table 1. Heat and electricity purchasing prices in Helsinki Metropolitan area

	Solar electricity price (c/kWh)	Traditional electricity (c/kWh)	Solar heat price (c/kWh) [4–100 m² collectors]	Fossil fuel price (c/kWh)	DH heat price (c/kWh)
Purchasing prices (excludes online purchase transfer fees, energy taxes, and value-added tax (24%))	6.67	4.5–5.8	3.7–9.2	Light fuel oil—6.9 Natural gas—4.6 Hard coal—3.9	5.81

Data sources: Helen Ltd. and Statics of Finland (Auvinen, 2016c; Helen, 2017; Tilastokeskus, 2017).

measurement for the cost of electricity produced by a solar plant in its lifetime which can be compared to the costs for different methods of electricity generations (Lai & McCulloch, 2016). It can be expressed as (Lai & McCulloch, 2016):

$$LCOE = \frac{Lifecycle\ cost(€)}{Lifetime\ energyproduction(kWh)} \qquad (4)$$

The average purchase price LCOE for utilities in Helsinki area is about 7.3–11.6 c/kWh and for small houses and condominium apartments are about 12–18 c/kWh (Auvinen, 2016c). According to the respondents, new solar enterprises have been established and solar promotions activities have been happening time to time. The City of Helsinki and HSY sometimes gather these enterprises to discuss about the solar energy investments.

In terms of investment cost per area for solar plant construction, larger solar plants require smaller investment costs in comparison to smaller plants (Equation 4). For instance, it costs approximately 300 €/m² for smaller and 200 €/m² for larger solar thermal plants constructions (Pasonen, Mäki, Alanen, & Sipilä, 2012). Similarly, system acquisition cost for residential solar plants reaches up to 7.3–11.6 c/kWh but it costs only 3.3–5.3 c/kWh for businesses and municipalities (Motiva, 2016b). The unit price covers the prices for panels, inverter, control, brackets, and wires and installation cost. Usually solar plants' major cost increases due to the secondary or system balance costs. For instance; about 53% of total cost increases due to the installation and additional costs such as cabling, permitting, financing, taxes, marketing, inspection, and interconnection costs (Pasonen et al., 2012). Maintenance cost usually arises between 5 and 10% of total investment cost. It depends on the size of the system, the smaller the system the higher the maintenance cost. For solar thermal plant, control unit, expansion vessel, and heat transfer fluids need to be replaced at least once in the lifetime of the system. The pump usually lasts for the full lifetime of a solar thermal plant (Motiva, 2016b).

The average price for district heat in Finland is about 5.5 c/kWh and it never reaches higher than 8 c/kWh (Pöyry, 2013) which is much lower than the cost for small-scale solar heat production in buildings. Therefore, solar heat production in the buildings is not seen as viable replacement of electricity and light fuel oil in the present pricing situation. Similarly, heat production from solar heating would be viable in the case of fuel crisis, tax exemption on solar energy, high subsidies for investment and installation cost, oil price increases, and price increases on CO_2 allowances (Auvinen, 2016c). However, heat production from larger solar collector systems connected to the district heat network would be beneficial because heat consumption from district heating network is enormous and spontaneous in comparison to typical solar heating network. This provides better return on solar investment but variable costs for replaced heat production also determines the return percentage (Pöyry, 2013). For example, Denmark has exploited its district heating network to integrate profitable heat production from solar (Hakkarainen et al., 2015). In fact, such integrations in the energy system are essential if the cities would like to generate significant energy from renewable sources to become carbon neutral. This will help to reduce further emissions from DH and helpful to fulfill the 20% renewable energy production target by 2020 and carbon neutral by 2050. At present, heat production from solar integrated combined heat and power (CHP) plant is also not feasible in dark climatic regions Finland.

Solar thermal is also profitable with other hybrid heating systems such as solar thermal collectors with electric heating, geothermal, air-water heat pump, radiant fireplace, and bioenergy (pellet boilers) (Hakkarainen et al., 2015). Hybrid solar energy production in Finland as well as Helsinki area is rare but promotions of such energy systems is essential to fulfill 20% renewable energy production target by 2020 and carbon neutral by 2050. It is also profitable for single family houses, especially when it replaces the DH because DH prices for single family houses in summer time becomes € 10/MWh higher in comparison to housing companies (Auvinen, 2016c). In addition, solar heating production businesses for corporate or municipal heating systems can be competitive with the other forms of heat production because such businesses can get investment subsidies and value added tax (VAT) is not liable for purchasing of heat production from solar heating systems (Auvinen & Liuksiala, 2016; Varonen & Myllymäki, 2014). Currently 20% subsidies are given to install solar heat collector plants which can rebate 5–10% on total investment cost during its lifetime production (Auvinen, 2016c). However, 20% subsidy is not available for small-scale solar heat production in building premises at present.

Subsidy systems play an important role in boosting the solar energy production. Several types of subsidy systems can be implemented, such as feed-in-tariff systems, renewable energy subsidies, tax exemption for solar energy production, investment cost subsidies, and subsidies in maintenance costs or inverter replacement costs. Feed-in-tariff system for renewable energy production is quite common in European Union (EU) member countries. Finland has investment support for solar energy productions, but the support is available only for the public sector and companies. High enough feed-in-tariff is required to promote solar energy production. If higher feed-in-tariff is applied, lower payback time is achieved and higher rate of PV installations will be possible. Feed-in-tariff for energy production from renewable energy sources (wind, biogas, and wood fuel) in Finland is 83.5 €/MWh but solar energy is excluded in the Finnish feed-in-tariff system at present (Energy Authority, n.d.).

Yet, some benefits are available to compensate the lack of investment subsidy and feed-in-tariff for small-scale solar energy production in Finland. For instance; energy taxes are not obligated to small-scale solar energy production. Electricity produced for own use (<100 kW or 800 MWh/year) are exempted from tax (Vero, 2016). The household production of PVs smaller than 50 kVA does not need to pay income tax if the maintenance and loss of system value are higher than the value of the electricity sales (Varonen & Myllymäki, 2014). Similarly, the PV power plant installed in the existing utilities and building roofs are not liable to property tax in Finland (Vero, 2016). In addition, small households can also get up to 50% domestic help tax subsidy for the home renovation and installations work which is helpful for solar plant installations (Vero, 2017). However, this domestic help subsidy is limited to € 2,400/per person per year for maximum two persons in a family house with 100 € deduction in it (Vero, 2017). Respondents say that this is, however, a good financial offset for home energy-efficiency improvements and small-scale solar energy production in the private buildings. If the surplus electricity needs to be sold, the producer earns about 2–6 c/kWh from surplus energy sells but it costs about 0.07 c/kWh for transfer fees at present (Auvinen, 2017). Even though such transfer fees are not much applicable in the small-scale solar energy production, it reduces income from the sold surplus electricity. This means the government has an option to revise policies to avoid such electricity transfer fees from small-scale renewable energy production.

Some regulations are different in different municipalities which also impacts on the cost of solar energy production. For example, operational licenses are not required to install the solar plants in Vantaa which reduces the costs and administrative work (MEAE, 2017; PKSRVA. n.d.). However, other municipalities have not eased such building regulations to enhance the production of solar energy.

4.3. Effects of current financial mechanisms and subsidies

Solar energy investment movements dependent on subsidies, energy taxations, and price development for other energy sources (Auvinen, 2016b; Pöyry, 2013). Higher subsidies and lower energy taxes can provide short payback time with higher investment rebate rate. Similarly, cheap system component prices can also reduce the investment cost and lower energy production price. Households, housing associations, and co-operatives are currently excluded from the scope of the Ministry of Employment and Economy's investment aids and feed-in-tariff systems (Pöyry, 2013). This means that all building premises owned by the housing cooperatives and private individuals in the Helsinki Metropolitan area are automatically limited from such energy subsidies. Respondents feel that these immature financial incentive systems discourage the production of household solar energy. If sufficient subsidies are available for them, small-scale solar energy productions would be cost-competitive. But instead, investment subsidy amount has been reduced from 30 to 25% for PV power plants (MEAE, 2017) which – according to the respondents – has discouraged solar energy development.

4.3.1. 25-year analysis for a small-scale PV system

For example, total investment cost for a small PV plant (3.5 kW) with 14 solar panels occupying 20 m^2 installed in a resident's building costs about 6,500 € which produces 2,500 kWh solar energy annually. All produced electricity is consumed in that building. As this small-scale solar energy production does not get investment subsidy but gets domestic help tax subsidy about 5%, current consumer electricity price including taxes is 15 c/kWh. Minimum bank interest rate in the loan is 2%. Assuming price of electricity increase by 1% every year, annual maintenance cost 0.1% and inverter replacing cost 10% for 25 years, payback time would be 17 years. The average production price of solar electricity [LCOE] for 25 years would be 23 c/kWh. Similarly, the solar electricity production price at 25th year becomes 10 c/kWh. Due to the maintenance costs and inverter needs to be replaced in 15th year, the system will bear loss of 1 310 € until 25th year. However, if the current 25% investment subsidy is provided, payback time would be 12 years. The average purchase price of electricity for 25 years would be 18 c/kWh. Similarly, the solar electricity price at 25th year becomes 8 c/kWh. The average profit during 25 years will be about € 1 323. This shows investment subsidies and possible subsidies for maintenance costs and inverter replacement plays important role to make the solar energy profitable. If the bank loan is not taken, the solar electricity cost would become more profitable.

Though income tax leverage is not large, this may influence negatively on installation of larger solar plants. If the domestic electricity is sold, income tax is liable over the value of the electricity sold after deduction of expenses on solar power plants but such expenses are higher than the sold electricity price in Finland at this moment (Vero, 2017). However, household production and consumption of heat and electricity do not need to pay energy taxes in the current legislation. This is a good scheme for enhancing small-scale and household solar energy production.

According to several respondents, a few domestic solar energy producers feel that maintenance cost became too expensive for them. For instance, the solar collector plant installed in 1998–2002 in the EcoViikki area in the City of Helsinki costed a lot for maintenance. The household solar energy production can cost up to 10% of total cost to replace inverters (Hakkarainen et al., 2015). Other annual maintenance cost can be about 1% annually. This increases significantly the life cycle cost of the solar power plant. If subsidies are available for maintenance cost or at least replace the inverter during 15-year period, this would reduce the LCOE as well as cost of the electricity and heat price. If the current 20% investment subsidy is also available for small-scale production, it will encourage building owners and housing associations to install solar thermal plants in their buildings. Providing investment or any other subsidies which are not available for hybrid solar systems currently would also ease the solar energy investments.

One respondent considered that if investment subsidies are provided to home solar installations, people do not want to obtain domestic help tax subsidy due to both subsidies falling under same subsidy mechanism. He also thought that instead of increasing subsidies amount, subsidies should be slowly discontinued as the LCOE price is getting aligned with the other types of energy systems.

There are several possible financial mechanisms to purchase solar power plants such as investment from own capital, loan or cash debt, hire-purchase, financial leasing, and operations lease financing (Auvinen, 2015). Own capital investment is more profitable in comparison to others while loan or cash debt easies on the investment on the installation of solar plants. Own capital evades the costs from interest on the loans. Respondents also thought that loans provided by banks and other finance companies for investment to solar energy can be helpful but they become expensive, especially for private individuals due to the higher end user prices. However, municipalities can be loan guarantees for solar energy investment at the household level. Hire purchases and leasing agreements have the advantages that a municipality or a company can make investments without the need for collaterals. Majority of buildings in the Helsinki Metropolitan area are commercial companies' spaces and municipal offices which can be the advantages for cities and companies in this regard.

Even though the government's energy policy measures for solar energy productions at house-holds and utilities is not looking attractive at present, residents and utility companies are quite satis-fied with their solar energy productions. This is not due to solar plant holders obtaining incentives for solar plant installation and the profits they may earn, but they are quite aware of the green energy systems. This is a good sign for the development of renewable energy production and realization of carbon neutral target.

4.4. Measures to boost solar energy in Helsinki Metropolitan area

With the current energy policies, subsidy system, and the development of investment costs and production rates, PV energy production in Helsinki area cannot reach the renewable energy produc-tion targets (Pasonen et al., 2012). Even though the solar energy production is growing, it is still in its infancy in the Helsinki Metropolitan area. This pace of solar energy production cannot contribute to fulfill renewable energy production targets by 2020 and to establish carbon neutrality in Helsinki Metropolitan area by 2050. However, several measures in terms of energy policies, awareness pro-grams, and attractive financial schemes can be implemented to boost solar energy. Such measures are, e.g. inclusion of solar energy in the real estate planning, profitable solar energy investment planning, municipalities' solar energy investment towards public interest favoring, cooperative par-ticipation of the municipalities in the acquisition of solar energy, and including zero energy building concept in energy policies (Auvinen, 2015). As many buildings and majority of the land are municipal properties in the Helsinki Metropolitan area, municipalities can provide their lands and roofs for solar energy production free of charge for a certain time.

Most respondents in this study were content with the current subsidies scheme but they also think that the government should increase the subsidies to boost solar energy production in Finland and Helsinki Metropolitan area. They think home renovation subsidies and investment subsidies both are essential to make solar energy profitable. Respondents also agree that subsidies are vital for the development of hybrid solar energy production and energy storage batteries both in the residential houses and utility premises. Few respondents think that cash payment for the investment finance is difficult for all the consumers but some energy producers invested in cash. Cash payment reduces the production cost of the solar energy and lowers the purchase price. They also consider easy loan access from banks or other finance companies can help to develop the solar energy productions at private and public buildings. Respondents reiterated that if the cities are obligated to significantly reduce emissions by certain deadlines or carbon neutrality by 2050, small-scale energy production in the public and private buildings is vital. They also agree that several packages of inspiring financial and policy measure are also essential to advance the small-scale solar energy productions in build-ings in the cities in the Helsinki Metropolitan area.

Some current energy policies can be revised to boost solar energy production in buildings. For in-stance, if the energy transmission fee is avoided, it encourages the building owners to produce sur-plus solar energy. The cities in Helsinki Metropolitan area can adopt an energy policy outlining that a certain percentage of energy should be produced from solar energy to achieve the 20% renewable energy targets by 2020. Several other policy measures can support solar energy development in the Helsinki area. For instance, adopting a serious carbon pricing policy can promote the solar energy production. Another way of promoting solar energy production is to regulate quota obligations for electricity distributors and industries so that they should obtain a certain amount of electricity from solar energy and other renewable energy sources. This helps to promote solar energy production in buildings.

Similarly, a respondent says that several solar businesses have been established offering solar panel installations as well as handling of administrative processes for the construction of solar plants in their customers' properties. This includes obtaining permissions, applications for subsidies, and training of solar plant owners for operation of the solar plants. Currently, installation costs vary from one company to another. If the cities direct all the solar businesses and energy companies through common guideline instructions to take minimum amount of installation costs acceptable for both the companies and users, this can be financially rewarding for both the parties. Likewise, municipalities, environmental administrations, and other concerned parties are sometimes organizing seminars and workshops for informing people about the advantages of the solar energy production (Finsolar, 2017; Smart Energy Transition, 2017) but such activities can be performed more frequently. Although the solar energy production is in its infancy, the respondents consider it is one of the positive things happening in the Helsinki Metropolitan area.

5. Discussion

The results show that solar energy production at household level in the Helsinki Metropolitan area can be improved through various incentives and financial measures. The solar energy production is growing progressively along with changes in perception of the residents towards solar energy. Although the current development is in its infancy, lower purchasing price of solar equipment has made solar energy more accessible to individuals and companies. Several utility companies and residents have shown positive attitude towards the production of solar energy in their buildings in recent years.

One reason for the increasing popularity of solar power is that individuals and companies are looking for more climate friendly sources of energy. For instance, several utilities and residents' houses in Helsinki Metropolitan area have started installing both PVs and solar thermal collectors. Solar PVs and thermal plants in present days are more advanced, efficient, and economical which lowers the production and maintenance costs of the solar energy during the entire life time of the solar plants.

It has been observed that solar energy production at private and public buildings can be economically beneficial with subsidies. In addition, as it is clean energy, it significantly reduces emissions which help to achieve renewable energy production goals and fulfill carbon neutrality targets of the cities. In the case of larger scale production, solar energy can also be integrated with other sources of renewable energy systems such as geothermal, bio-energy or heat pump systems. It can also be integrated with the DH system. Such integrated systems are economically viable, techno friendly, and complementary to mainstream energy system (Hakkarainen et al., 2015). It will be helpful in fulfilling the carbon neutral target of the cities. Furthermore, retrofit efforts, such as improving energy efficiency and adding solar PVs and solar thermal systems, need to be expanded in buildings.

Nowadays some solar PVs and collectors as well as auxiliary equipment for installations are manufactured also in Finland (Auvinen, 2016a). Moreover, utilizing local production of community-based digital services which would work without changes in wiring can reduce the total investment cost for solar energy production. Currently, a few institutions and associations have started to promote solar energy production in Finland including Helsinki Metropolitan area (Hakkarainen et al., 2015). A few solar technology providers in Helsinki Metropolitan area are in the process for acquiring the permission from local energy network and municipal permissions for solar plant installations. All these activities are vital to the development of solar energy productions. Yet, solar energy promotion activities are happening at slow pace.

Subsidies are vital for the promotion of solar energy. With the current subsidies and current state of investment costs and production rates, small-scale PV energy production in residential houses and building corporations is not economically competitive as compared to energy production from fossil fuels. However, larger production of solar energy with current 25% investment subsidies have made it inexpensive as compared to small-scale production and comparable with the other sources of energy production. The government can bring attractive subsidy mechanisms such as feed-in-tariff-system as in Germany (12.31 c/kWh for larger production and 6.26 c/kWh for smaller production for 20 years) (Wirth, 2017) to boost small-scale solar energy production in Finland. If the currently distributed 25% investment subsidy is also available for small-scale solar energy production in building premises, it will definitely attract more citizens to install solar plants in their buildings as it becomes profitable energy production. However, solar energy is excluded from the Finnish feed-in-tariff system at present which discourages solar energy investors and citizens to utilize solar energy. Subsidies can also be provided to rental housing which are more often occupied by low-income families (Fuller et al., 2009). This will encourage all income levels of people to produce and consume solar energy from the buildings where they reside. These subsidies are not required for longer period, just to boost the current solar installations pace.

Even though the property tax is exempted for solar plants installations at existing buildings and utilities, it can also be exempted for solar plants installations at the permitted lands to increase the solar energy production. At the moment, household solar electricity is not sold much due to the smaller production but selling of large quantity of household energy require to pay income tax as per the legislation. Taxable income from sold household solar energy is calculated after the deduction of expenses on solar power plants (Varonen & Myllymäki, 2014). This can be a barrier for installation of large solar plants in buildings of housing companies and associations. In addition, subsidies can be available also for maintenance costs because such costs are more than 10% of total investment costs. Private houses can get domestic help tax subsidy for repairing and installations of equipment at buildings. If the investment subsidies are provided to solar plant installations in the private buildings, the fear of losing domestic aid arises among citizens. Home renovation subsidies and investment subsidies both are important to make the solar energy profitable. Thus, the government can provide both investment subsidies as well as domestic aid for renovations and installations to make solar plant installations at private buildings attractive and affordable to residents.

In fact, promoting solar energy productions in housing buildings can also increase the real estate business profits. If the government desires to promote solar energy productions at housing companies and residential property, it can upgrade current energy subsidies systems. It can also revise the energy tax amount so that solar energy along with other renewable energy productions will be increased. As in Germany (Wirth, 2017), government and cities can regulate quota obligations for electricity distributors and other industries to boost the solar energy production in private and public buildings. Due to possessing a substantial solar potential and falling prices of PVs in recent years, solar energy promotion in the Helsinki area becomes a strong alternative.

There are also other financial measures that can be adopted by various entities within the cities and by national government. One such measure can be financial providers such as banks can offer easy loan schemes for solar energy producers. They can bring special financial measures as in Germany and many cities in California and Colorado (USA). Germany provides soft loans up to 2000 €/kW in addition to capital grant up to 25% for the eligible solar PV panels (PV Magazine, 2016). The City of Boulder in Colorado has issued up to $40 million in special assessment bonds at lower costs to finance clean energy improvements (Fuller et al., 2009). The central government or city administrations can also provide capital grants and personal loans at lower costs. Such mechanisms can include paying for the improvements up front, refinancing their mortgages or securing home equity lines of credit, and acquiring personal loans (Fuller et al., 2009). Additional financial measures such as on-bill financing, specialized unsecured bank loans for solar installations and energy-efficiency retrofits, mortgages designed to reward investments in energy efficiency, and other funds such as second mortgages and unsecured personal loans (Fuller et al., 2009). These financial measures are also applicable in the cities in Helsinki Metropolitan area.

If the cities are to significantly reduce emissions by set deadlines and aim at carbon neutrality, small-scale energy productions are vital and cities should adopt innovative financial mechanisms for small-scale clean energy development. For instance, City of Berkeley's program "Berkeley FIRST (Financing Initiative for Renewable and Solar Technology)" has facilitated the residents to generate small-scale solar energy production through various cost-effective financial measures (Fuller et al., 2009). Similarly, City of London has prioritized developing off-grid energy systems which include small-scale solar energy production (Bulkeley et al., 2014). City's cooperation to approve loans to citizens can drive the residential solar energy productions in the right direction. Choosing partners and formulating an appropriate marketing strategy when launching a new solar energy development program are extremely important. Such partnerships between an academic or other analysis group, city councils, and central government officials can facilitate the team needed to overcome the diverse issues that can arise (Fuller et al., 2009).

In addition to subsidies, financial and policy measures, issues related to attitudes are challenging. Although there is increasing interest towards solar energy production, implementation of attitude changing programs can help to enhance solar energy production in the Helsinki area. For instance, city administrations and environmental institutions can repeatedly organize workshops, seminars, webinars, and solar energy demonstrating programs. They can also organize visit packages for citizens to familiarize themselves with existing solar plants. Similarly, such organizations can organize participatory programs for citizens to establish community solar plant installations. The metropolitan cities can also collaborate with other neighboring towns to establish new community solar plants. They can share their knowledge, technology, and other resources with towns.

Proper solar energy production targets have not been set in the cities in the Helsinki area which is vital to achieving the low carbon or carbon neutral goals. For instance, development of solar energy production in London has been one of the key strategies to the emerging low-carbon regime (Bulkeley et al., 2014). The four Metropolitan cities in Finland can adopt a joint climate strategy outlining that a certain percentage (e.g. 20% of total renewable energy production) of energy should be produced from solar energy to achieve the 20% renewable energy targets by 2020 and carbon neutrality by 2050. If the strict climate strategies to promote renewable energy productions are applied, the electricity price will be increased and feasibility of solar PVs will also be improved. PVs will also be feasible option outside the electricity grid as its consumption demand will fits to the small-scale electricity production from PVs.

6. Conclusions

This study shows that solar energy development in the Helsinki Metropolitan area is progressing steadily but current development is in its infancy in comparison to its production capacity. Thus, there is potential for solar energy production to be expanded in the future, which will have positive impact to achieving 20% renewable energy production by 2020 and carbon neutrality goal of the cities in the Helsinki Metropolitan area. This expansion is already taking place despite lack of attractive energy subsidies. Residents and utility companies who have installed solar plants in their buildings are positive towards their investments on solar energy production. Yet, the current financial measures and subsidy mechanisms are not strategic enough and supportive for the development of solar energy production and carbon neutrality target in the Helsinki Metropolitan area. The current subsidy policies restrict the financial incentives to small-scale energy production in the public and private buildings. Investment and production cost for household solar energy is quite high and it is not competitive at this time. In addition, exclusion of investment subsidy and feed-in-tariff for solar energy production has discouraged energy producers and investors to invest into the solar energy development. Without such motivating financial incentives, real estate owners have a significant financial burden to install solar PVs and thermal plants. Current income taxes, lower surplus electricity price and fees for transmission are barrier for the production of larger solar plants in the public and private buildings. For that reason, policies for energy subsidies, taxation system, and electricity purchasing regulations should be revised to promote small-scale solar energy productions in the buildings in the Helsinki Metropolitan area.

Financial assistance for solar power installations should be more appealing and accessible to private and public building owners. The system balances and maintenance costs become larger for small-scale solar energy plants. Thus, financial aids can be available also for both the system balances and maintenance costs. The government can provide both investment subsidies, as well as domestic tax subsidies simultaneously for renovations and installations of solar plants at private buildings, which promote small-scale solar energy productions. The attractive financial schemes such as easy loans, cities' guarantee for loans approval, and establishment of funds for financing to solar plants are very essential. Cities can also adopt a solar energy production target through strategy outlining a certain proportion of energy (e.g. 20%) should be produced from solar energy to achieve renewable energy and carbon neutral targets of the cities. City authorities and the government can provide awareness through workshops, seminars, webinars, and street climate demonstrations to enhance the solar energy production and change the attitudes of people towards solar energy.

Author Contributions
Karna Dahal collected the data, performed the analyses, and drafted the manuscript, Jari Niemelä provided supervision of all stages, edited, and commented on the manuscript, and Sirkku Juhola edited and commented on the manuscript.

Acknowledgments
We would like to thank all the personnel at HSY and City of Helsinki Environmental Centre for assisting us to collect the data. We also thank all the interviewees who provided their lengthy time and useful data for this research work. Similarly, we are so grateful to anonymous reviewers who contributed to improve the quality of this paper.

Funding
The authors received no direct funding for this research.

Competing Interests
The authors declare no competing interest

Author details
Karna Dahal[1]
E-mail: karna.dahal@helsinki.fi
ORCID ID: http://orcid.org/0000-0002-1747-382X
Jari Niemelä[1]
E-mail: jari.niemela@helsinki.fi
Sirkku Juhola[1,2]
E-mail: sirkku.juhola@helsinki.fi
[1] Department of Environmental Sciences, University of Helsinki, P.O. Box 65, Viikinkaari 2, Helsinki FI-00014, Finland.
[2] Department of Built Environment, Aalto University, P.O. Box 11000, Aalto, Espoo FI-00076, Finland.

References
Auvinen, K. (2015). *Rahoitusmallit aurinkoenergiainvestoinneille.* Retrieved April 5, 2017, from http://www.finsolar.net/aurinkoenergian-hankintaohjeita/kuntien-aurinkoenergian-hankinta-ja-rahoitusmallit/
Auvinen, K. (2016a). *Aurinkolämpöjärjestelmien hintatasot ja kannattavuus. Finsolar project.* Retrieved March 19, 2017, from http://www.finsolar.net/kannattavuus/aurinkolampojarjestelmien hintatasot-ja-kannattavuus-suomessa/
Auvinen, K. (2016b). *Romuttaako siirtohinnoittelun tuleva muutos aurinkosähkön kannattavuuden?* Finsolar. Retrieved February 23, 2017, from http://www.finsolar.net/romuttaako-siirtohinnoittelun-tuleva-muutos-aurinkosahkon-kannattavuuden/
Auvinen, K. (2016c). *Aurinkoenergiainvestointien kannattavuuden haasteet.* Retrieved March 13, 2017, from http://www.finsolar.net/aurinkoenergian-hankintaohjeita/aurinkoenergian-tuotantohintoja/
Auvinen, K. (2017). *Aurinkosähkön hyvityslaskentamalli.* Retrieved April 15, 2017, from http://www.finsolar.net/taloyhtiot/hyvityslaskentamalli/
Auvinen, K., & Liuksiala, L. (2016). *Aurinkoenergiainvestointien tuet.* Retrieved March 24, 2017, from http://www.finsolar.net/aurinkoenergian-hankintaohjeita/lait-ja-saadokset/haettavat-tuet-aurinkoenergialle/

Bhandari, K. P., Collier, J. M., Ellingson, R. J., & Apul, D. S. (2015). Energy payback time (EPBT) and energy return on energy invested (EROI) of solar photovoltaic systems: A systematic review and meta-analysis. *Renewable and Sustainable Energy Reviews, 47,* 133–141. https://doi.org/10.1016/j.rser.2015.02.057
Bridge, G., Bouzarovski, S., Bradshaw, M., & Eyre, N. (2013). Geographies of energy transition: Space, place and the low-carbon economy. *Energy Policy, 53,* 331–340. https://doi.org/10.1016/j.enpol.2012.10.066
Bulkeley, H., Castán Broto, V., & Maassen, A. (2014). Low-carbon transitions and the reconfiguration of urban infrastructure. *Urban Studies, 51*(7), 1471–1486. https://doi.org/10.1177/0042098013500089
Campbell, M. (2017). The economics of PV system. In I. A. Reinders, P. Verlinden, W. V. Sark, & A. Freundlich (Eds.), *Photovoltaic solar energy: From fundamentals to applications* (pp. 623–649). New York, NY: Wiley.
Dahal, K., & Niemelä, J. (2016). Initiatives towards carbon neutrality in the Helsinki Metropolitan Area. *Climate, 4*(3), 36. https://doi.org/10.3390/cli4030036
Energy Authority. (n.d.). *Renewable energy. Feed-in-tariff.* Retrieved March 25, 2017, from https://www.energiavirasto.fi/web/energy-authority/feed-in-tariff
European Commission. (2012). *Photovoltaic geographical information system (PVGIS).* Ispra: Joint research Centre (JRC). Institute for Energy, Renewable Energy Unit. Retrieved July 10, from, http://re.jrc.ec.europa.eu/pvgis/
Finsolar. (2017). *Kutsu Smart Solar Growth -seminaarin 18.2.2016.* Retrieved April 8, 2017, from http://www.finsolar.net/tag/seminaari/
Fuller, M. C., Portis, S. C., & Kammen, D. M. (2009). Towards a low-carbon economy: Municipal financing for energy efficiency and solar power. *Environment: Science and Policy for Sustainable Development, 51*(1), 22–33.
Galkin-Aalto, M. (2015). *Kivikko solar panels in high demand.* Retrieved March 6, 2017, https://www.helen.fi/en/news/2015/kivikko-solar-panels-are-in-high-demand/
Gill, P., Stewart, K., Treasure, E., & Chadwick, B. (2008). Methods of data collection in qualitative research: Interviews and focus groups. *British Dental Journal, 204,* 291–295. https://doi.org/10.1038/bdj.2008.192
Glenn, B. A. (2009). Document analysis as a qualitative research method. *Qualitative Research Journal, 9*(2), 27–40.
Hakkarainen, T., Tsupari, E., Hakkarainen, E., & Ikäheimo, J. (2015). *The role and opportunities for solar energy in Finland and Europe* Espoo: VTT Technical Research Centre of Finland Ltd. Retrieved March 3, 2017, from http://www.vtt.fi/inf/pdf/technology/2015/T217.pdf
Helen. (2017). *Electricity products and prices.* Retrieved March 12, 2017, from https://www.helen.fi/en/electricity/homes/electricity-products-and-prices/
International Energy Agency. (2013). *Energy Policies of IEA Countries. Finland. Review.* Retrieved July 7, 2017, from https://www.iea.org/publications/freepublications/publication/Finland2013_free.pdf
International Energy Agency. (2014). *Technological roadmap. Solar photovoltaic energy. 2014 Edition.* Paris: Author. Retrieved February 27, 2017, from https://www.iea.org/publications/freepublications/publication/TechnologyRoadmapSolarPhotovoltaicEnergy_2014edition.pdf
International Renewable Energy Agency. (2016). *The power to change: Solar and wind cost reduction potential to 2025.* Retrieved July 10, 2017, from http://www.irena.org/DocumentDownloads/Publications/IRENA_Power_to_Change_2016.pdf

Jung, N., Moula, M. E., Fang, T., Hamdy, M., & Lahdelma, R. (2016). Social acceptance of renewable energy technologies for buildings in the Helsinki Metropolitan Area of Finland. *Renewable Energy, 99*, 813–824. https://doi.org/10.1016/j.renene.2016.07.006

Kammen, D. M., & Sunter, D. A. (2016). City-integrated renewable energy for urban sustainability. *Science, 352*(6288), 922–928. https://doi.org/10.1126/science.aad9302

Kesäniemi, O., & Räsänen, H.-K. (2017). *Unpulished solar data obtained from Helsinki region Environmental Services Authority (HSY)*. Finland: Helsinki.

Lai, C. S., & McCulloch, M. D. (2016). Levelized cost of energy for PV and grid scale energy storage systems. *Systems and Controls, V1*, 1–1.

Laleman, R., Albrecht, J., & Dewulf, J. (2011). Life cycle analysis to estimate the environmental impact of residential photovoltaic systems in regions with a low solar irradiation. *Renewable & Sustainable Energy Reviews, 15*(1), 267–281. https://doi.org/10.1016/j.rser.2010.09.025

MEAE. (2017). *Ministry of economic affairs and employment. The maximum applicable support (Tuen enimmäismäärät)*. Retrieved April 3, 2017, from http://tem.fi/tuen-enimmaismaarat

Motiva. (2016a). *Toimialueet. Auringonsäteilyn määrä Suomessa*. Retrieved March 5, 2017, from http://www.motiva.fi/toimialueet/uusiutuva_energia/aurinkoenergia/aurinkosahko/aurinkosahkon_perusteet/auringonsateilyn_maara_suomessa

Motiva. (2016b). *Aurinkosähköjärjestelmien hinta*. Retrieved March 20, 2017, from http://www.motiva.fi/toimialueet/uusiutuva_energia/aurinkoenergia/aurinkosahko/jarjestelman_valinta/aurinkosahkojarjestelmien_hinta

MTV. (2016). *Uutiset. Lähes 3000 aurinkopaneelia - Suomen suurin aurinkovoimala käyttöön Helsingin Kivikossa*. Retrieved March 1, 2017, from http://www.mtv.fi/uutiset/kotimaa/artikkeli/lahes-3000-aurinkopaneelia-suomen-suurin-aurinkovoimala-kayttoon-helsingin-kivikossa/5842674

Pasonen, R., Mäki, K., Alanen, R., & Sipilä, K. (2012). *Arctic solar energy solutions*. Kuopio: VTT. Retrieved March 18, 2017 from http://www.vtt.fi/inf/pdf/technology/2012/T15.pdf

PKSRVA. (n.d.). *Rakennusvalvonta Helsinki-Espoo-Vantaa-Kauniainen yhteneäiset htenäiset käytännöt. The City of Vantaa building order*. Retrieved March 29, 2017, from http://www.pksrava.fi/doc/yleiset/rivi_236.pdf

Pöyry. (2013). *Aurinkolämmön liiketoimintamahdollisuudet kaukolämmön yhteydessä Suomessa*. Vantaa: Pöyry Management Consulting Oy. Retrieved March 22, 2017,
from https://tem.fi/documents/1410877/2872337/Aurinkolämmön+liiketoimintamahdollisuudet+kaukol\ämmön+yhteydessä+Suomessa+05072013.pdf

PV Magazine. (2016). *Germany's solar + storage subsidy extended to 2018*. Retrieved April 9, 2017, from https://www.pv-magazine.com/2016/02/22/germanys-solarstorage-subsidy-extended-to-2018_100023314/

Rohracher, H., & Späth, P. (2014). The interplay of urban energy policy and socio-technical transitions: The eco-cities of Graz and Freiburg in retrospect. *Urban Studies, 51*(7), 1415–1431. https://doi.org/10.1177/0042098013500360

Smart Energy Transition. (2017). *Esitykset ja puheenvuorot*. Retrieved April 6, 2017, from http://www.smartenergytransition.fi/fi/hanke/esitykset-ja-luennot/

Tilastokeskus. (2017). *Statistics Finland's PX-Web databases*. Retrieved July 8, 2017, from http://pxnet2.stat.fi/PXWeb/pxweb/en/StatFin/StatFin__ene__ehi/?tablelist=true#_ga=2.198358127.1311833222.1499601717-864856679.1499349423

Uusitalo, S. (2015). *Solar power production started in Helsinki District of Suvilahti*. Retrieved March 7, 2017, from https://www.helen.fi/en/news/2015/solar-power-production-started-in-helsinki-district-of-suvilahti/

Varonen, S., & Myllymäki, J. (2014). *Finnish tax administration (VERO)*. Retrieved March 25, 2017, from https://www.vero.fi/fi-FI/Syventavat_veroohjeet/Henkiloasiakkaan_tuloverotus/Kotitalouden_sahkontuotannon_tuloverotus

Vero. (2016). *Finnish tax administration Energiaverutus*. Retrieved April 4, 2017, from https://www.vero.fi/fi-FI/Syventavat_veroohjeet/Valmisteverotus/Energiaverotus(41357)

Vero. (2017). *Finnish tax administration. Henkiloasiakkaat. Kotitalousvähennys*. Retrieved April 1, 2017, from https://www.vero.fi/fi-FI/Henkiloasiakkaat/Kotitalousvahennys

Wang, Z., Yang, W., Qiu, F., Zhang, X., & Zhao, X. (2015). Solar water heating: From theory, application, marketing and research. *Renewable and Sustainable Energy Reviews, 41*, 68–84. https://doi.org/10.1016/j.rser.2014.08.026

Wirth, H. (2017). *Recent Facts about Photovoltaics in Germany*. Retrieved February 26, 2017, from https://www.ise.fraunhofer.de/content/dam/ise/en/documents/publications/studies/recent-facts-about-photovoltaics-in-germany.pdf

Zalpyte, J., Work Efficiency Institute (TTS), BEF-Latvia, & Hovi, M. (2013). In I. Brēmere, D. Indriksone, & I. Aleksejeva (Eds.), *Energy efficient and ecological housing in Finland, Estonia and Latvia: Current experiences and future perspectives* (pp. 1–108). Riga, Latvia: Energy Efficient and Ecological Housing.

4

Developing CBDR-RC indices for fair allocation of emission reduction responsibilities and capabilities across countries

Izzet Ari[1]* and Ramazan Sari[2,3]

*Corresponding author: Izzet Ari, Ministry of Development, Necatibey Cad., No: 110/A 06100, Yucetepe, Cankaya, Ankara, Turkey
E-mails: iari@dpt.gov.tr, izzetari@gmail.com, izzet.ari@kalkinma.gov.tr
Reviewing editor: Md Younus, University of Adelaide, Australia

Abstract: The aim of this paper is to develop two indices for quantifying common but differentiated responsibilities (CBDR) and respective capabilities (RC) of countries in mitigating climate change. These composite indices can help facilitate fair allocation of GHG emission reduction responsibilities across countries. Indices are formulated by taking into account the economic, environmental, social, and technical indicators of a given country. These indicators are usually highly correlated. An index using these indicators must take this high correlation into account, otherwise it will either over or underestimate the responsibility and the capability of a country. This study takes the correlation between the indicators into account in developing the CBDR and RC indices via the principal components method. However, the novelty of this study arises from measuring and using economic, social, technical, and environmental indicators together in creating the composite indices. The CBDR and RC are constructed for 50 countries that are responsible for at least 81% of global GHG emissions, including OECD countries and emerging economies. The Cluster Analyses

ABOUT THE AUTHORS

Izzet Ari is head of Department of Environment and Sustainable Development at the Ministry of Development of Turkey. He received his PhD from Earth System Science in the international climate change policies. He also received MSc in Climate Change and Policy degree from University of Sussex. He participates in the UNFCCC negotiations on behalf of the Ministry of Development. He also followed post-2015 development regime and actively participates in discussions on SDGs since Rio+20. He currently coordinates the implementation of SDGs in Turkey.

Ramazan Sari is a professor of Economics at Middle East Technical University in the Department of Business Administration, Ankara Turkey. He received his PhD in Economics at Texas Tech University in 2000. His area of interest is Energy Economics, Environmental Economics, Forecasting, and Labor Economics, Economic Growth. His recent works are on social acceptance, energy justice, and consumer behavior issues in energy policy development. He is a co-editor of forthcoming The Routledge Handbook of Energy Economics.

PUBLIC INTEREST STATEMENT

The Paris agreement entered into force in 2016. The Agreement aims to strengthen the global response to the threat of climate change, in the context of sustainable development. The Agreement will be implemented to reflect equity and the principle of common but differentiated responsibilities (CBDR) and respective capabilities (RC). Responsibilities and capabilities are determined as developed and developing countries in the Agreement. However, there is no clear classification for those countries. This article creates a tool to understand the level of responsibility (CBDR) and capability (RC). It was found that definitions of developed and developing countries should not be based on static, already done within the United Nations Framework Convention on Climate Change and historical emissions. Usage of economic, social, technical, and environmental dimensions of development can improve understanding of definitions on developed and developing countries.

are employed to classify the countries according to their CBDR and RC scores. The results suggest revision of current responsibility and capability classifications of the UNFCCC.

Subjects: Development Studies; Sustainable Development; Environment & the Developing World

Keywords: CBDR and RC principle; principal component analysis; cluster analysis

1. Introduction

According to the UNFCCC, countries should protect the global climate system on the basis of equity and in accordance with their common but differentiated responsibilities (CBDR) and respective capabilities (RC) (UNFCCC, 1992). Thus, all countries share a common set of responsibility frame, but the responsibilities are differentiated according to the differing capabilities of each country (Basic, 2007; UNFCCC, 1992). The UNFCCC suggests that emission mitigation efforts should also be taken by non-Annex I countries in line with their capabilities. However, non-Annex I countries have not taken any quantified emission mitigation targets, according to their responsibilities and respective capabilities. While developing countries have blamed the developed countries for their past emissions in the context of historical responsibilities, developed countries have invited developing countries to take serious actions in accordance with CBDR (Dubash, 2009). However, this polarized situation and tension between developed and developing countries hinders reaching the objective of the UNFCCC, and leads to failed climate agreement attempts such as the Copenhagen Accord in 2009 (Dubash, 2009).

At the Conference of Parties (COP) of the UNFCCC in 2011 (COP-17), a new negotiation platform, the *Ad Hoc Working Group on the Durban Platform for Enhanced Action (ADP)*, was established in order to construct a post-2020 international climate regime. ADP opens a venue for developing countries to be involved in emission mitigation, since the ADP's mandate encourages inclusion of all parties without Annex-I/non-Annex I, developed/developing separations (UNFCCC, 2012). It is expected that non-Annex I countries will contribute to emission reduction in the post-2020 climate regime through the ADP process significantly (Aldy & Stavins, 2012).

In 2015, 21st Conference of Parties (COP-21) of the UNFCCC has adopted a new legal document called the Paris Agreement. This Agreement categorizes countries' responsibilities as developed and developing countries rather than Annexes system of the UNFCCC. However, developed and developing countries are not defined in the agreement. This undefined classification might cause problems during the implementation of the Paris Agreement. One of the problem might be about countries' resistance on keeping *state que* rather than ensuring fair burden sharing on emission reduction. Copenhagen Climate Summit in 2009 was a well known example for the similar problem. Lesson learned from the failure of Copenhagen Summit showed that non-Annex countries resisted to take emission mitigation targets even many of them have higher capabilities and responsibilities than Annex-I countries (Ari & Sari, 2015). Besides, they blamed Annex-I countries for their insufficient efforts as required in the UNFCCC and the Kyoto Protocol, and prefer to keep the existing classifications of the UNFCCC. This resistance caused problems of unfair allocation of responsibilities among countries. Thus, the resistance from non-Annex countries and insufficient efforts of Annex-I have led to a failure progress in the climate change negotiations for the post-2012 regime (Dubash, 2009).

In terms of ensuring equity and operationalization of the principle of CBDR-RC, all countries ratification of the Paris Agreement will be an opportunity, but allocation of emission reduction among all countries particularly developed and developing countries should be fair. For instance, in the Article 4 of the Paris Agreement, while developed countries should continue taking the lead by absolute emission reduction targets, developing country should reduce their emissions in the light of different national circumstances (UNFCCC, 2015). In this case, the undefined terms of "developed" and "developing" countries might result in referring back to the Annexes of the UNFCCC (Stavins & Stowe, 2016). Thus, the persisting in the Annex system of the UNFCCC, while implementing the Paris

Agreement, might break the deal of the adopted Agreement even entered into force in 2016. This might cause another problem. It is clear that before the implementation of the Paris Agreement, there is a need for a guidance, rules, and procedure for definition of developed and developing countries.

Besides the necessity of updating the UNFCCC's annexes, the main principles of the UNFCCC, such as CBDR and RC, are needed to be analyzed to find a common and objective ground for revision of the current allocation of responsibilities (Bodansky, 2012; Heyward, 2007). Although many differentiation proposals have been submitted from countries or scholars to find a true position for countries (Bodansky, Chou, & Jorge-tresolini, 2004; Karousakis, Guay, & Philibert, 2008; Wei, Zou, Wang, Yi, & Wang, 2013), none of them directly use the operationalized and quantified CBDR and RC in their differentiation proposals. To better represent and utilize country-specific characteristics, CBDR and RC must be quantified. For the purpose of quantification, composite indices need to be constructed. The main reason behind creating separated responsibility (CBDR) and capability (RC) indices is to determine countries' true responsibility and capability level. For example, a big emitter country is expected to be one of the most responsible party, but the country might not have capabilities to overcome to reduce its emissions. Similarly, a country with low responsibility might assist to other countries according to its capability level.

The aim of this paper is to construct, for first time, composite indices for CBDR and RC in order to differentiate countries' positions. These indices are formed based on selected indicators that represent different dimensions of responsibility and capability. In addition to the commonly used economic and environmental indicators, social and technical indicators are also used in the construction of indices. Social and technical indicators were not quantitatively considered in the literature previously. Technical variables indicate whether countries are capable of climate change mitigation without hindering their economic growth, and social variables indicate perceptions and abilities of societies' for the mitigation policy shift. As emphasized in the latest IPCC report, the causes and effects of climate change are varying in developed and developing countries depending on social capital such as education and awareness level of society, customs and cultural dimensions for social acceptability, ability for collective action and coordination in a country, and degree of equality (Kolstad et al., 2014). Social concerns and opportunities such as education, health care, sustainable environment, and social lives, that are complementary to economic issues, are related to capability, level of development, and quality of life in a country (Sen, 1999). Social variables can explain public perceptions and social transformation abilities in response to mitigation policies. These social variables should be included in the set of determinants of effective environmental policies. Economic and technical indicators alone do not sufficiently reflect mitigation capabilities of a country. Social indicators must be accounted for in assessing potential national contribution to climate change mitigation.

The challenge is the determination of appropriate indicators and methods for computing CBDR and RC indexes. The candidate indicators are highly correlated, thus, a methodology that can overcome this problem should be employed. The Principles Component Analysis (PCA) is employed to eliminate the effects of highly correlated indicators when accounting for the total variation covered by all indicators. The Cluster Analyses (CLA) technique will then serve to classify countries into natural groups based on CBDR and RC index scores. The data are limited to 50 most emitting countries, including all OECD member states, emerging economies such as China, Brazil, South Africa, India, Indonesia, and Malaysia as well as economies in transition countries. The selected 50 countries are responsible for at least 81% of global GHG emissions according to 2010 (WRI, 2014).

2. Conceptual framework

The climate change is a tragedy of the commons problem (Hardin, 1968) since a country's emissions affects the whole globe, and an emission reduction by a country benefits all countries (Kolstad et al., 2014). Besides, climate change is still an important global problem that has not been solved through the allocation of responsibilities based on current annexes of the UNFCCC. To reach the ultimate objective of the UNFCCC, more efforts are required. But the question of "how should effort or burden sharing among countries be done?" raises the difficult concerns of climate equity, justice, fairness, and rights (Kolstad et al., 2014).

2.1. Responsibility

Allocation of responsibility is the most important point of equity in the context of climate change (Ashton & Wang, 2003). The responsibility for climate change has scientific, political, and ethical dimensions (Wei et al., 2013). However, the determination of responsibilities in terms of scientific and moral concerns is quite difficult (Heyward, 2007). Although several proposals for quantifying countries' causal or moral responsibilities have been made (Füssel, 2010), there is no consensus on how to define responsibility (Weisbach, 2010).

Responsibility theories deal with protecting sufferers from others' wrong actions (Weisbach, 2010). The principle of polluter pays, which is based on addressing climate change problem in line with accountability for the consequences of the problem, distributes efforts as a compensation tool by sharing responsibilities in accordance with contribution to the problem (Ashton & Wang, 2003; Heyward, 2007; Rive & Fuglestvedt, 2008). Polluter pays principle is based on the ethical notion that *thou shalt not harm others* or at least not harm others "knowingly" (Mattoo & Subramanian, 2012). In other words, any polluter who is conscious about the damage he/she has caused should have greater responsibility than others who are not aware of the consequences of their actions (Ashton & Wang, 2003).

2.2. Capability

Capability is highly related to the welfare, technology, institutions, skills, information, and opportunities in a society. It is expected that developed countries have more capacity to address the problem (Ashton & Wang, 2003). During the negotiations under the ADP, for instance, many countries' and NGOs' stressed the relationship between commitments and RC (IISD, 2012; Müller & Mahadeva, 2013). The concerns were based on the impossibility of success in the climate change negotiations without appropriately and fairly considering the RC. When considering the importance of GDP in the determination of capability, increasing capacity in terms of per capita GDP is strongly associated with high emissions and high per capita emissions (Aslam, 2002). However, capability is not only related to per capita GDP and per capita emission, but also related to social, technological, and institutional factors (Winkler, Baumert, Blanchard, Burch, & Robinson, 2007). For capability, Müller and Mahadeva (2013) propose a new framework, known as "The Oxford Approach", to measure countries' ability to pay for climate change on the basis of income tax paradigm. Müller and Mahadeva (2013) modify GDP in terms of relative prosperity levels, which are determined by the taxable income, and obtain the gross taxable income based on the general ability to pay. However, they only handle the economic dimension of the RC. The technological and social dimensions should also be integrated into economic dimensions holistically, because un-advanced technological level may hinder the economic growth and/or negative attitude may prevent the necessary social support during the implementation of environmental policies.

The differentiation of commitments and contributions in the light of national circumstances, the principle of CBDR and RC were reiterated in the Paris Agreement. Therefore, the representation of the countries' national circumstances through quantification of related indicators gains a significant importance for reflecting equity and the principle of the CBDR and RC in nationally determined contributions including emission reductions and providing finance. Thus, there is an urgent need to accurately use CBDR and RC to assess the INDC of each country. Therefore, objectively quantified CBDR and RC criteria can be useful for policy-makers and negotiators.

3. Data and methodology

To quantify CBDR and RC and to assess countries' emission mitigation responsibilities, a set of economic, social, technological, and environmental variables are used. By utilizing these indictors, CBDR Index and RC Index are created to compare responsibilities and capabilities of countries. Additionally, countries are classified into groups using the calculated CBDR and RC indices. The scope of the analysis is limited to 50 countries because of availability of data for these countries.

3.1. Creating composite indices

Composite indices provide simple comparisons of countries by creating an index in a wide-ranging field, e.g. environment, economy, society, or technology (Nardo et al., 2008). In the climate change literature, different types of indicators have been selected to explain countries' emissions profile, responsibilities, abilities, and performances to combat climate change. Without considering multidimensional aspects, only emission related and economic indicators might not be sufficient to reflect countries' national circumstances. Therefore, all relevant aspects of responsibility and capability such as economic, emission, social, and technical indicators are attempted to include in this study (Table 1). This section consists of two parts: variables for CBDR and variables for RC.

3.1.1. Variables for CBDR index

Total GHG emissions, per capita GHG emissions, change of per capita GHG emissions, change of GHG emissions per unit GDP, cumulative CO_2 since 1850, cumulative GHG since 1990, consumption-based per capita CO_2 emissions, and sectoral per capita GHG emissions are used as indicators. Besides basic emission indicators such as annual, historical emissions and per capita emissions, sectoral and consumption-based per capita emission indicators are also used. It is assumed that sectoral emissions can provide countries' sector-specific emission intensities. In climate change negotiations, not only economy-wide emission reduction, but also sectoral emission reduction targets for countries have being discussed. For this reason, energy, industry, agriculture, and waste per capita emissions are considered in the calculation of CBDR index. Besides domestic or production-based GHG emissions in a country, some studies emphasize on relations among consumption, trade, and national GHG emissions budgets (Wiedmann, 2009). In the climate policy literature, it is suggested that consumption-based GHG emission accounting should be concerned as a necessary tool (Peters, 2008). Since emissions embodied in trade and consumption are rapidly increasing, there is a growing gap

Table 1. CBDR and RC indicators		
CBDR	Responsibility	GHG emissions
		Per capita GHG emissions
		Consumption-based per capita CO_2 emissions
		Sectoral per capita emissions
		Cumulative emissions since 1850 and 1990
		Change of per capita emissions
		Change of GHG emissions per GDP
		GHG emissions per GDP
RC	Economic	GDP
		Gross savings
		Per capita GDP
	Social	Literacy rate
		Unemployment rate
		Income inequality
		Poverty rate
		Secondary school enrollment
		Number of hospital beds
	Technical	High technological export
		Research and development expenditure
		Renewable energy
		Patent numbers

between production emissions and the emissions associated with consumption (Barrett et al., 2013). The coverage of consumption-based GHG emissions is the inclusion of GHG emission associated with products and services in a country where they are consumed, and exclusion of GHG emissions related to exported products (Csutora & Vetőné mózner, 2014). Therefore, the needs for adoption of consumption-based GHG emissions accounting has started to be discussed in the climate policy domain (Kokoni & Skea, 2013). While creating CBDR index, emission intensity indicators such as change of emission intensity of economy are also used. The change of GHG emissions per unit GDP for a country gives information about the tendency of emission intensity of the economy. For instance, some countries can follow more climate-friendly or less carbon intense growth path; as the change is negative, the country's efforts positively affects to reduce its responsibility. When the intensity is high, it means that the economy has emission intense structure or production style; so its responsibility can increase due to the emission intense economy.

3.1.2. Variables for RC index

RC has economic, technological, and social capability dimensions. Economic capabilities can be represented by GDP, per capita GDP, and the percentage of gross savings in the GDP. Technical capabilities can be based on the share of research and development (R&D) expenditure in GDP, patents number in one million people, the share of high technology exported products in total manufactured products, and the usage of renewable energy sources as primary energy sources in the total energy supplies. For social capabilities, literacy rate, unemployment rate, Gini Coefficient to measure the income inequality, poverty rate, and hospital bed for per 10,000 people can be taken into account. Economic indicators might be considered as the main indictors for determination of capability, however, social and technical dimensions should also be included to ensure multi-dimensionality of capability issue.

As a technical capability indicator, technological and scientific research and development (R&D) are necessary for innovative solutions to combat climate change. According to the IPCC's report, it is almost impossible to stabilize GHG emissions in the atmosphere without R&D (Edenhofer et al., 2014). Related to R&D expenditure and other technical infrastructure of a country, patent number can show the efficiency of R&D expenditure, technological and scientific institutions for being ready to deployment of technologies in the context of mitigation of GHG emissions. To reduce emissions, innovative and high-tech products can play a critical role. Increasing output efficiency in manufacturing industries, replacing old-fashioned production style with more climate-friendly and sustainable production styles, and solving the complicated problems depend on owning of high technological products in a country. High technological production capabilities in any country can also indicate countries' intellectual knowledge and innovations to overcome climate problems. In terms of technical capabilities, there are many types of technologies reducing GHG emissions through alternating fossil fuels with renewable energy sources (Edenhofer et al., 2014). In fact, many studies consider the renewable energy sources as one of the main solutions to mitigate GHG emissions (IEA, 2013; IPCC, 2011).

As it is needed to expand emission mitigation activities to all societies, consumer behavior and basic knowledge are highly critical to follow countries' emission reduction policies at the individual level. Literacy rate in a society can be used as a proxy for countries' capacity deficiency in terms of general knowledge on emission reduction policies. Education is an essential ingredient of a society to combat climate change to raise awareness and realize long-term targets. Transformation of societies from traditional to new development pathway is positively related to the level of education. Education level is a part of social capital (Fleurbaey et al., 2014). In this study, total enrollment in secondary school education is used as one of the measurements of social capability. Improved education level of a country can enlarge the capacity of mitigation to combat climate change (UN, 2014). Similar to education level, the employment rate can be considered as a social capital in the national circumstances; so, the unemployment rate can provide information about country's general socioeconomic situation as well. Number of hospital beds as an indicator can represent the different aspects of social capabilities. This variable, as a component of the health care system, can provide information about the existence and sufficiency of social opportunity in a country (Sen, 1999). A rise in the temperature has negative impacts on human health, water resources, agricultural productivity, and energy use (USGCRP, 2017). Younus (2017a) quantifies impact of extreme climatic events on communities. Creating a weighted index for vulnerability and adaptation levels of societies can assist to evaluate the impact of extreme weather events on local communities and to come up with new policies for reducing future vulnerability (Younus, 2017a). Adaptation to climate change is a very important subject for developing countries because the actions involved are highly essential in improving the resilience of a developing society (Younus, 2017b). Developed countries should assist developing countries in the context of adaptation through proving financial resources, transferring technology, and capacity-building (Younus, 2017b).

Social capabilities and resilience in case of any natural disaster can depend on ability to recover the impacts of the disaster. For instance, heat waves in Europe in 2003 caused many health, agricultural, and infrastructure problems (Schar et al., 2004; Stott, Stone, & Allen, 2004). Income inequality can be considered as a measure of the development level of a country (Baer, Kartha, Athanasiou, & Kemp-Benedict, 2008). It can also provide information about the emission distribution within a country between different income levels (Fleurbaey et al., 2014). Social capability is a part of the development issue. In designating a development threshold, each country's income distribution is an essential indicator. The Gini coefficient indicates the level of social welfare linked to the income distribution, which may have an impact on the capacity of effective integration to environmental policies. Besides, decreasing income inequality is among the Sustainable Development Goals (SDG-10: Reduce inequality within and among countries) to sustain social prosperity. Additionally, poverty which is a kind of shortage of income, causes deprivation of basic capabilities (Sen, 1999). Because climate change is a multi-dimensional issue, and any country with high poverty rate can be less capable to reduce GHG emissions.

3.2. Normalization of indicators

Indicators can have different measuring units and scales, so they should be normalized in a common scale before starting the weighting and aggregation steps of index development (Nardo et al., 2008). There are different types of normalization methods such as min-max, standardization (z-scores), ranking, and categorical scales. Standardization (z-scores) technique is used in this study. In order to obtain each normalized indicator, I_k, the average of X_k, μ, and standard deviation, σ, are used as shown below.

$$I_k = \frac{X_k - \mu}{\sigma} \tag{1}$$

3.3. Principal component analysis

After the normalization of the indicators, the following step is the employment of PCA. The aim of the PCA is to reduce the large number of variables into a smaller number of components that represents most of the variance in the variables (Verma, 2013). PCA attempts to maximize correlation between original variables and new components, and among new components ensures non-correlation and orthogonality (Ul-Saufie, Yahaya, Ramli, Rosaida, & Hamid, 2013). After new components are obtained from the original data, PCA provides a new formula with a new source of variation as formalized below:

$$PC_i = I_{1i}X_1 + I_{2i}X_2 + \ldots + I_{ni}X_n \tag{2}$$

where PC_i is i^{th} principal component and I_{ji} is the loading of the observed variable X_j.

In the PCA, new components are linear combinations of the original variables, and the levels of these new variables are known as eigenvalues. An eigenvalue is a ratio between the common variance and the specific variance explained by a specific factor extracted (Ho, 2006). When all eigenvalues are summed, they are equal to the number of variables (Verma, 2013). Kaiser's Criteria are used to determine the number of eigenvalues. This Criteria suggests that the eigenvalues greater than 1.0 should stay in the analysis (Abson, Dougill, & Stringer, 2012; Verma, 2013). While determining new components, there are a bunch of techniques for rotation including varimax, quartimax, equamax, direct oblimin, and promax. The varimax rotation technique is commonly used in the PCA (Verma, 2013). It is an orthogonal rotation of the component axes to maximize the variance of the squared loadings of a component on all variables in a component matrix (Verma, 2013). The criteria to check the variables in terms of suitability of PCA are The Kaiser–Mayer–Olkin (KMO) sampling adequacy test and Bartlett's sphericity tests (Abson et al., 2012). KMO test result should be higher than 0.500 and Bartlett's sphericity tests should be less than 0.05 for all PCA analyses (Abson et al., 2012).

All CBDR indicators are utilized by PCA to find a number of principal components for calculations of CBDR index. Similar steps are also followed for RC index development. The results of the PCA are expected to provide a number of principal components and weights of each principal component as follows:

$$I_{CBDR} = \sum_{j=1}^{m} \gamma_j PC_j \tag{3a}$$

$$I_{RC} = \sum_{j=1}^{m} \gamma_j PC_j \tag{3b}$$

where,

$$\gamma_j = \frac{eigen\,value_j}{\sum_{j=1}^{m} eigenvalue_j} \tag{4}$$

I_{CBDR} is the CBDR index, I_{RC} is the RC index, γ_j is the weight of jth principal components, and PC is the principal component.

3.4. Cluster analysis

Cluster analysis (CLA) is employed to classify 50 countries based on CBDR and RC index scores. The aim of CLA is to group countries by the similarity of index values, so a group or cluster consists of homogenous and similar variables or subjects (Anderberg, 1973). The similarity is measured by the distance among the cases or objects. The most common and simple measurement of distance method is the Euclidean distance. After the measurement of distances, the linkage procedure of the cases should be selected. There are three different procedures to do this: hierarchical clustering, non-hierarchical clustering, and two-step clustering (Verma, 2013). The hierarchical classification, which can be presented via tree clustering in CLA, is based on distance measurement among cases (Nardo et al., 2008). In order to form a cluster within the hierarchical clustering procedure; there are centroid, variance (Ward), and linkage methods. In the centroid method, the cluster is formed on the basis of the Euclidian distance among cluster centroids; in the variance method, the cluster is formed in order to minimize the within-cluster-variance; and in the linkage method the cluster is formed on the basis of minimum distance between closest/farthest/average of member of clusters (Verma, 2013). In this study, Ward's minimum variance method with squared Euclidian distance measurement technique is used to ensure the most probable homogeneous clusters. Classification and grouping of all 50 countries using CLA can provide appropriate segregation among groups without losing any information in index scores.

4. Results

4.1. Results for CBDR index

Before creating CBDR index, KMO and Bartlett's sphericity tests are conducted for PCA. The findings suggest that the KMO is 0.525, and Barlett's test is smaller than 0.05, so PCA can be applied. The next step is to determine the number of principal components. The PCA results indicate that there are

Table 2. The principal components for CBDR index		
Principal components	Eigenvalues	γ
1	2.912	0.503
2	1.764	0.305
3	1.111	0.192

three principal components, explaining 82.7% of total variation.[1] In order to find weights of each principal component in the CBDR index, ratio of each eigenvalues to total eigenvalues are calculated (Equation (3a)). Thus, the weights of each principal component symbolized by "γ" are listed in Table 2. The sum of principal component weights must be 1.0.

To find CBDR index, each principal component is multiplied by its corresponding weights (γ) (Equation (3a)). The calculated index scores are provided in Table 3. Once an index for each country is calculated, then countries are grouped using CLA methodology (Column 4). The CBDR indices indicate that the USA is the most responsible country, and Philippines is the least responsible country for climate change among the 50 countries. The CLA results suggest that countries in Group-A have the highest, and Group-E the lowest responsibility to mitigate emissions.

Table 3. CBDR index scores and country groups			
Rank	Country	CBDR index	Group
1	USA	2.447	A
2	Australia	1.501	B
3	Canada	1.098	B
4	Russia	0.887	B
5	Saudi Arabia	0.865	B
6	China	0.818	B
7	Estonia	0.548	C
8	New Zealand	0.535	C
9	South Korea	0.426	C
10	Finland	0.405	C
11	Israel	0.348	C
12	Ireland	0.222	C
13	South Afr.	0.165	C
14	Germany	0.162	C
15	Czech Rep	0.147	C
16	Belgium	0.142	C
17	Malaysia	0.104	C
18	Japan	0.066	C
19	Netherlands	0.052	C
20	Singapore	0.039	C
21	Greece	0.031	C
22	Argentina	0.013	C
23	UK	−0.054	D
24	Norway	−0.085	D
25	Denmark	−0.088	D
26	Poland	−0.132	D

(Continued)

Table 3. *(Continued)*			
Rank	**Country**	**CBDR index**	**Group**
27	Austria	−0.138	D
28	Slovenia	−0.163	D
29	Italy	−0.197	D
30	Portugal	−0.225	D
31	Iceland	−0.229	D
32	Bulgaria	−0.230	D
33	France	−0.269	D
34	Mexico	−0.280	D
35	Spain	−0.289	D
36	Slovakia	−0.375	E
37	Malta	−0.389	E
38	Lithuania	−0.437	E
39	Hungary	−0.466	E
40	Brazil	−0.474	E
41	Turkey	−0.485	E
42	Switzerland	−0.505	E
43	Romania	−0.556	E
44	Thailand	−0.572	E
45	Sweden	−0.580	E
46	Latvia	−0.621	E
47	India	−0.708	E
48	Chile	−0.719	E
49	Indonesia	−0.723	E
50	Philippines	−1.035	E

The map illustrated in Figure 1 presents the distribution of countries based on CBDR index scores. There is no visible convergence in terms of institutional perspective such as being an OECD member state or not. Likewise, there is no clear distinction between Annex I and non-Annex I countries of the UNFCCC. Yet, except China and Saudi Arabia, the North–South division still exists in the map. There is no country from Southern part located in Group A. Nevertheless, the Northern countries are not as homogenous as the Southern countries in terms of responsibility issue. For instance, the European countries can be observed in Group B, C, D, and E. Further differentiation among the Northern and the Southern countries are required. It seems that the results indicate a new group emerging in the same cluster, consisting of Canada, Russia, Saudi Arabia, and Australia. The index of these countries is closer to each other, and yet they have different economic levels but they still should take similar emission mitigation commitments. All these results indicate that the current classification based on Annexes or institutional criterion according to the early 1990s may fail to be used in determination of countries' emission mitigation responsibilities. Also, any differentiation referring to the existing grouping criterion such as being an OECD member states or G/77 + China countries highly likely leads to incorrect classification.

4.2. Results for RC index

The test statistics of 0.683 for KMO and 0.05 for Barlett's test indicate the applicability of the PCA. Results suggest that four principal components explain 67.67% of total variation. In order to determine weights of each principal component in the RC index, ratio of each eigenvalues to total eigenvalues are calculated as in Equation (3b). The weights of each principal's coefficient shown with "γ" are presented in Table 4.

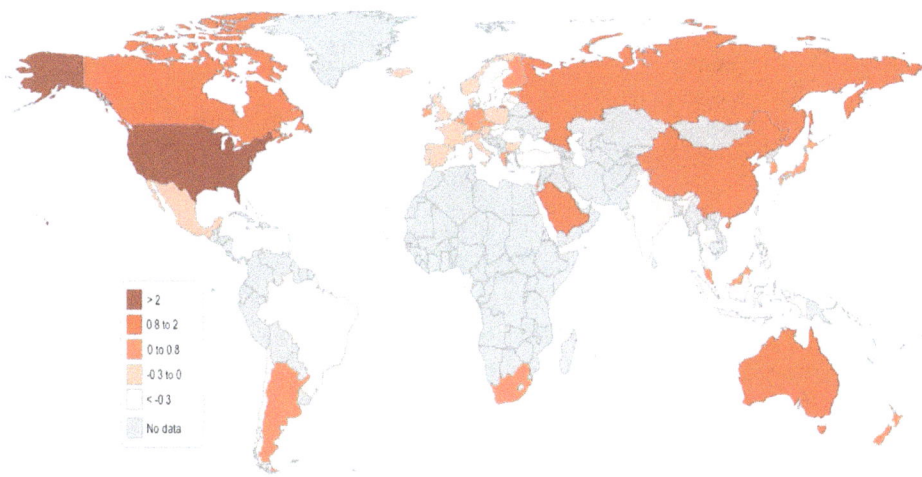

Figure 1. The world map based on CBDR index.

Table 4. The principal component for RC index		
Principal component	**Eigenvalues**	**γ**
1	3.591	0.442
2	2.192	0.267
3	1.320	0.163
4	1.018	0.125

Using Equation (3b), RC indices are calculated and then the CLA methodology is employed to determine country groups, as shown in Table 5. The RC index scores indicate that Singapore is the most capable and South Africa is the least capable country to combat climate change among the 50 countries. The CLA results suggest that there are five groups. In Group A, there are five countries and two of them are non-Annex I countries. On the other hand, in Group-E, there are five countries and only two of them are Annex-I countries (Turkey and Greece).

The world map in Figure 2 illustrates countries based on RC index scores. In terms of continental perspective, while North American Countries such as Canada and the USA are in the same group, it is not the similar case among Latin American countries. The EU countries donot have same capability level. In terms of institutional perspective, the OECD member states do not resemble each other. Also, there is no clear divergence between Annex-I and non-Annex I countries. Except, South Korea, all the Southern countries analyzed in this study have lower capabilities than many Northern countries. It seems that the current classifications of countries may address to the climate change problem when it comes to emission mitigation capabilities of countries.

There are four regions in the Figure 3. Countries in the *first* area with low RC and CBDR index values are not only less responsible, but also have low capability for emission mitigation. Whatever these countries' classifications are in the UNFCCC, their reduction targets should be the lowest among 50 countries selected. Although India is one of the biggest emitter, it has low CBDR and RC index scores. The position of India becomes clearer when compared to first method's indices scores. Countries in the *second* area with high RC and low CBDR are less responsible, but highly prosperous countries. Countries in this area do not need to take high reduction commitments, but they should assist to other countries through providing finance and technology transfer. The *third* area represents countries with high RC and high CBDR. Countries in this area should commit for both emission reduction and finance, technology transfer and capacity-building to other countries. They should also lead in combating climate change. Countries in the fourth area with low RC and high CBDR have high

Handbook of Environmental Science

Table 5. RC index scores and country groups			
Rank	Country	RC index	Group
1	Singapore	1.296	A
2	Japan	1.100	A
3	South Korea	1.005	A
4	Norway	0.794	A
5	Australia	0.748	A
6	Switzerland	0.612	B
7	Finland	0.569	B
8	Iceland	0.543	B
9	Sweden	0.532	B
10	Austria	0.499	B
11	USA	0.490	B
12	Netherlands	0.461	B
13	Germany	0.446	B
14	New Zealand	0.380	B
15	Denmark	0.353	B
16	Canada	0.351	B
17	France	0.350	B
18	Israel	0.220	C
19	Belgium	0.184	C
20	China	0.162	C
21	Ireland	0.127	C
22	Malaysia	0.079	C
23	UK	0.076	C
24	Czech Rep	0.053	C
25	Malta	0.047	C
26	Slovenia	−0.004	C
27	Russia	−0.029	C
28	Hungary	−0.085	C
29	Estonia	−0.151	D
30	Italy	−0.174	D
31	Thailand	−0.213	D
32	Poland	−0.278	D
33	Lithuania	−0.293	D
34	Romania	−0.360	D
35	Latvia	−0.374	D
36	Philippines	−0.401	D
37	Saudi Arabia	−0.417	D
38	Slovakia	−0.453	D
39	Portugal	−0.495	D
40	Spain	−0.498	D
41	Indonesia	−0.510	D
42	Chile	−0.514	D
43	Brazil	−0.534	D
44	Argentina	−0.602	D
45	Bulgaria	−0.629	E
46	Turkey	−0.693	E
47	Greece	−0.747	E
48	Mexico	−0.826	E
49	India	−0.989	E
50	South Afr.	−1.207	E

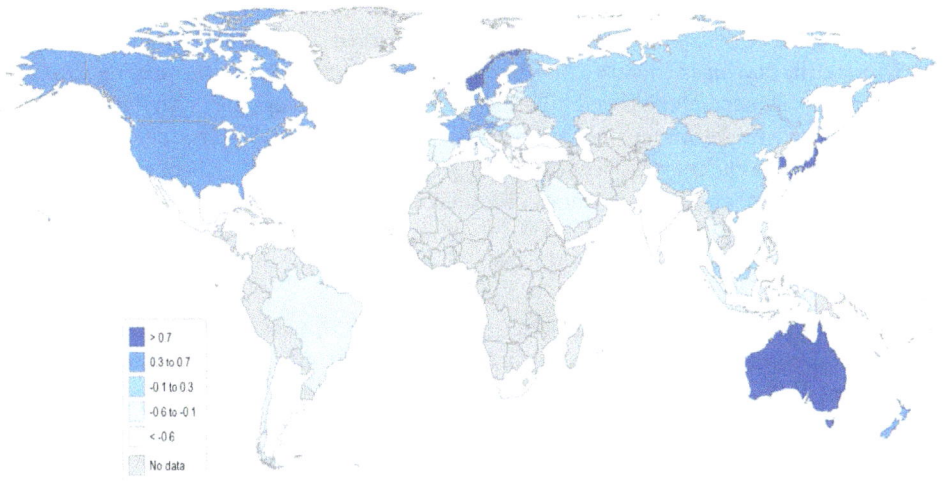

Figure 2. The world map based on RC index.

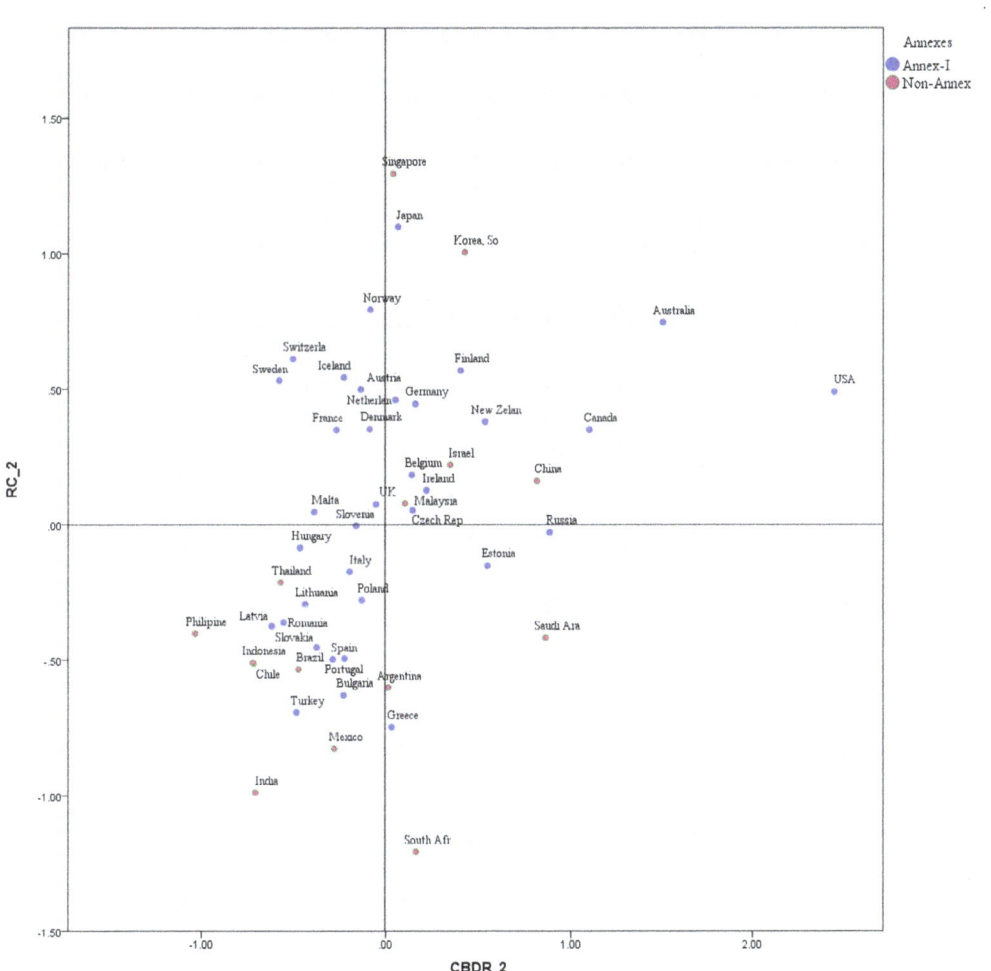

Figure 3. CBDR index vs. RC index.

responsibility and low prosperity. It can be suggested that countries in this group should need external assistance to reduce their emission intensity and emission level until reaching the same prosperity level as owned by developed countries. These countries should also voluntarily make contributions in line with nationally appropriate policies. Besides, all countries in four regions should implement emission reduction activities without compromising their sustainable development needs. As a conclusion, in terms of the current classification of countries under the UNFCCC, there are no concrete differences between Annex-I and non-Annex countries. Thus, there is a need for revisiting the classification of countries in the UNFCCC.

5. Discussions

At the Conference of Parties (COP) of the UNFCCC in 2011 (COP-17), a new negotiation platform, the *Ad Hoc Working Group on the Durban Platform for Enhanced Action (ADP)*, was established in order to construct a post-2020 international climate regime. ADP opens a venue for developing countries to be involved in emissions mitigation, because the ADP's mandate encourages to include all parties without diverging as "Annex-I", "non-Annex I"(UNFCCC, 2012). It is expected that non-Annex I countries will contribute a significant emission reduction in the post-2020 climate regime through the ADP process (Aldy & Stavins, 2012). Between 2011 and 2015, climate change negotiations were discussed in order to reach a comprehensive and inclusive agreement applicable for all countries. Finally, a new legal document called Paris Agreement has been adopted in COP-21 Paris in 2015.

With important and historic moment in the ongoing process of climate negotiations (Bodansky, 2015), according to the Vienna Convention, the Paris Agreement has legal force and provides on the degree of political will behind it (Day et al., 2015). The Agreement also provides an evolution in the global climate change policy and regime in terms of shifting away from categorical binary approach (annex-1 and non-annexes in the UNFCCC) to forms of differentiation as referencing developed and developing countries (C2ES, 2015). Besides, the Agreement creates a flexible framework which covers all countries for burden sharing in emission mitigation responsibilities according to the principle of common but differentiated responsibilities and respective capabilities in light of different national circumstances. In the Agreement, though definitions and classifications of developed and developing countries are not explicitly determined, these country groups' responsibilities are differentiated in operational articles particularly in mitigation and finance (Iddri, 2015). Besides, escaping from Annex-I and non-Annex classification of the UNFCCC during preparation of the Paris Agreement, the Agreement has a paradigm shift from a closed to an open architecture (Iddri, 2015). The closed architecture which had been seen in the Kyoto Protocol, means "top-down" approaches including precise targets and fixed lists of countries (Iddri, 2015). On the other hand, the Paris Agreement is based on "bottom-up" approaches in terms of commitments types including the nationally determined contributions (NDCs), voluntarily participation of all countries, and implementation (C2ES, 2015; Iddri, 2015). In addition, there are no international enforcement mechanisms and any form of penalization for non-compliance (Day et al., 2015), it is seen that the Agreement's particular articles including transparency reflects "top-down rules". Thus, establishment process of the Agreement can be also classified as a "hybrid approach".

In order to reach long-term goals including explicit targets such as keeping the increase in the global average temperature to well below 2°C above pre-industrial levels and efforts to limit the temperature increase to 1.5°C above pre-industrial levels and climate resilience (UNFCCC, 2015), implicitly deep de-carbonization, international and regional cooperation among country parties should be enhanced. For this reason, the success of the Agreement depends on country parties ability to work, evolve, and improve, and strong policy cooperation (Iddri, 2015).

The global cooperation among all countries on the basis of equity and in accordance with their CBDR and RC has already been agreed in the UNFCCC. This study attempts to introduce and develop differentiation tools for both capabilities and responsibilities of countries in the UNFCCC and the Paris Agreement. The tools are the CBDR and RC indices. Having index scores, any country can compare its commitment level with other countries through comparison of index values. Both index values can also be used by countries as a criterion for the determination of commitment efforts.

It is advantageous to build an innovative classification of countries that will allow more inclusive and flexible sharing of responsibilities with a dynamic nature. The differentiation of countries based on quantitative analyses with ethical concerns, particularly the equity dimensions of climate change rather than the political dimension of climate change, is perhaps one way to go. It may also help the construction of common science-based understanding. Creating a composite index involves subjective judgments regarding the selection of indicators, utilization of all data, treatment of missing data, selection of model weights, assessment of uncertainties, transparency, and identification of countries (Dobbie & Dail, 2013; Nardo et al., 2008). In this study, possible sources of uncertainty include quality of indicators, missing data, normalization method, and weighting and aggregation. All these possible sources are taken into consideration for the robustness of both CBDR and RC indices. For normalization, we employed various normalization methods, such as z-scores (standardization), min-max, and ranking. We used z-scores in our analysis to prevent significant variations in rankings of countries. In order to minimize the missing data problem, we collected data from various sources. Data used in this study are acquired from the OECD, the World Bank, the United Nations, and statistics offices of countries. This is intended to reduce the concerns on missing data and the quality of data sources. Selection of indicators for the index might be subjective particularly due to some of the social indicators in the economic capability, such as the unemployment rate. Because we conduct social capability dimension of respective capabilities in the identification of a society's resilience level, we prefer to include unemployment rate to social pillar rather than economic dimensions. The weighting and aggregations are highly subjective while creating an index. Determination of any weighting and aggregation schemes are highly challenging issues. To tackle this challenge, we utilize the principal component analysis (PCA) and use the weights from the principal components of CBDR and RC. For aggregation, additive aggregation (arithmetic mean) methods (Equations (3a) and (3b)) are applied. There is only a small difference between additive aggregation and geometric aggregation results. On the other hand, there is a significant difference between PCA and equal weights. Equal weighting is likely to lead to double counting and misinterpretation of indicators (Freudenberg, 2003), hence we applied PCA method to overcome this problem. There are different base years for selected data. For example, for cumulative greenhouse gas emissions data, there are two base years, namely, 1850 and 1990. The reason for the two different base years is related to discussions on "historical contribution" and "historical responsibility". An indicator having 1850 as base year refers to "historical contribution", while having 1990 refers to "historical responsibility". A strong correlation between "historical contribution" and "historical responsibility" might lead to biased results. The PCA method is expected to eliminate this problem.

6. Recommendation and conclusion
The Paris Agreement includes all countries for emission mitigation. This is a significant opportunity to change developing countries' emission intense production and consumption style. It is known that economic activities of major developing countries and related emissions have been increasing, and they are converging toward Western style consumption, so their exclusion from the emission mitigation responsibility is not in line with the climate justice (Harris, 2010). Until the Paris Agreement, developing countries gained a comparative advantage for being out of the quantified emission targets and hide behind their countries' low per capita income without making any contribution to global cooperation (Harris, 2010). However, after the agreement, developing countries should reduce their emissions in accordance with the principle of CBDR-RC.

After the evaluation of the various differentiation proposals, it can be concluded that the differentiation of countries for developed and developing countries should be based on quantitative analyses with dynamic review. Countries should be periodically updated so that the static position of countries can be prevented. In this manner, essential emission reduction can be globally achieved by including both developed and developing countries. Thus, Paris Agreement's classifications such as developed and developing countries cannot be based on the current annexes of the UNFCCC.

Thus, scores based on CBDR and RC indices suggest that updated annexes of the UNFCCC or new classification of countries might be one of the solution to the equity-based burden sharing. The revision should be considered not only on the mitigation commitments point of view, but also on the provision of necessary support to combat climate change.

The first recommendation is based on the ranking of countries according to the results of CBDR and RC indices. The point is that the higher a country's CBDR rank, the greater its level of responsibility. In this manner, desired and required emissions reduction can be achieved by globally including both developed and developing countries. In this case, all contributions, even very small ones, will be inside the system. This ranking system can be periodically updated.

The second alternative is based on all countries' efforts according to self-differentiation. According to the Paris Agreement, all parties' contributions are expected to combat climate change. This agreed approach can be an opportunity to include all countries' intended nationally determined contributions (INDCs) without losing their economic development. Aggregated contributions from all countries should ensure to expect an emissions reduction explained by the IPCC in the global carbon budget. CBDR and RC indices should be applied to allocate equity-based emissions reduction.

Funding

The authors received no direct funding for this research.

Author details

Izzet Ari[1]
E-mails: iari@dpt.gov.tr, izzetari@gmail.com, izzet.ari@kalkinma.gov.tr
ORCID ID: http://orcid.org/0000-0002-6117-3605
Ramazan Sari[2,3]
E-mail: rsari@metu.edu.tr
[1] Ministry of Development, Necatibey Cad., No: 110/A 06100, Yucetepe, Cankaya, Ankara, Turkey.
[2] Department of Business Administration, Middle East Technical Universities, Cankaya 06800, Ankara, Turkey.
[3] Department of Earth System Sciences, Middle East Technical Universities, Cankaya 06800, Ankara, Turkey.

Note

1. When we increase the number of components the results do not change and the rankings of the countries remain the same.

References

Abson, D. J., Dougill, A. J., & Stringer, L. C. (2012). Using principal component analysis for information-rich socio-ecological vulnerability mapping in Southern Africa. *Applied Geography, 35*, 515–524. doi:10.1016/j.apgeog.2012.08.004

Aldy, J. E., & Stavins, R. N. (2012). *Climate negotiations open a window: Key implications of the durban platform for enhanced action*. Cambridge, MA: Harvard Kennedy School.

Anderberg, M. R. (1973). *Cluster analysis for applications*. New York, NY: Academic Press.

Ari, I., & Sari, R. (2015). The comparison of G-20 countries in terms of climate change indicators. *Energy Dipl Journal, 1*, 14–33.

Ashton, J., & Wang, X. (2003). Equity and climate. In *Principle and practice* (pp. 61–84). Washington, DC: Pew Center on Global Climate Change.

Aslam, M. A. (2002). Equal per capita entitlements: A Key to Global Participation on Climate Change?, In K. Baumert, O. Blanchard, S. Llosa, & J. F. P. (Eds.), *Building on the Kyoto protocol: Options for protecting the climate* (pp. 175–201). Washington, DC: World Resource Institute.

Baer, P., Kartha, S., Athanasiou, T., & Kemp-Benedict, E. (2008). *The greenhouse development rights framework, climate and development*. Heinrich Böll Foundation, Christian Aid, EcoEquity and the Stockholm Environment Institute. Retrieved from https://doi.org/10.3763/cdev.2009.0010

Barrett, J., Peters, G., Wiedmann, T., Scott, K., Lenzen, M., Roelich, K., & Le Quéré, C. (2013). Consumption-based GHG emission accounting: A UK case study. *Climate Policy, 13*, 451–470. doi:10.1080/14693062.2013.788858

Basic. (2007). *The Sao Paulo proposal for an agreement on future climate policy*.

Bodansky, D. (2012). *The durban platform negotiations: Work stream one*.

Bodansky, D. (2015). *Reflections on the Paris conference* [WWW Document]. Retrieved December 23, 2015 from http://opiniojuris.org/2015/12/15/reflections-on-the-paris-conference/

Bodansky, D., Chou, S., & Jorge-tresolini, C. (2004). *International climate efforrts beyond 2012: A survey of approaches*. Washington, DC: Pew Center on Global Climate Change.

C2ES. (2015). *Outcomes of the UN climate change conference in Paris* [WWW Document]. Retrieved December 20, 2015 from http://www.c2es.org/international/negotiations/cop21-paris/summary

Csutora, M., & Vetőné mózner, Z. (2014). Proposing a beneficiary-based shared responsibility approach for calculating national carbon accounts during the post-Kyoto era. *Climate Policy, 14*, 599–616. doi:10.1080/14693062.2014.905442

Day, T., Röser, F., Tewari, R., Kuramochi, T., Warnecke, C., Hagemann, M., … Höhne, N. (2015). *What the Paris agreement means for global climate change mitigation* [WWW Document]. Retrieved December 22, 2015 from http://newclimate.org/2015/12/14/what-the-paris-agreement-means-for-global-climate-change-mitigation/

Dobbie, M. J., & Dail, D. (2013). Robustness and sensitivity of weighting and aggregation in constructing composite indices. *Ecological Indicators, 29*, 270–277. doi:10.1016/j.ecolind.2012.12.025

Dubash, N. K. (2009). Copenhagen: Climate of mistrust. *Economic and Political Weekly, xliv*, 8–11.

Edenhofer, O., Pichs-Madruga, R., Sokona, Y., Kadner, S., Minx, J. C., Brunner, S., ... Zwickel, T. (2014). Technical summary. In: Climate change 2014: Mitigation of climate change. Contribution of working group III to the Fifth assessment report of the intergovernmental panel on climate change. In *Climate change 2014: Mitigation of climate change*. Cambridge: Cambridge University Press.

Fleurbaey, M., Kartha, S., Bolwig, S., Chee, Y. L., Chen, Y., Corbera, E., ... Sagar, A. (2014). Sustainable development and equity. In *Climate change 2014: Mitigation of climate change. Contribution of working group III to the fifth assessment report of the intergovernmental panel on climate change, in: climate change 2014: mitigation of climate change* (pp. 283–350). Cambridge: Cambridge University Press.

Freudenberg, M. (2003). *Composite indicators of country performance: A critical assessment.* https://doi.org/10.1787/18151965

Füssel, H.-M. (2010). How inequitable is the global distribution of responsibility, capability, and vulnerability to climate change: A comprehensive indicator-based assessment. *Global Environmental Change, 20*, 597–611. doi:10.1016/j.gloenvcha.2010.07.009

Hardin, G. (1968). The tragedy of the commons. *Science (80-.), 162*, 1243–1248. https://doi.org/10.1126/science.162.3859.1243

Harris, P. G. (2010). *World ethics and climate change: From international to Glbal justice.* Edinburg: Edinburg University Press.

Heyward, M. (2007). Equity and international climate change negotiations: A matter of perspective. *Climate Policy, 7*, 518–534. https://doi.org/10.1080/14693062.2007.9685674

Ho, R. (2006). *Handbook of univariate and multivariate data analysis and interpretation with SPSS.* New York, NY: Taylor & Francis. https://doi.org/10.1201/9781420011111

Iddri. (2015). *The Paris agreement: Historic! what's next?* [WWW Document]. Retrieved December 22, 2015 from http://www.blog-iddri.org/2015/12/14/the-paris-agreement-historic-but-whats-next/

IEA. (2013). *Renewables information-2013.* Paris Cedex.

IISD. (2012). Summary of the Doha climate change conference: 26 November–8 December 2012.

IPCC. (2011). *IPCC special report on renewable energy sources and climate change mitigation, choice reviews online.* Cambridge: The Intergovernmental Panel on Climate Change (IPCC). https://doi.org/10.5860/CHOICE.49-6309

Karousakis, K., Guay, B., & Philibert, C. (2008). *Differentiating countries in terms of mitigation commitments, actions and support.* IEA: OECD.

Kokoni, S., & Skea, J. (2013). Input-output and life-cycle emissions accounting: Applications in the real world. *Climate Policy, 14*, 1–25. doi:10.1080/14693062.2014.864190

Kolstad, C., Urama, K., Broome, J., Bruvoll, A., Olvera, M.C., Fullerton, D., ... Mundaca, L., 2014. Social, economic and ethical concepts and methods. In *Climate change 2014: Mitigation of climate change. Contribution of working group III to the fifth assessment report of the intergovernmental panel on climate change, in: Climate change 2014: Mitigation of climate change* (pp. 207–282). Cambridge: Cambridge University Press.

Mattoo, A., & Subramanian, A. (2012). Equity in climate change: An analytical review. *World Development, 40*, 1083–1097. doi:10.1016/j.worlddev.2011.11.007

Müller, B., & Mahadeva, L. (2013). *The oxford approach: Operationalising "respective capabilities".* European Capacity Building Initiative. https://doi.org/10.26889/9781907555688

Nardo, M., Saisana, M., Saltelli, A., Tarantola, S., Hoffmann, A., Giovanni, E. (2008). *Handbook on constructing composite indicators.*

Peters, G. P. (2008). From production-based to consumption-based national emission inventories. *Ecological Economics, 65*, 13–23. doi:10.1016/j.ecolecon.2007.10.014

Rive, N., & Fuglestvedt, J. (2008). Introducing population-adjusted historical contributions to global warming. *Global Environmental Change, 18*, 142–152. doi:10.1016/j.gloenvcha.2007.09.004

Schar, C., Vidale, P. L., Luthi, D., Frei, C., Haberli, C., Liniger, M. A., & Appenzeller, C. (2004). The role of increasing temperature variability in European summer heatwaves. *Nature, 427*, 3926–3928. doi:10.1038/nature02230.1

Sen, A. (1999). *Development as freedom*, 1st ed. Anchor Press.

Stavins, R. N., & Stowe, R. C. (2016). *The Paris agreement and beyond: International climate change policy post-2020.* Cambridge: Harvard Project on Climate.

Stott, P. A., Stone, D. A., & Allen, M. R. (2004). Human contribution to the European heatwave of 2003. *Nature, 432*, 2–6. doi:10.1029/2001JB001029

Ul-Saufie, A. Z., Yahaya, A. S., Ramli, N. A., Rosaida, N., & Hamid, H. A. (2013). Future daily PM10 concentrations prediction by combining regression models and feedforward backpropagation models with principle component analysis (PCA). *Atmospheric Environment, 77*, 621–630. doi:10.1016/j.atmosenv.2013.05.017

UN. (2014). *Open working group proposal for sustainable development goals.*

UNFCCC. (1992). *United nations framework convention.*

UNFCCC. (2012). *Decisions adopted by the COP-17.* Author.

UNFCCC. (2015). *Adoption of the Paris agreement* [WWW Document]. Retrieved December 14, 2015 from http://unfccc.int/resource/docs/2015/cop21/eng/l09r01.pdf

USGCRP. (2017). Climate Science Special Report: Fourth National Climate Assessment. D. J. Wuebbles, D. W. Fahey, K. A. Hibbard, D. J. Dokken, B. C. Stewart, & T. K. Maycock (Eds.), *U.S. Global Change Research Program* (vol. I, p. 470). Washington, DC, USA. doi:10.7930/J0J964J6

Verma, J. P. (2013). Data analysis in management with SPSS software. Springer India. doi:10.1007/978-81-322-0786-3

Wei, Y.-M., Zou, L., Wang, K., Yi, W., & Wang, L. (2013). Review of proposals for an agreement on future climate policy : Perspectives from the responsibilities for GHG reduction. *Energy Strategy Reviews, 2*, 161–168. doi:10.1016/j.esr.2013.02.007

Weisbach, D. (2010). *Negligence, strict liability, and responsibility for climate change.* Cambridge, MA: Harvard Kennedy School.

Wiedmann, T. (2009). Editorial: Carbon footprint and input-output analysis – An introduction. *Economic Systems Research, 21*, 175–186. doi:10.1080/09535310903541256

Winkler, H , Baumert, K., Blanchard, O., Burch, S., & Robinson, J. (2007). What factors influence mitigative capacity? *Energy Policy, 35*, 692–703. https://doi.org/10.1016/j.enpol.2006.01.009

WRI. (2014). *Climate analysis indicators tool (CAIT 2.0)* [WWW Document]. GHG Emissions. Retrieved December 27, 2014 from http://cait2.wri.org/wri/Country

Younus, M. A. F. (2017a). An assessment of vulnerability and adaptation to cyclones through impact assessment guidelines: A bottom-up case study from Bangladesh coast. *Natural Hazards, 89*, 1437–1459. doi:10.1007/s11069-017-3027-8

Younus, M. A. F. (2017b). Adapting to climate change in the coastal regions of Bangladesh : Proposal for the formation of community-based adaptation committees. *Environmental Hazards, 16*, 21–49. doi:10.1080/17477891.2016.1211984

5

Wastewater treatment potential of *Moringa stenopetala* over *Moringa olifera* as a natural coagulant, antimicrobial agent and heavy metal removals

Asaminew Abiyu[1]*, Denghua Yan[2], Abel Girma[1], Xinshan Song[1] and Hao Wang[2]

*Corresponding author: Asaminew Abiyu, College of Environmental Science and Engineering, Donghua University, Shanghai, China
E-mail: asaminewab@yahoo.com

Reviewing editor: Murat Eyvaz, Gebze Technical University, Turkey

Abstract: Moringa is a multipurpose tree with considerable economic and social potential and its cultivation is currently being actively promoted in many developing countries. Seeds of this tropical tree contain water-soluble, positively charged proteins that act as an effective coagulant for water and wastewater treatment. This study evaluated the effectiveness of *Moringa oleifera* and *Moringa stenopetala* seed powder in water purification as a replacement coagulant. Water treatment with *M. stenopetala* was found to be more effective for water purification than treatment with *M. oleifera* seed. Indeed, it has been given little research and development attention. Unlike *M. oleifera*, little scientific research has been conducted on the properties and potential uses of *M. stenopetala* in general and its seeds in particular. However, the method should be encouraged in communities without safe water supply.

Subjects: Freshwater Biology; Water Quality; Water Science

Keywords: Moringa; *Moringa oleifera*; *Moringa stenopetala*; coagulant; purification water

ABOUT THE AUTHORS

Asaminew Abiyu is a Moringa Program Coordinator in the Hunger project Ethiopia International based NGO for the last 2 years. Asaminew received the Mission for change and development association in Ethiopia (MCDAE) award for his contribution and successful completion of the management and leadership courses in the year 2015. Asaminew holds an MSc degree in Environmental Science form Addis Abeba University Ethiopia, in the year 2013 and a BSc degree in Production Forestry form Hawassa University, Wondo Genet College of Forestry and Natural Resources, 2006 and BTH in Theology from Holy Trinity Theological University College, 2013 Addis Ababa, Ethiopia. Asaminew's objective is to seamlessly merge different projects with applied research in the service of communities. His research interest focuses on the complex connectivity of human and environmental relations, addressing indigenous ways of knowing, Green Energy, waste treatment, climate change and Hydrology. And now he is a PhD student at Donghua University, School of Environmental Science and engineering.

PUBLIC INTEREST STATEMENT

Moringa stenopetala is one of the facultative deciduous trees having multipurpose use. *M. stenopetala* has potential benefits for many purposes such as ecological, nutritional, medical, and industrial. It is also a good source of generating income to the rural poor. Various findings on the medicinal use of *M. stenopetala* support the traditional claim of the plant particularly its indigenous use for chronic diseases. The leaves are nutritious due to the presence of important micronutrients.

This review has focuses to address wastewater treatment potential of *M. Stenopetala* over *Moringa oleifera* as a natural coagulant, antimicrobial agent and heavy metal removals. Besides this review evaluates the effectiveness of *M. oleifera* and *M. stenopetala* seed powder in water purification as a potential replacement of coagulants. Water treatment with *M. stenopetala* is found to be more effective and efficient for water purification than *M. oleifera* seed powder.

1. Introduction

Water is one of the most vital natural resources for all life on Earth. The availability and quality of water have always played an important part in determining not only where people can live, but also their quality of life. Even though there always has been plenty of fresh water on Earth, water has not always been available when and where it is needed, nor is it always of suitable quality for all uses. Water must be considered as a finite resource that has limits and boundaries to its availability and suitability for use.

Lack of potable water is a huge problem and a major cause of death and disease in the world. According to WHO/UNICEF (2012), 783 million people worldwide are without improved drinking water, and the World Health Organization estimates that lack of proper drinking water causes 1.6 million deaths each year from diarrheal and parasitic diseases. In many parts of the world river water which can be highly turbid is used for drinking purposes. This turbidity is conventionally removed by treating the water with expensive chemicals; many countries must import expensive chemicals to clarify the water, limiting the amount they can afford to produce these imported chemicals with a great expense.

Conventional drinking water treatment includes, but is not limited to: coagulation, flocculation, sedimentation, filtration, and disinfection. Coagulation and filtration are the most critical unit processes (other than disinfection) determining success or failure of the whole system and they are the bottlenecks for upgrading treatment plants. The two units are so closely linked that the design of one affects the other. When they are well designed and operated, other units, such as flocculation and sedimentation, may not be required (Conley, 1961) and the burden on disinfection is significantly reduced.

Based on this, purification of drinking water is the process of removing undesirable chemicals, biological contaminants, suspended solids and gases from contaminated water. The goal is to produce water fit for a specific purpose. Most water is disinfected for human consumption (drinking water), but water purification may also be designed for a variety of other purposes, including fulfilling the requirements of medical, pharmacological, chemical and industrial applications. The methods used include physical processes such as filtration, sedimentation, and distillation; biological processes such as slow sand filters or biologically active carbon; chemical processes such as flocculation and chlorination and the use of electromagnetic radiation such as ultraviolet light.

Purifying water may reduce the concentration of particulate matter including suspended particles, parasites, bacteria, algae, viruses, fungi, as well as reducing the amount of a range of dissolved and particulate material derived from the surfaces that come from runoff due to rain. Chemical coagulants like Aluminum sulfate (alum), $FeCl_2$ are used in municipal drinking water treatment plant for purification process. This excess use of amount of chemical coagulants can affect human health e.g. Aluminum has also been indicated to be a causative agent in neurological diseases such as pre-senile dementia.

Indeed, wastewater treatment is carried out to remove turbidity, chemicals, and microbiological pollutants that may constitute health hazards by series of unique processes. The most important stage of wastewater treatment is disinfection. Hence, natural coagulants have been used for centuries in traditional water treatment practices throughout certain areas of the world. The new study compares protein from the seeds of different varieties of Moringa trees that are grown in different countries. It also allows estimates of the optimum amount of seed extract that should be used to minimize residues in treated water. Powerful research tools such as those for neutron scattering are important to tackle challenges facing developing countries as well as industrialized regions. Use of Moringa seeds, both to substitute conventional materials in large water treatment plants and in small scale units.

2. Facts about Moringa tree

Moringa is multipurpose drought-resistant and salt tolerant farm tree cultivated predominantly in South Ethiopia (Jiru, 2016). According to Dechasa there are 13 species which are grouped into three morphological types (stem structural forms) namely the slender, the bottle and shrubs/bushes or tree. *Moringa stenopetala* is native to East Africa linking South Ethiopia, North Kenya and West Somalia—a region that is characterized by unreliable rain and affected repeatedly by drought. The tree has special attributes in reclaiming salt affected farms in the Rift Valley. The massive Moringa is a cabbage and camel tree is nutritious, rich in all vitamins grows when most dry land trees shed their leaves. It is a good source of sustainable food and feed (including bee forage) supplier. The cabbage tree is a small tree up to 12 m (39 ft), with a many-branched crown and sometimes with multiple trunks.

The leaves are bipinnate or tripinnate, with about five pairs of pinnae and three to nine elliptic or ovate leaflets on each pinna. The fragrant flowers have creamy-pink sepals, white or yellow petals, and white stamens. The fruits are long reddish pods with a greyish bloom. The cabbage tree was planted by agriculturalists on the complex system of terraces built high up in the Ethiopian Highlands, where they became domesticated and were bred to improve productivity, the taste of their leaves, and the size of their seeds. Since then, the improved trees have been introduced into other areas such as the Rift Valley. Multidimensional direct and indirect role of the societies has been reviewed for economic, social acceptability and environmental privilege has been presented from research results of the presenter and secondary information sources. Its role in the hydrological function compares to other major land use, land cover is compared and addressed from secondary and primary data.

M. stenopetala is mostly known for its importance as a nutritious vegetable food crop in the terraced fields of Konso, Ethiopia. In this way, it is similar to its Indian relative, *Moringa oleifera* (Jiru, 2016). It is also used for shading of capsicum and sorghum crops, as a companion plant; and additionally in folk medicine. Another use is the clarification and purification of water to make it potable. A powder made by grinding the seeds is found to be more effective at coagulating substances in suspension than the seeds of the closely related horseradish tree (*M. oleifera*), which is used for this purpose in India (Jiru, 2016).

M. stenopetala is a massive tree; it is native to Ethiopia and has been used for food and feeds in the south Ethiopia for centuries by the Konso, Darachie and other Nation and Nationalities Peoples of the South. In recent years, it has been distributed in almost all parts of the lowland and the midlands of Ethiopia. This important multi-purpose vegetable, fruit tree and oil crop has not been tapped in relation to its potential (Wakjira, Jiru, & Asaminew, 2016).

M. stenopetala is a drought-resistant multipurpose food and feed that can sustain when used under dry condition. It is rich in nutrition and multivitamins. The seed cake purifies water where turbidity and unsafe water is a common phenomenon in the lowland in general and the lowest and hottest prefer in particular. The tree originated in South Ethiopia around eastern and northern Turkan Lake, North Kenya and West Somali (Olson, 2001) cited as Wakjira et al. (2016). These spots are both the lowest and driest where conflict arises between the three neighboring countries due to shortage of animal feed. Moringa being a good quality, sustainable fodder provider can be termed as conflict resolver.

According to Keay, Moringa is a tree which belongs to the family Moringaceae, consisting of only one genus with about 13 species of deciduous trees (Keay, 1989) cited as Anjulo (2016). Among these, there are two popular and widely grown species of moringa as food for human consumption. *M. oleifera* is native of India, introduced into other countries in the tropics which is easily recognized by the compound pinnate leaves (2 or 3 times pinnate) and the long, narrow angular fruits containing large, usually winged seeds (Hutchinson & Dalziel, 1966) cited as Anjulo (2016). On the other hand *M. stenopetala* is native of East Africa, which is widely grown for its palatable leaves cooked for human consumption in southern Ethiopia and northern Kenya. The compound leaves of this species are wider than *M. oleifera* and are preferred as cooked vegetables by the local communities in south and southwestern Ethiopia.

Reports showed that Moringa is extremely rich in vital nutrients and, can grow very fast in dry and impoverished areas of the world, where food is scarce. Moringa has been in use as a medicinal plant since time immemorial to heal and ease many diseases, starting from various inflammations to parasitic diseases, diabetes and cancers, shown by various reports. In this line, the recent woks of the Ethiopian Public Health Institute (EPHI) are encouraging. Moringa has been gaining much popularity, as a very nutritious plant that can feed the needy people and may save various lives. Its leaves or leaf powder can be used successfully as a supplemental food to nourish children, pregnant or nursing women, and everyone else. Report showed that, Moringa seed contains up to 42% oil, which possess the highest quality (up to 72% oleic) and hence possess tremendous market values. The oil is required as cooking oil and as industrial oil, specific ingredient for medicines and detergents. In short, the role and prospects of Moringa in farming system apparently look bright, and thus need to be well understood and be given high attention in order to further improve and utilize its productivity and profitability in a sustainable manner.

2.1. The power of Moringa as natural coagulant

Use of natural coagulants for treatment of water and wastewater in developing countries is an area that is gaining interest. Tropical plants of the family of Moringa, are among some of the natural co-agulants that have been studied for clarification of turbid water. Both *M. oleifera* and *M. stenopetala* is the most widely distributed, well-known and studied species of the family Moringaceae because of its previous economic importance as a source of the commercially important and more recently, as a multipurpose tree for arid lands and a source of water-purifying agents for developing countries (Morton, 1991).

M. oleifera is native to sub-Himalayan North-Western India and Pakistan but the plant was distributed to other areas of tropical Asia in prehistoric times and to other parts of the world including Malawi during the British colonial era. *M. stenopetala*, often referred to as the African Moringa tree, originates from southern Ethiopia and Kenya (Jahn, 1991) cited as Seifu (2015). The food, fodder, water clarifying and medicinal uses of the Moringaceae, especially *oleifera* are well documented and the trees are recommended for live fencing, inter-cropping, and pollution control (Moges, 2004; Morton, 1991).

The water-soluble Moringa seed proteins possess coagulating properties similar to those of alum and synthetic cationic polymers. The use of Moringa species for water clarification is a part of African indigenous knowledge. Jahn (1991) first studied and confirmed the coagulating properties of Moringa seeds after observing women in Sudan use the seeds to clarify the turbid Nile waters. *M. stenopetala* is less widely distributed than *M. oleifera* but *stenopetala* is reportedly more resistant to insect pests than other members of the family and its seeds are larger and easier to process than those of *oleifera* (Kayambazinthu, D., Forestry Research Institute of Malawi, personal communication). Although the water clarifying properties of *M. stenopetala* have not been as extensively studied as those of *M. oleifera*, Jahn, Musnad, and Burgstaller (1986) cited as Seifu (2015) reported that 100–150 mg/L of *M. stenopetala* was as effective in water clarification as 200 mg/L of *M. oleifera* which indicates that *stenopetala* is more effective than *oleifera*. The mechanism of coagulation by Moringa is not well understood and different authors have attributed it to existence of proteins and non-protein flocculating agents (Gassenschmidt, Jany, Tauscher, & Niebergall, 1995; Ndabigengesere, Narasiah, & Talbot, 1995; Okuda et al., 2001).

Although a number of papers are found on the water clarification properties of Moringa, only a few deal with heavy metal removal potential of Moringa. Metal ion removal ranging from 70 to 89% for lead, 66 to 92% for iron and 44 to 47% for cadmium using *M. oleifera* seed kernels and ram press cakes with initial metal concentration of 7 ppm and sorbent dose of 120 mg/L has been reported by our group (Sajidu, Henry, Kwamdera, & Mataka, 2005). Study showed that, the removal and recovery of arsenic from aqueous system using shelled *M. oleifera*. The study revealed removal capacities of 60.21% As 3 + and 85.6% As 5 + at a biomass dosage of 2.0 g in 200 ml of 25 mg/L of the metal. Sharma, Kumari, Srivastava, and Srivastava (2006) reported favorable performance of bio-sorption of Cd^{2+}, Cr^{3+} and Ni^{2+} on shelled *M. oleifera* seeds. The study reported that *M. stenopetala* is better than *M. oleifera* at removing lead from contaminated water (Mataka, Henry, Masamba, & Sajidu, 2006). Lead removal of 96% from initial concentration of 7 mg/L at a dosage of 2.4 g of seed powder in 100 ml of the metal solution had been reported. In continuation of the work, the present investigation reports the pH dependence of sorption of Cd^{2+}, Zn^{2+}, Cu^{2+} and Cr^{2+} cations on pure water and sodium chloride extracts of *M. stenopetala* and *M. oleifera*. The removal of organic and inorganic material from raw water is essential before it can be disinfected for human consumption. In a water treatment works, this clarification stage is normally achieved by the application of chemical coagulants which change the water from a liquid to a semi-solid state. This is usually followed by flocculation, the process of gentle and continuous stirring of coagulated water, which encourages the formation of "flocs" through the aggregation of the minute particles present in the water. Flocs can be easily removed by settling or filtration. For many communities in developing countries, however, the use of coagulation, flocculation and sedimentation is inappropriate because of the high cost and low availability of chemical coagulants, such as aluminum sulfate and ferric salts. The study reported that Moringa seeds treat water on two levels, acting both as a coagulant and an antimicrobial agent. It is generally accepted that Moringa works as a coagulant due to positively charged, water-soluble proteins, which bind with negatively charged particles (silt, clay, bacteria, toxin, etc.) allowing the resulting "flocs" to settle to the bottom or be removed by filtration. The antimicrobial aspects of Moringa continue to be researched. Findings support recombinant proteins both removing microorganisms by coagulation as well as acting directly as growth inhibitors of the microorganisms. While there is ongoing research being conducted on the nature and characteristics of these components, it is accepted that treatments with Moringa solutions will remove 90–99.9% of the impurities in water (Figure 1).

2.2. Water treatment with Moringa seeds

According to Jennifer (2015), solutions of Moringa seeds for water treatment may be prepared from seed kernels or from the solid residue left over after oil extraction (press cake). Moringa seeds, seed kernels or dried press cake can be stored for long periods but Moringa solutions for treating water should be prepared fresh each time. In general, 1 seed kernel will treat l L (1.056 qt) of water.

Dirty Water + Moringa Seeds Powder = Clean Water

Figure 1. Water treatment using Moringa Seed.

- Low turbidity NTU < 50 1 seed per 4 L (4.225 qt) water
- Medium turbidity NTU 50–150 1 seed per 2 L (2.112 qt) water
- High turbidity NTU 150–250 1 seed per 1 L (1.056 qt) water
- Extreme turbidity NTU > 250 2 seeds per 1 L (1.056 qt) water

Jennifer puts the following instructions to clean water with Moringa seeds

(1) Collect mature Moringa seed pods and remove seeds from pods
(2) Shell seeds (remove seed coat) to obtain clean seed kernels; discard discolored seeds.
(3) Determine quantity of kernels needed based on amount and turbidity of water in general 1seed kernel will treat 1 L (1.056 qt) of water.
(4) Crush appropriate number of seed kernels (using grinder, mortar & pestle, etc.) to obtain a fine powder and sift the powder through a screen or small mesh.
(5) Mix seed powder with a small amount of clean water to form a paste.
(6) Mix the paste and 250 ml (1 cup) of clean water into a bottle and shake for 1 min to activate the coagulant properties and form a solution.
(7) Filter this solution through a muslin cloth or mesh screen (to remove insoluble materials) and into the water to be treated.
(8) Stir treated water rapidly for at least 1 min then slowly (15–20 rotations per minute) for 5–10 min.
(9) Let the treated water sit without disturbing for at least 1–2 h.
(10) When the particles and contaminates have settled to the bottom, the clean water can be carefully poured off.
(11) This clean water can then be filtered or sterilized to make it completely safe for drinking.

The report puts the following sentence as dangers:

Secondary infection: The process of shaking and stirring must be followed closely to activate the coagulant properties; in the flocculation process takes too long, there is a risk of secondary bacteria growth during flocculation.

Recontamination: The process of settling is important. The sediment at the bottom contains the impurities so care must be taken to use only the clear water off the top and not allow the sediment to recontaminate the cleared water.

In general, water purified with Moringa seeds is acceptable for drinking only where people are currently drinking untreated, contaminated water. *M. oleifera* The *M. oleifera* is native to the sub-Himalayun tracts of north-west India, Pakistan, Bangladesh and Afghanistan (Makkar & Becker, 1997). This multipurpose tree has been introduced to Ethiopia over the last few years and is grown on nursery sites parallel to *M. stenopetala* in southern parts of the country. Both *M. stenopetala* and *M. oleifera* trees are the most commonly cultivated Moringa species in the tropics and subtropics which have the potential as alternative animal feed resources during dry periods of the tropics. However, the removal of heavy metals from wastewater of both Moringa species in raw water under Ethiopian conditions is hardly documented. Scholars indicated that *M. oleifera*, reduce turbidity of water. The reduction efficiency is higher for more turbid waters. Turbidity reduction exceeding 90% was achieved for all the three extracts on shallow well water with an initial turbidity of about 50 NTU. *M. oleifera* exhibited the most favorable results followed by G. gum and lastly J. curcas. The results indicated that *M. oleifera* can reduce turbidity of shallow well water. *M. oleifera* results for more turbid water (200 NTU) was better than less turbid water. It was noted that pH of the samples increased as the concentration of the extracts increased. There was, in general, an overall reduction in the number of coliforms and E. coli after the water had been treated with M. oleifera (Abatneh, Sahu, & Yimer, 2014).

Similarly, in Sudan dry M. oleifera seeds are used in place of alum by rural women to treat highly turbid Nile water (Jahn et al., 1986) cited as Seifu (2015). In Northern Nigeria, the fresh leaves are used as a vegetable, roots for medicinal purposes and branches for demarcation of property boundaries and fencing. Studies by Eilert, Wolters, and Nahrstedt (1981) cited as Lürling and Beekman (2009) identified the presence of an active antimicrobial agent in M. oleifera seeds. The active agent isolated was found to be 4a Lrhamnosyloxy-benzyl isothiocyanate, at present the only known glycosidic mustard oil. Madsen, Schlundt, and Omer (1987) carried out coagulation and bacterial reduction studies on turbid Nile water in the Sudan using M. oleifera seeds and observed turbidity reduction of 80–99.5% paralleled by a bacterial reduction of 1–4 log units (90–99.9%) within the first one to two hours of treatment, the bacteria being concentrated in the coagulated sediment. Other scholars at Asia said that M. oleifera, native to India, grows in the tropical and subtropical regions of the world. It is commonly known as "drumstick tree" or "horseradish tree" (Moringa can withstand both severe drought and mild frost conditions and hence widely cultivated across the world. With its high nutritive values, every part of the tree is suitable for either nutritional or commercial purposes. The leaves are rich in minerals, vitamins and other essential phytochemicals. Extracts from the leaves are used to treat malnutrition, augment breast milk in lactating mothers. It is used as potential antioxidant, anticancer, anti-inflammatory, antidiabetic and antimicrobial agent. M. oleifera seed, a natural coagulant is extensively used in water treatment. The scientific effort of this research provides insights on the use of moringa as a cure for diabetes and cancer and fortification of moringa in commercial products. This review explores the use of moringa across disciplines for its medicinal value and deals with cultivation, nutrition, commercial and prominent pharmacological properties of this "Miracle Tree".

2.3. M. stenopetala

M. stenopetala known as African Moringa has a wide range of adaptation from the arid to humid climates and can be grown in a various land use patterns. It grows in the lowlands of west of the Great Rift Valley Lakes from arid to semi-humid areas altitudinal ranging from 390 m to about 2200 masl. It is a strategic multi-purpose tree plant in being a unique food tree in drought prone areas and has recently been distributed to other regions of Ethiopia, beyond its place of origin. Leaves are used for human consumption and animal feed (Aberra, Workinesh, & Tegene, 2011). A study conducted by Aberra, Bulang, and Kluth (2009) indicated that the leaves of M. stenopetala are rich in crude protein (28.2%) and contain reasonable amounts of essential amino acids. M. stenopetala is native to Ethiopia, and it is known by various vernacular names. It is called "Haleko" in Gofa areas, "Shelagda" in the Konso language, and "Shiferaw" in Amharic (Engels & Goettsch, 1991; Jahn, 1991; Teketay, 2010) cited as Seifu (2015). M. stenopetala is particularly important as human food because the leaves, which have high nutritional value (Abuye et al., 2003), appear toward the end of the dry season when few other sources of green vegetables are available (Figure 2). The leaves contain high amounts of essential amino acids and vitamins A and C (Abuye et al., 2003) cited as Seifu (2015).

Scholars said that M. stenopetala has also antimicrobial properties; the seeds of the tree are used to clarify muddy water (ICRAF, 2006b). The seeds of M. stenopetala have natural flocculating and antimicrobial properties (ICRAF, 2006b; Jahn, 1991) cited as Seifu (2015). Earlier experiments showed that whole crushed seeds of M. stenopetala were effective in removing turbidity from waters with high initial turbidity, and bacterial contamination was reduced by 90 to 99.9%. M. stenopetala seeds have better water purifying properties than M. oleifera (HDRA, 2002). The active coagulating substances are found in the cotyledons of the seeds (ICRAF, 2006a). A recent study (Hellsing et al., 2014) also indicated that proteins extracted from the seeds of the M. stenopetala tree are effective flocculent for particles dispersed in water and are attractive as a natural and sustainable product for use in water purification. M. stenopetala seed powder could be used to remove heavy metals and

Figure 2. *Moringa stenopetala* plant in the backyard of a farmer in southern Ethiopia.

industrial wastes from water. A study by Mataka et al. (2006) aimed at investigating the potential of *M. stenopetala* and *M. oleifera* for the removal of cadmium(II) ions from water indicated that *M. stenopetala* seed powder, at a dose of 2.50 g/100 ml, reduced the concentration of cadmium by 53.8%.

The leaves, roots, and seeds of *M. stenopetala* and *M. oleifera* have a long tradition of use in folk medicine. Various parts of the *M. stenopetala* tree are claimed to contain disease-preventing chemicals (Endeshaw, 2003). People with high blood pressure boil the leaves and drink the water to get relief from their ailment (Endeshaw, 2003). A recent report by Mengistu, Abebe, Mekonnen, and Tolessa (2012) indicated that *M. stenopetala* has blood pressure lowering effects. These researchers showed that crude aqueous leaf extract of *M. stenopetala* caused a significant drop in systolic blood pressure, diastolic blood pressure, and mean arterial blood pressure in normotensive anesthetized guinea pigs. Leaf extracts of *M. stenopetala* are used to lower blood glucose and cholesterol levels.

Ghebreselassie, Mekonnen, Gebru, Ergete, and Huruy (2011) reported that aqueous leaf extract of *M. stenopetala* is shown to increase body weight and reduce serum glucose and cholesterol levels in mice. Serum glucose and serum cholesterol levels decreased significantly after six weeks of treatment. They indicated the need for further studies in order to fractionate the active economic and social importance of *M. stenopetala*.

Moringa is a multipurpose tree of significant economic and social importance, as it has vital nutritional, industrial, and medicinal applications (Jahn, 1991; NRC, 2006). On the other hand, the rest of the species of moringa have not been studied in detail, and their potential uses have not been fully understood. *M. stenopetala* was domesticated in the east African lowlands and is indigenous to southern Ethiopia. Many different ecotypes and varieties of *M. stenopetala* are found in Ethiopia. *M. stenopetala* is often called "cabbage tree" and is an important indigenous vegetable in south western Ethiopia where it is cultivated as a food crop. The Gofa, Konso, Burji, and Gamo tribes consume its leaves as a vegetable, especially during the dry season (Abuye et al., 2003; Demeulenaere, 2001; Jahn, 1991) cite as Seifu (2015).

The economic status of an individual in low lands of southern Ethiopia is closely associated with the number of moringa trees they have in their backyard. For example, when a young man proposes marriage in the former administrative region of Gamo Goffa of the South Ethiopia, the girl's (bride) families enquire whether or not the would-be husband has Haleko trees in his farm (Endeshaw, 2003). Every part of *M. stenopetala* tree is important; different parts of the tree such as leaves, seeds, steam, and root of the tree are used for different purposes. Apart from being consumed as a vegetable, *M. stenopetala* is also marketed as a source of income in different countries. *M. stenopetala* leaves and pods are used as fodder for animals (ICRAF, 2006a).

Melesse, Getye, Berihun, and Banerjee (2013) reported that M. stenopetala leaf meal could be used as an alternative and inexpensive source of protein in the diets of grower Koekoek chicken breeds. They indicated that replacing roasted soybean with M. stenopetala leaf meal has resulted in general improvement of the growth performance, feed efficiency, and carcass yield of Koekoek chickens without affecting the vital organs. Gebregiorgis, Negesse, and Nurfeta (2012) reported that supplementing a basal diet of Rhodes grass hay with dried Moringa leaves improved dry matter intake, body weight gain, and nitrogen retention in sheep indicating that M. stenopetala can serve as a protein supplement to low-quality grass during the dry season under a smallholder sheep production system.

The leaves, roots, and seeds of M. stenopetala and M. oleifera have a long tradition of use in folk medicine. Various parts of the M. stenopetala tree are claimed to contain disease-preventing chemicals (Endeshaw, 2003). People with high blood pressure boil the leaves and drink the water to get relief from their ailment (Endeshaw, 2003). Mengistu et al. (2012) indicated that M. stenopetala has blood pressure lowering effects. This research showed that crude aqueous leaf extract of M. stenopetala caused a significant drop in systolic blood pressure, diastolic blood pressure, and mean arterial blood pressure in normotensive anesthetized guinea pigs. Leaf extracts of M. stenopetala are used to lower blood glucose and cholesterol levels. Ghebreselassie et al. (2011) reported that aqueous leaf extract of M. stenopetala is shown to increase body weight and reduce serum glucose and cholesterol levels in mice. Serum glucose and serum cholesterol levels decreased significantly after six weeks of treatment. They indicated the need for further studies in order to fractionate the active. Demeulenaere (2001) observed in some parts of southern Ethiopia, especially among the Konso people, that the abundance of Moringa species in the garden or on farmland was an indication of the social status of the owner among the society. The one with many Moringa tree in the garden or on farmland had a higher social status and was also considered as a prosperous person. In Konso, in the course of marriage engagement, a common query posed to the to-be husband is the abundance of Moringa trees in his garden. Their belief is that if the husband has many Moringa trees in his garden or farmland then to-be wife will have no problem to feed her babies even when drought occurs. For this reason Konso people especially young men are encouraged to plant Moringa in their garden as well as on their farmlands. A lesson to learn from this practice is that culture in itself has a great role in conservation and sustainable utilization of locally important tree species.

M. stenopetala is planted together with fruit trees in the cropped fields In southern Ethiopia and many other east African countries. Sometimes the trees are also used to provide partial shade for crops like sorghum in the southern Ethiopia. Whole plants have been used as hedges and fences. M. stenopetala can also be planted as a windbreak. As soon as the upper branches of the tree grew broader, the tree can be pruned to stimulate more profuse growth of their lower branches, thus thickening the hedge. Vegetables cultivated behind it profited from this protection. The species can also be grown as an ornamental tree in private gardens and home compounds.

2.4. Removal capacities of M. stenopetala and oleifera
The present study explores heavy metal sorption property of crude water and sodium chloride extracts of M. oleifera and M. stenopetala seed powder. The metals studied are cadmium, zinc, copper and chromium. Cadmium is one of the most toxic heavy metals without any known biological function. Zinc and copper, which are essential nutrients for plants and animals, can be toxic at very high concentrations. Chromium occurs in two redox states, Cr(III) and Cr(VI) with the former occurring as a cation and the later as an anion (chromate). Cr(III) is less toxic than Cr(VI). Because of the toxicities of cadmium and chromium the World Health Organization's maximum contaminant levels in drinking water are 0.003 and 0.05 mg/L, respectively (WHO and Parsons, 2004).

M. oleifera and M. stenopetala are common species among the species of the Moringa family, both species have many characteristics in common. They are commonly grown, but oleifera is widely cultivated and got research development attention (Seifu, 2015).

Comparison of removal capacities between *M. stenopetala* and *oleifera*; research indicated that *M. stenopetala* was more effective than *M. oleifera* in removing cadmium from water. Study results showed that Moringa seeds could be used as a less expensive biosorbent for the removal of cadmium (Cd) from polluted water. Reports indicated that *M. stenopetala* seed powder could remove lead from contaminated water (Mataka et al., 2006). Mataka reported that *M. stenopetala* was more effective in removing lead from water than *M. oleifera*.

Another study indicated that *M. stenopetala* seed powder could be used to remove chromium (Cr) from tannery effluent (Gatew & Mersha, 2013). The results showed that *M. stenopetala* seed powder at a dose of 1 g/100 ml and pH of 9.5 decreased the concentration of Cr in tannery waste by 99.86%. A similar study by Degefu and Dawit showed that the seed powder of *M. stenopetala* was found to be effective in the removal of chromium from tannery wastewater. *M. stenopetala* seed powder resulted in a 99.74% removal of chromium from tannery waste. The use of bioadsorbents like *M. stenopetala* seed powder that are easily available and effective for removal of heavy metals could be an innovative and economical approach for treatment of industrial wastewater. The seed of *M. stenopetala* is an important source of oil that could be used for cooking or for different industrial applications. A recent study reported that *M. stenopetala* seed oil could be used as a potential feedstock for biodiesel production. The study indicated that *M. stenopetala* seeds yield 45% w/w of oil. The oil contains 78% mono-unsaturated fatty acid and 22% saturated fatty acid. Oleic acid is the dominant fatty acid and accounts for about 76%. When mixtures of alcohols were used, the amount of ethyl ester formed was 30% that of methyl ester. The recommended way to use the oil as a fuel is as a mixture of esters. This study showed that *M. stenopetala* has a number of advantages compared to biodiesel fuels derived from other vegetable oils. Moreover, *M. stenopetala* seed cake obtained after oil extraction contains high protein content and can be used as an important protein supplement in animal feed. A research finding also indicated that *M. stenopetala* seed cake powder can be used for biogas production (Mekete, 2008).

In addition to the seeds, the roots of *M. stenopetala* can also be used to clarify dirty water. Nomadic peoples in the Omo Valley of Ethiopia apparently use the roots of wild *M. stenopetala* to clarify muddy water (Demeulenaere, 2001). The root is also used in traditional medicine to treat different ailments.

3. Way forward

To compare both moringa trees that is between *M. oleifera* and *M. stenopetala*; not only *M. oleifera* but using *M. stenopetala* also as a replacement coagulant for proprietary coagulants meets the need for water and wastewater technology in developing countries which is simple to use, robust and cheap to both install and maintain. Water purified with Moringa seeds, is acceptable for drinking only where people are currently drinking untreated, contaminated water.

Based on the reports, nutritional value, drought tolerance, fust growth, and many of its potential uses, *M. stenopetala* should be given due attention by all concerned bodies and considered a priority crop to alleviate malnutrition and reduce poverty. Despite the enormous economic and social values *M.stenopetala* has among the rural community it has been given little research and development attention. Unlike *M. oleifera*, little scientific research has been conducted on the properties and potential uses of *M. stenopetala* in general and its seeds in particular. This calls for detailed and rigorous scientific study on *M. stenopetala* in order to fully understand the characteristics of the plant and exploit its full potential.

Despite the enormous economic and social values *M. stenopetala* has among the rural community in southern Ethiopia, it has been given little research and development attention. Unlike *M. oleifera*, little scientific research has been conducted on the properties and potential uses of *M. stenopetala* in general and its seeds in particular.

4. Conclusion

For conclusion using *M. stenopetala* as a natural coagulant has advantage to many countries of the developing world. It could be viewed as sustainable, appropriate, effective and robust water treatment means. The effective enhancement of particular wastewater treatment processes can decrease reliance on the importation and distribution of treatment chemicals, creating a new cash crop for farmers and employment opportunities for the rural dwellers in particular.

In addition, the benefits derivable from Moringa tree are inexhaustible if all its usefulness is to be considered. No wonder some researchers regarded it as a miracle tree in backyards that have all nutritional needs, take care of medicinally, and purify water. The benefits endowed with this tree actually need to be explored in all ramifications especially its economic importance going by its profitability index in the study area as shown above. This study therefore recommends that more technologically improved methods of processing, packaging, and preservations should be adopted and encouraged for upward economic efficiency.

Funding

This work was supported by Donghua University and IWHR [grant numbers 91547209, 41571037].

Competing interests

The authors declare no competing interest

Author details

Asaminew Abiyu[1]
E-mail: asaminewab@yahoo.com
ORCID ID: http://orcid.org/0000-0002-2412-1507
Denghua Yan[2]
E-mail: yandh@iwhr.com
Abel Girma[1]
E-mail: abelethiopia@yahoo.com
Xinshan Song[1]
E-mail: newmountain@163.com
Hao Wang[2]

[1] College of Environmental Science and Engineering, Donghua University, Shanghai, China.
[2] State Key Laboratory of Simulation and Regulation of Water Cycle in River Basin, China Institute of Water Resources and Hydropower Research, Beijing, China.

References

Abatneh, Y., Sahu, O., & Yimer, S. (2014). Purification of drinking water by low cost method in Ethiopia. *Applied Water Science, 4*(4), 357–362. https://doi.org/10.1007/s13201-013-0151-9

Aberra, M., Bulang, M., & Kluth, H. (2009). Evaluating the nutritive values and in vitro degradability characteristics of leaves, seeds and seedpods from *Moringa stenopetala. Journal of the Science of Food and Agriculture, 89*(2), 281–287.

Aberra, M., Workinesh, T., & Tegene, N. (2011). Effects of feeding *Moringa stenopetala* leaf meal on nutrient intake and growth performance of Rhode Island Red chicks under tropical climate. *Tropical and Subtropical Agroecosystems, 14*(2), 485–492.

Abuye, C., Urga, K., Knapp, H., Selmar, D., Omwega, A. M., Imungi, J. K., & Winterhalter, P. (2003). A compositional study of *Moringa stenopetala* leaves. *East African Medical Journal, 80*(5), 247–252.

Anjulo, A. (2016). Review of insect pests and diseases on moringa with possible management options (un pub.). In *Proceedings of the workshop on moringa with moringa societies and experts*. Held at Gullele Botanical Garden, Addis Ababa, Ethiopia.

Conley, W. R. (1961). Experience with anthracite-sand filters. *Journal American Water Works Association, 53*(12), 1473–1483.

Demeulenaere, E. (2001, October 29–November 2). *Moringa stenopetala*, a subsistence resource in the Konso district. In *Proceedings of the international workshop development potential for moringa products* (pp. 2–29). Dar-Es-Salaam, Tanzania.

Eilert, U., Wolters, B., & Nahrstedt, A. (1981). The antibiotic principle of seeds of *Moringa oleifera* and *Moringa stenopetala. Planta Medica, 42*, 55–61. https://doi.org/10.1055/s-2007-971546

Endeshaw, H. (2003, July 19), Promoting the miracle tree of hope. *Ethiopian Herald*.

Engels, J. M. M., & Goettsch, E. (1991), *Konso agriculture and its plant genetic resources: Moringa stenopetala*. Retrieved from http://www.moringanews.org/documents/Konsogene

Gassenschmidt, U., Jany, K. K., Tauscher, B., & Niebergall, H. (1995). Isolation and characterization of a flocculating protein from *Moringa oleifera* Lam. *Biochimica et Biophysica Acta (BBA) - General Subjects, 1243*, 477–481. https://doi.org/10.1016/0304-4165(94)00176-X

Gatew, S., & Mersha, W. (2013). Tannery wastes water treatment using *Moringa stenopetala* seed extract. *Global Journal of Environmental Sciences, 12*(1), 29–39.

Gebregiorgis, F., Negesse, T., & Nurfeta, A. (2012). Feed intake and utilization in sheep fed graded levels of dried moringa (*Moringa stenopetala*) leaf as a supplement to Rhodes grass hay. *Tropical Animal Health and Production, 44*(3), 511–517. https://doi.org/10.1007/s11250-011-9927-9

Ghebreselassie, D., Mekonnen, Y., Gebru, G., Ergete, W., & Huruy, K. (2011). The effects of *Moringa stenopetala* on blood parameters and histopathology of liver and kidney in mice. *Ethiopian Journal of Health Development, 25*(1), 51–57.

HDRA. (2002). *Moringa oleifera: A multi-purpose tree*. HDRA – The Organic Organization.

Hellsing, M. S., Kwaambwa, H. M., Nermark, F. M., Nkoane, B. B., Jackson, A. J., Wasbrough, M. J., ... Rennie, A. R. (2014). Structure of flocs of latex particles formed by addition of protein from Moringa seeds. *Colloids and Surfaces A: Physicochemical and Engineering Aspects, 460*, 460–467.

Hutchinson, J., & Dalziel, J. M. (1966). *Flora of tropical Africa* (Vol. 1).

ICRAF. (2006a). *Moringa stenopetala*. Retrieved from http://www.worldagroforestry.org/Sea/Prod

ICRAF. (2006b). *Moringa stenopetala* (Vol. 3, No. 4, pp. 8–19). ISSN 1929-0969. Retrieved from http://www.worldagroforestry.org/Sea/Products/AFDbases/AF/asp/SpeciesInfo.asp

Jahn, S. A. A. (1991). The traditional domestication of a multipurpose tree *Moringa stenopetala. Ambio, 1*, 244–247.

Jahn, S. A., Musnad, H. A., & Burgstaller, H. (1986). The tree that purifies water: Cultivating multipurpose Moringaceae in the Sudan. *Unasylva, 38*(152), 1–6.

Jennifer, S. (2015). *Researchers study inexpensive process to clean water in developing nations*. Nigeria: Development Council in Abuja.

Jiru, D. (2016). Review of multidimensional role of the Miraculous (un pub.). In *Proceedings of the workshop on moringa with moringa societies and experts*. Held at Gullele Botanical Garden Ababa, Ethiopia.

Keay, R. W. J. (1989). *Trees of Nigeria*. Oxford: Oxford University Press.

Lürling, M., & Beekman, W. (2009). Anti-cyanobacterial activity of *Moringa oleifera* seeds. *Journal of Applied Phycology, 22*(4), 503–510.

Madsen, M., Schlundt, J., & Omer, E. F (1987). Effect of water coagulation by seeds of *Moringa oleifera* on bacterial concentration. *Journal of Tropical Medicine and Hygiene, 90*, 101–109.

Makkar, H. P. S. & Becker, K. (1997). Nutrients and antiquality factors in different morphological parts of the Moringa oleifera tree. *Journal of Agricultural Science, 128*(3), 311–322.

Mataka, L. M., Henry, E. M. T., Masamba, W. R. L., & Sajidu, S. M. (2006). Lead remediation of contaminated water using *Moringa Stenopetala* and *Moringa oleifera* seed powder. *International Journal of Environmental Science and Technology, 3*, 131–139. https://doi.org/10.1007/BF03325916

Melesse, A., Getye, Y., Berihun, K., & Banerjee, S. (2013). Effect of feeding graded levels of *Moringa stenopetala* leaf meal on growth performance, carcass traits and some serum biochemical parameters of Koekoek chickens. *Livestock Science, 157*(2–3), 498–505. https://doi.org/10.1016/j.livsci.2013.08.012

Mengistu, M., Abebe, Y., Mekonnen, Y., & Tolessa, T. (2012). In vivo and *in vitro* hypotensive effect of aqueous extract of *Moringa stenopetala. African Health Sciences, 12*(4), 545–551.

Mekete, E. (2008). *Moringa stenopetala seed cake powder: a potential for biogas production and brewery wastewater treatment through coagulation*. Ehiopia: Addis Ababa University.

Moges, Y. (2004). Recommended agroforestry/multipurpose trees for borana lowlands/midlands and their production techniques. Sub-report No. 4, FARM Moringa Tree: A focus on sustainable food security (un pub.). In *Proceedings of the Workshop on Moringa with Moringa Societies and Experts*. Held at Gullele Botanical Garden, Addis Ababa, Ethiopia.

Morton, J. F. (1991). The horseradish tree, Moringa pterygosperma (Moringaceae): A boon to arid lands? *Economic Botany, 45*(3), 318–333. https://doi.org/10.1007/BF02887070

Ndabigengesere, A., Narasiah, K. S., & Talbot, B. G. (1995). Active agents and mechanism of coagulation of turbid waters using *Moringa oleifera. Water Research, 29*(2), 703–710. https://doi.org/10.1016/0043-1354(94)00161-Y

NRC. (2006). *Lost crops of Africa, vegetables* (Vol. II). Washington, DC: The National Academies Press.

Okuda, T., Baes, A. U., Nishijima, W., & Okada, M. (2001). Isolation and characterization of coagulant extracted from *Moringa oleifera* seed by salt solution. *Water Research, 35*(2), 405–410. https://doi.org/10.1016/S0043-1354(00)00290-6

Olson, M. E. (2001). *Introduction to the Moringa family* (L. J. Fuglie, Ed.). London: The White Friars Press Ltd.

Sajidu, S. M., Henry, E. M. T., Kwamdera, G., & Mataka, L. (2005). Removal of lead, iron and cadmium ions by means of polyelectrolytes of the *Moringa oleifera* whole seed kernel. *WIT Transactions on Ecology and the Environment, 80*, 1–8.

Seifu, E. (2015). Actual and potential applications of *Moringa stenopetala*, underutilized indigenous vegetable of Southern Ethiopia: A review. *International Journal of Agricultural and Food Research, 3*(4).

Sharma, P., Kumari, P., Srivastava, M. M., & Srivastava, S. (2006). Removal of cadmium from aqueous system by shelled *Moringa oleifera* Lam. seed powder. *Bioresource Technology, 97*(2), 299–305.

Teketay, D. (Ed.). (2010). *Edible wild plants in Ethiopia* (pp. 260–262). Addis Ababa: Addis Ababa University Press.

Wakjira, A., Jiru, D., & Asaminew, G. (2016). The potential role of moringa in oil production under rainfed and irrigated arid and semiarid farm (un pub.). In *Proceedings of the workshop on Moringa with Moringa societies and experts*. Held at Gullele Botanical Garden Ababa, Ethiopia.

WHO/UNICEF. (2012). *Fast facts; water sanitation hygiene, WHO/UNICEF joint monitoring report 2012*. Retrieved from http://www.who.int/water_sanitation_health/monitoring/jmp2012/fast_facts/en/

World Health Organization, & Parsons, E. (2004). *Review of assessing microbial safety of drinking water: Improving approaches and methods*. North-West University.

6

Environmental sanitation unleashed: Effectiveness and challenges of the National Sanitation Day as a community sanitation participatory approach in Aboabo, Ghana

Emmanuel Mawuli Abalo[1]*, Seth Agyemang[1], Samuel Atio[1], Derrick Ofosu-Bosompem[1], Prince Peprah[1] and Rita Ampomah-Sarpong[2]

*Corresponding author: Emmanuel Mawuli Abalo, Department of Geography and Rural Development, PMB, Kwame Nkrumah University of Science and Technology (KNUST), Kumasi, Ghana
E-mail: eabalo92@gmail.com

Reviewing editor: Terry Tudor, University of Northampton, UK

Abstract: Investment in technical facilities by government alone is not enough to meeting the challenge of providing adequate sanitation services in communities. This study aimed at examining the effectiveness and challenges of the National Sanitation Day (NSD) as a community participatory module towards environmental sanitation in Aboabo, Ghana. Adopting a descriptive and inferential mixed method design, 10 key informants were interviewed using purposive sampling and 180 study participants using systematic sampling techniques. Result indicates an encouraging number of participation during the NSD. Statistically, higher educational attainment was a predictor of respondents' participation in the exercise. Using respondents' perception about the extent of environmental sanitation before the NSD exercise and their perception about the current environmental sanitation as benchmarks for assessing the programme's effectiveness, the exercise was found to be ineffective. In recommendation, if needs be that prominent people are invited to intermittently grace the NSD exercise, their invitations should be to empower the local people to

ABOUT THE AUTHOR

Emmanuel Mawuli Abalo holds a BA in Geography and Rural Development from the Kwame University of Science and Technology, Ghana. He is a member of the Rural Research and Advocacy Group (RRAG-Ghana). His research interests include Land use and Land cover changes, Climate Change issues, Poverty studies and Food Security, Environmental Sanitation and Management and health. The authors work contributes to studies on the importance of community members' involvement in the management, sustainability and sanitary of the environment. Specifically, this study examines the effectiveness of the National Sanitation Day Module as a community participatory approach in environmental sanitation in Ghana, with an emphasis on the Aboabo community.

PUBLIC INTEREST STATEMENT

Investment in sanitation facilities by government alone is not enough to meet the challenge of waste management hence, the need to involve the community. Though community participation efforts were an integral component of Ghana's national rural programme in the 1990s, the initiative waned over time. However, the cholera outbreak in 2014, and other diseases resulting from poor sanitation led to the declaration of the first Saturday of every month as a National Sanitation Day (NSD) by the government. Using interviewer–administered questionnaire, the inclusiveness, effectiveness and challenges of the NSD were assessed. The knowledge base and participation of respondents during the NSD were high, though economic activities restricted others. Given the importance of a clean environment on socio-economic life and health, the author(s) argue that community participation should be recognised as an interaction, rather than a coercion, which should be undertaken willingly so as to reap its full potential.

esteem the importance of the programme by making it their own rather than depending on their presence to increase patronage.

Subjects: Environmental Management; Motivation; Urban Geography; Environmental Geography

Keywords: environment; sanitation; environmental sanitation; community participation; National Sanitation Day; Aboabo

1. Introduction

Environmental sanitation has been a topical issue drawing different views on ways to improve and maintain proper sanitation in communities. The need for improving sanitation has been necessitated by population increase, industrialization, urbanization, the alteration of urban consumption pattern towards packaging and economic growth, resulting in an increase in solid waste (SW) generation in developing countries (Dhokhikah, Trihadiningrum, & Sunaryo, 2015; Wahabu, Oduro-Kwarteng, Monney, & Kotoka, 2014; Zhu, Asnani, Zurbrügg, Anapolsky, & Mani, 2008; Zurbrügg, 2002).

Poor sanitation resulting from the practice of widespread open defecation and indiscriminate dumping of refuse have negative health and social impacts on communities with consequence of diarrhoea and cholera (Kar, 2005). For a long time, planning of sanitation service provision consisted of what came to be known as a "Top-Down" approach where the needs of communities were determined by well-meaning officials or political representatives at the central, regional, district and/or municipal levels to the neglect of community members who are beneficiaries of the project. This resulted in a poor maintenance of the services provided (Eawag, 2005).

Community participation was born out of the need of placing community members at the centre of the planning process with regard to environmental sanitation. Minkler (2005) defines community participation as a process brought about through social interactions expressed collectively, embedded in a community of place, and directed to the achievement of a specific task that is perceived to lead to the betterment of the community. Communities remain economically and socially viable when members of the community are involved in community projects (Taylor, Wilkinson, & Cheers, 2006).

Some notable community participatory approaches adopted to improve sanitation include the household-centred environmental sanitation approach (HCES) and the community-led total sanitation approach (CLTS) which focused on community empowerment, rather than negotiated development (Eawag, 2005; Kar, 2005; Lüthi & Tilley, 2008).

Developed by the Environmental Sanitation Working Group of the Water Supply and Sanitation Collaborative Council (WSSCC) in 2002, the HCES was targeted at providing stakeholders at every level, but particularly at the household level and neighbourhood level, to have the opportunity to participate in the planning, implementation and operation of the urban environmental sanitation services (UESS) in an effort to achieving the Sustainable Development Goals. This process was adopted to help community members live healthy and productive lives and also to protect and restore the natural environment (Eawag, 2005; Morel, Luethi, & Schertenleib, 2008).

Similarly, the CLTS involves facilitating a process to inspire and empower rural communities to stop open defecation and to build and use latrines, without offering external subsidies to purchase hardware such as pans and pipes (Kar, 2005). Countries where these approaches have been implemented and proved effective include Costa Rica, Burkina Faso, Kenya, Tanzania, Laos, Nepal and Bangladesh (Kar, 2005; Morel et al., 2008).

Caring for the environment and ensuring cleanliness has been a traditional communal affair in many sub-Saharan African countries, particularly Ghana, where all farming and other economic activities were halted and community members collectively weeded and cleaned both public places as toilets, streets, markets, footpaths, as well as their private homes and compounds. These "communal labour" days were taken very serious by the rank and file, and even children had their part to play by running errands and also helping to clean the "drop down" public toilets. Severe punishment was meted out to members who deliberately refused to participate, including public ridicule (naming and shaming), being made to pay a fine, or prohibiting access to a community facility (such as toilet and water source) for a prescribed number of days.

However, the dawn of modernity has seen a roll back or erosion of most of these traditional mechanisms for safeguarding the environment, opening the floodgates for indiscriminate littering and mountains of filth that we see all around us. It is estimated that the cost of poor sanitation to Ghana every year is $290 million, representing 1.6% of the country's Gross Domestic Product (Water & Sanitation Program [WSP], 2012). Approximately 13,900 Ghanaian adults and 5,100 children under five, die each year from diarrhoea, with nearly 90% of the deaths directly attributed to poor sanitation and hygiene. Although these figures are worrying and poses threat to the well-being of citizens in the country, not much effort have been channelled to prevent the menace of sanitation related diseases until the 2014 cholera outbreak where 17,000 cases and 150 deaths were recorded, the worst ever, after the 1982 incidence in Ghana (Adubofour, Obiri-Danso, & Quansah, 2013; WSP, 2012).

It is against this backdrop that efforts to bring back the communal spirit and inculcate a sense of environmental awareness into the citizenry should be lauded by all. The government of Ghana, through the Ministry of Local Government and Rural Development (MIGRD), invoked the traditional communal spirit of collectively taking care of the environment through the institution of the National Sanitation Day (NSD) as a means of ensuring environmental cleanliness and dealing with the filth that has engulfed most parts of the country, especially urban areas. The NSD was earmarked on the first Saturday of every month to remove heaps of garbage at all refuse dumping sites across the country, and to educate the populace on sorting techniques (Ministry of Local Government & Rural Development [MLGRD], 2014).

In the initial stages of its implementation, the NSD were held up as a "game changer" in the ongoing sanitation albatross, amid much fun fare and publicity. Community members enthusiastically teamed up with their elected representatives, waste management companies and opinion leaders to rid their surroundings of filth. This led to a cautious conclusion that the nation's challenge with environmental sanitation would soon be over if the momentum of the community approach was sustained. However, what started on a very bright note soon began losing its steam. The enthusiasm and fervor that greeted the exercise has waned down considerably, while participation and commitment have dwindled, pointing to an imminent dissipation of the spirit and purpose of the activity.

In the Aboabo Township, the outbreak of sanitation related diseases such as 171 typhoid cases in 2015 and diarrhoea, malaria and intestinal worm infections; 163, 174 and 47 cases, respectively, in the first quarter of 2016 are worrying and needs addressing (Asokore Mampong Municipal Health Directorate Report, 2016). These infections may have a negative toll on the economic activities of the Municipality, specifically, the study communities whose active labour force are involved in petty trading and other commercial activities.

Although various works have been undertaken with regard to sanitation improvement in the Ghana (Adubofour, 2010; Adubofour et al., 2013; Dakpallah, 2011), none of these ascertained the effectiveness of the NSD as a community participation module towards environmental sanitation. It is in this regard that this study seeks to assess the effectiveness of the NSD as a community participatory approach towards environmental sanitation and its associated challenges using the case of the Aboabo communities within the Asokore Mampong Municipality. Within the broad objective of the study, the author(s) seeks to find out whether respondents' demographic characteristics influence their participation during the NSD.

In as much as government institutions are tasked with keeping the environment clean, the role of community members towards environmental sanitation, with regard to their attitude, cannot be underestimated. A person's attitude towards the environment has a significant influence on his or her behaviour. The interaction between man's behaviour and the social environment creates a conflict referred to as cognitive dissonance theory (Festinger, 1957). The theory explains the perceived inconsistency between one's behaviour and attitudes. It looks at why individuals get involved in negative activities amidst their knowledge of its consequences (Eagly & Chaiken, 2007; Festinger, 1957; Kassarjian & Cohen, 1965). Applying this theory to examine why people continue to smoke amidst their knowledge on hazards of smoking, Kassarjian and Cohen (1965) found that smokers justify their continued smoking by eliminating their responsibility for their behaviour; denying, distorting, misperceiving or minimizing the degree of health hazard involved and selectively drawing out information that reduces the inconsistency of the smoker's behaviour. To have a systematic and inclusive framework that articulates an in-depth understanding of how personal attitude impact on participation towards environmental sanitation, Festinger (1957) theory of cognitive dissonance was adopted to guide the study.

1.1. Literature review

The world's cities continue to experience population growth at a rate that far exceeds their absorptive capacity in terms of conventional sanitation infrastructure and environmental protection (United Nations Population Fund [UNFPA], 2007). This situation is compounded by the inability of governmental agencies to meet the corresponding waste generation associated with population growth through the provision of basic environmental services (Tukahirwa, Mol, & Oosterveer, 2010). Moreover, the challenge of sanitation service delivery results from the fact that many poor urban residents live in the unplanned and underserved informal settlements commonly known as slums. The United Nations Joint Monitoring Programme (UN JMP) report exemplifies the situation of poor sanitation services among the world's urban population. The report indicated that the urban population without access to improved sanitation will increase from 661 million (2006) to 898 million (2015) (UN JMP, 2008).

Though poor urban sanitation issue continues to be difficult for policy-makers, and presents great challenge in the development of integrated solutions for managing a variety of waste streams (Tilley, Atwater, & Mavinic, 2008), most of the interventions undertaken to reduce this canker are often implemented without consultation or participation of stakeholders and beneficiaries (Eawag, 2005; Rosemarin et al., 2008). Despite this trend, a number of recent initiatives have been embarked upon, focusing on motivating community involvement and encouraging appropriate technology geared towards environmental sanitation (Atkinson, 2007; Smith, 2006). These initiatives are collectively termed as community participation. It is argued that communities are able to raise more resources, achieve more result and develop a more holistic and more beneficial way when their citizens and partners engage in developmental activities (Ton & Patrick, 2003). Nance (2004) described the concept of community participation as the involvement in and contributions to a project by individual residents, households and the community as a whole. Undeniably, a high level of involvement in mobilization often signals a strong demand for a project and environmental sanitation is no exception (Nance & Ortolano, 2007). Bill (2007) articulates that participation carries with the feelings of ownership, and builds a strong base for intervention in the community.

Consequently, Nance and Ortolano (2007) discovered in their Brazilian study that the form of participation as well as community influence are key roles to ensuring the smooth implementation and execution of sanitation services. Commenting on how this is achieved, they outlined four forms of participation which are associated with various project outcomes and also helps to efficiently manage the task involved in taking a project from conception to operation. These forms of participation are: planning, design, construction and maintenance. The planning process is also known as the mobilization stage. The focus of this stage is to bring community members together so that the cost and benefit, in terms of improved health and sanitation, of the project could be explained to them. Hence, a high turn-out at this stage is considered as an indication of goodwill for the project (Sara &

Katz, 1997). In effect, this stage focuses primarily on fostering support for the project by creating awareness among community members and also provides a platform for them to air their needs and preferences concerning the project (Nance & Ortolano, 2007). The next form of participation, design, incorporates the insights and local knowledge of community members into the project. Thus, this stage is described as the participation in decision-making. Specifically, this stage examines the receptive capacity of residents to the project implementation. Since each community participation initiative is targeted at making community members identify with the project, incorporating their local knowledge in the decision-making process helps to strengthen the projects' success (Nance, 2004). Significantly too, whereas the participation in construction contributes to reducing project cost and brings about a sense of ownership among the participants, participation in maintenance looks at the regular and timely preservation of the project to ensure its success. Together, participation in construction and maintenance is assessed based on the amount of time, money, tools, materials and labour contributed by the beneficiaries of the project. Moreover, at these stages, the role played by sanitation agencies can potentially lead to better cost recovery, higher connection rates and reduce the risk of ill health since the quality of construction and maintenance work performed by residents may not be high (Nance & Ortolano, 2007).

The concept of participation in all developmental activities is certainly not a new one. The concept emerged out of the recognition of the limitations of top-down development approaches. The limitation led to a shift towards participatory research and planning methods by communities. Proponents in the participatory approach are Fritz Schumacher with his seminal work "Small is Beautiful" (1976) and Robert Chambers, the father of participatory rural appraisal (Chambers, 1983). That notwithstanding, the past decade witnessed the benefits associated with participation in development which was described as participatory development (Lawrence, 2006). Different studies have examined the concept of community participation from varying angles. For instance, Nance and Ortolano (2007) examined the concept of participation more carefully by distinguishing among forms of participation in Northeastern Brazil, community participation in rural water projects (Ishaku & Majid, 2010; Isham & Kahkonen, 1999; Narayan, 1995; Ostrom, 1992; Prokopy, 2002, 2005; Sara & Katz, 1997; Wade, 1988) and from other public service sectors where the concept of participation may be viewed as a form of "co-production" (Brudney & England, 1983; Levine, 1984; Percy, 1984; Sundeen, 1985). Community participation has also been identified as a key factor for attaining the goal of solid waste management (SWM) (Chung & Poon, 2001; Kamara, 2006; Sukhor, Mohammed, Sani, & Awang, 2011; World Bank, 2004) and plays an important role in achieving SW management in developing countries (Dhokhikah & Trihadlningrum, 2012; Mongkolnchaiarunya, 2005; Zurbrügg, Drescher, Patel, & Sharatchandra, 2004). In these countries, the human resource holds potential for development and is useful in sanitation services (Dhokhikah et al., 2015).

Aside the above studies, the notable community participatory modules towards environmental sanitation are the HCES and CLTS approaches which have been examined thoroughly in other studies (Kar, 2005; Lüthi, McConville, & Kvarnström, 2010). Hamdi and Goethert (1997) describes the HCES approach as a communicative planning framework that focuses on participatory, bottom-up methodologies where planners solicit the participation of a variety of stakeholders in a democratic planning process. The approach proffer solutions from storage to transport to treatment and disposal/re-use of municipal waste (Lüthi, Schertenleib, & Tilley, 2007). Notable countries where this approach have been implemented and proved successful include the Hatsady Tai Village, Vientiane and Tanzania (Lüthi et al., 2010). The CLTS was initiated in Bangladesh in 1999, as an innovative methodology for eliminating open defecation (Evans, Colin, & Jones, 2009; Kar, 2005; MoLG, 2005). CLTS uses a participatory approach to empower local communities to stop open defecation and promote the building and use of latrines through community-led action instead of subsidies. The approach has been lauded for its especially in rural communities (Kar & Chambers, 2008). Areas where the approach has been successful include rural South and South-east Asian, Southern Region of Ethiopia (WSP, 2007) and Kolkata, India (Stockholm Environment Institute [SEI], 2008).

In the Ghanaian context, different responses have been initiated by individuals and governments at keeping its cities clean. Government agencies in charge of sanitation activities provide skips and dustbins at vantage points so that refuse would be disposed of by community members. Later, the pay-as-you-dump (PAYD) policy was introduced. Repetto, Dower, Jenkins, and Geoghegan (1992) stated that: "fixed fee charged for waste disposal would serve as an incentive for respondents to reduce the amount of waste they generate". Skumatz (1996) also described the PAYD policy as a mechanism put in place by city governments to make residents responsible for the waste they generate by charging them based on the quantity of waste they disposed of. However, the rapid springing up and sprawling of urban communities has reduced the effectiveness of these waste management policies. This is evidenced from the mounting filth on the streets of urban communities, left to be attended to by sanitation agencies in the community. The recent response to the fight against poor environmental sanitation is the advent of the NSD which was declared in response to the 2014 cholera outbreak in the country. In line with the forms of participation outlined by Nance & Ortolano, it can be said that the NSD adheres to all the four forms. Though the programme had not been in existence for long, the Millennium Development Goal (MDG) report in 2015 revealed that the country ranked very poorly on the global environmental sanitation ladder: the country was pronounced the seventh dirtiest in the world, having slipped from the tenth position (MDG, 2015). This results from the mounting garbage or filth disposal and poor sanitation services in especially the urban areas (WHO & UNICEF, 2015).

Though the benefits of community participation in developmental activities have been largely lauded, the concept is not without some challenges. Dukeshire and Thurlow (2002) indicated that, it is important for communities to understand that government also faces barriers and challenges that can hinder its progress in responding to and recognizing the priorities of developmental issues in communities, and environmental sanitation is no exception. Some of the challenges outlined by Dukeshire include; lack of understanding of the policy process; lack of community resources and lack of access to information. Also, Nour (2011) posited that community participation in densely populated areas with little social consistency and a low level of popular organisation could pose as a challenge to participation efforts.

2. Materials and methods

2.1. Study area context
The study was conducted at Aboabo, an urban slum settlement within the Asokore Mampong Municipal Assembly in the Ashanti Region of Ghana. Aboabo, which is actually made up of two communities, Aboabo No.1 and No. 2 (with the two communities being contiguous with each other) used to be part of the Kumasi Metropolitan Assembly until 2012. Other notable communities that fall within Asokore Mampong Municipality are Adukrom, Akurem, Asokore Mampong, Sawaba, Asawasi, New Zongo, Sepe-Tinpom and Akwatialine. Figure shows the study area in the municipality (Ghana Statistical Service [GSS], 2010).

The Municipality covers a total land area of 23.91 km² and lies within latitude 6°42′North: 1°34′West: 6.700°North and 1.567°West. The municipality shares boundaries with the Kumasi Metropolitan Assembly to the East, South and West, Kwabre East Municipal to the North-West and Ejisu-Juabeng Municipal Assembly to the South-East. The Municipality has a total population of 304,815 (GSS, 2010). Aboabo was chosen because of it being a beneficiary of the Urban Environmental Sanitation Project I and Component II of the same project (UESP I and II), as well as the Urban Poverty Reduction Project (UPRP) which sort to address sanitation infrastructure challenges in the communities (Dakpallah, 2011). However, though the community had benefited immensely from these projects, Aboabo is still beset with poor waste management practices due to indiscriminate disposal of refuse into gutters, on the road side and backyards which breeds sanitation related diseases such as diarrhoea, malaria, intestinal worm and typhoid among others (Asokore Mampong Municipal Health Directorate Report, 2016) (See Figure 1 for the Map of the study area within the Municipality).

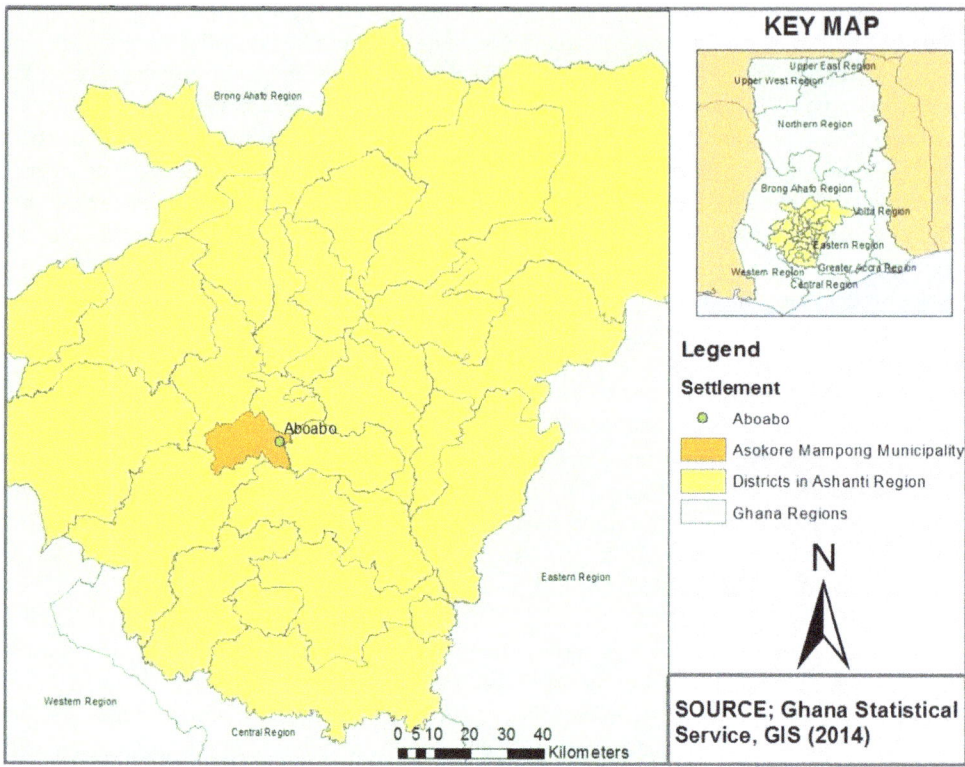

Figure 1. Map of study area in the Asokore Mampong municipal assembly.

2.2. Sampling design, and data collection

The study was undertaken from September 2015 to June 2016. However, the period of data collection spanned from January to March 2016 on the days when the sanitation exercise was embarked upon (first Saturday of each Month). Prior to the period of data collection, the authors visited the communities for familiarization, and interacted with the Assemblymen of the two communities. This was necessary as the authors needed the permission of these prominent personnel's for entry into the two communities. During the period of familiarization, the authors also took transient walks to ascertain the extent of cleanliness of the communities. This exercise was instrumental in designing the research instrument which was used during the data collection process.

During the three month period, both probability and non-probability sampling methods were used to select a total of 190 respondents for the study. From the 190 participants, 180 household respondents were chosen from Aboabo No. 1 and No. 2, whereas the remaining 10 participants constituted the key informants in the study prefecture. Being one of the first empirical studies focused on the effectiveness of the NSD as a community participatory approach since its inception in 2014, the authors arbitrarily sampled 180 respondents from different households to serve as a basis for future studies on the NSD programme. It is important to note that, the homogeneity of the study participants allows for smaller sample sizes which are representative of the total population since it assumes equal expected value and variance (Business Advocay Network, n.d.; Chambers & Clark, 2012). Moreover, the sample was representative as the χ^2 test of homogeneity conducted between gender and the other demographic variables was significant ($p < 0.05$). That notwithstanding, a minimum sample of 70 is considered large enough for any statistical analysis (Bryman, 2001). The absence of a coherent house numbering system made it difficult to employ simple random sampling technique. Thus, on the field, systematic sampling was adopted, whereby every fifth house was entered and one adult (male or female aged 18 years and above) was contacted. In a situation where an identified person in a particular house refused to participate, another person in the same house was chosen for replacement. In each community, the procedure continued until the total number allocated for it was exhausted. Thus, 90 participants from Aboabo No. 1 and No. 2, making a total of 180 respondents, took part in the study.

In addition to the survey sample, purposive sampling was used to select the 10 key informants from key sectors in the Municipality. These were: Environmental Health and Sanitation Officer of the Asokore-Mampong Municipal Assembly (AMMA)-1, one official from Zoomlion Ghana Limited, the Assemblymen of the two Aboabo communities, three Unit Committee members each from the two communities.

Standardized questionnaires were used to elicit responses from the survey sample. To avoid the problem of non-response created by low reading and writing abilities, the questionnaires were interviewer–administrated, making the process pass as structured interview or interview schedule. Thus, the questions were read in the local dialect, Asante Twi, and the responses recorded on the question paper. However, in a few instances, respondents who could read and write insisted on filling the questionnaires themselves. Issues covered included personal and household characteristics, awareness of the NSD, extent of participation, and consequences and reasons for non-participation. As regards the extent of participation, respondents were queried: "Have you ever heard of the NSDs?" The options provided were: (a) yes (b) no. Those who have heard about the NSD were asked to explain what they know about it. Respondents were further quizzed: "Have you ever participated in any of the NSDs in the last 12 months?" with the same options: (a) yes (b) no. Those who do not participate were asked: "What prevents you from partaking part?" The options provided were: (a) busy work schedule (b) not aware of such exercises (c) sickness (d) apathy on the part of the community members (e) inadequate tools to work with (f) inadequate communication and information (g) insufficient equipment for the cleanups. In an attempt to ascertain the seriousness attached to the NSD, the researchers asked whether any punishment were meted out to non-participants: "What normally happens to people who refuse to participate in the community clean up exercises?" The options provided were: (a) nothing is done to them (b) they are punished/pay fines (c) they are arrested and cautioned.

Data from the key informants was obtained by means of semi-structured interviews to avoid time loss and the problem of call-backs. Data from the final leg of respondents, the Assemblymen and the unit committee members, was obtained by means of in-depth interviews using a checklist. The list contained items spanning community attitudes towards environmental cleanliness and sanitation. The use of these qualitatively inclined approaches afforded the opportunity to clarify some of the issues raised in the survey. Interviews were conducted with the Environmental Health and Sanitation Officer and the Sanitation officer of Zoomlion Ghana in their respective offices at their convenience. The interviews were audio-taped and lasted for 40 min. Responses from the interviews were transcribed verbatim and presented through direct quotes.

2.3. Measure of NSD programme effectiveness

In order to estimate the effectiveness of the NSD, respondents' perception about the extent of environmental sanitation before the NSD exercise and their perception about the current environmental sanitation were used as benchmarks. For respondents' perception about the extent of environmental sanitation before the NSD exercise, the options provided were: "don't know, very clean, clean, filthy, very filthy". Respondents were asked: What is your perception about the sanitary condition of the community before the NSD exercise? As regards the respondents' perception about the current environmental sanitation, the following variables were used as indicators: "highly improved, improved, worsened, significantly worsened, no change". The specific question asked for this indicator was: What is your idea about the current environmental sanitation in the community? For the two indicators, a minimum higher rating of either "very filthy, filthy or both" for the sanitary condition of the community before the NSD exercise, and a further minimum higher rating of either "highly improved, improved or both" for the current sanitary condition of the community were regarded as proof of the programme's effectiveness. However, a higher response for description of the post-sanitary condition before the NSD exercise as either "filthy, very filthy or both" and either "worsened, significantly worsened or both, and no change" for the current sanitary condition of the community were considered as indicators of the programme's ineffectiveness

2.4. Data analysis

The field data were first subjected to checks and editing for consistency and validation. The quantitative data was entered into the Statistical Package for Social Sciences (SPSS) version 16.0 and analysed with the aid of descriptive analysis using counts, frequencies, tables and percentages from which patterns and trends were inferred and used to form the basis of the analysis. Also binary logistic regression model was used to tease out the influencing demographic factors on respondents' participation during the NSD exercise. In this case, participation during the NSD exercise was the dependent variable, whereas respondents' demographic characteristics were the independent variables. Participation during the NSD in the last 12 months preceding the survey was captured in the dependent variable box of the binary logit model, whereas the demographic variables were captured in the covariates box. All the demographic variables were captured in the covariate section and the first responses selected under the categorical section as the reference point with which other responses were compared to. The 95% confidence interval [CI] and the Hosmer and Lemeshow test were selected in the options section, whereas "enter" was maintained under the Method option to allow the step-by-step outcome of each variable been tested. A value of $p < 0.05$ means the measure is not a good fit, whereas a value of $p > 0.05$ is an indication of a good fit for the Hosmer and Lemeshow test. The qualitative data were transcribed verbatim and presented through direct quotes and linked to the requisite quantitative analysis, for better enlightenment. Besides, the primary data was supplemented with secondary information in the form of journal articles, unpublished thesis, and other official documents provided for by the Asokore Mampong Municipal Assembly.

3. Results

3.1. Sample characteristics

The background characteristics of the respondents are presented in Table 1. There were more male (58%) than female (42%) participants. In all, most respondents (86%) were youthful in nature, within the age cohort of ≤ 41, schooled up to the basic level (48%) and 13% having no education at all. The average household size was 2, unemployed (12%), students (28%) and employed (60%), with (50%) in the informal sector comprising of petty traders and artisanal economic ventures. This low level of educational attainment is actually mirrored by the low percentage of the study sample working in the formal sector (11%).

3.2. Community participation

3.2.1. Knowledge and awareness of the NSD

Generally, respondents were aware of the NSD as a module for improving sanitation in their communities. In view of this, 98% of the respondents described the NSD as the first Saturday of every month set aside for community members to clean their community in order to avoid diseases. Meanwhile, some 2% of the sample had no idea of the existence of the exercise. These responses show that with the exception of a few study participants who were not aware of the NSD, the majority are adequately informed about the exercise.

On the mode of organization for the exercise, majority (72%) of the respondents mentioned that announcement was made on the local radio station days before the exercise comes off. Others (22%) indicated that they were mostly informed through word of mouth by the assemblymen and members of the unit committee, whilst 7% participated in the exercise based on their own initiative (see Table 2). This indicates that without the involvement of the local radio station in disseminating information on the NSD to the participants, indulgence during the exercise would be low.

Table 1. Background characteristics of study participants		
Variables	N = 180	Percent (%)
Gender		
Male	104	57.8
Female	76	42.2
Age		
18–25	40	22.2
26–33	81	45
34–41	34	18.9
42–49	17	9.4
50+	8	4.5
Level of education		
None	24	13.3
Primary	31	17.2
Middle/J.S.S/J.H.S	56	31.1
Secondary/Technical/Vocational	42	23.3
Training College/University	27	15.0
Household size		
Mean	2.3	
Minimum	1	
Maximum	7	
Occupation		
Unemployed	72	40
Petty trader	55	30.5
Artisan/Craftsmanship	34	18.9
Civil servant/Public servant	19	10.6

Table 2. NSD related knowledge and mode of organisation		
Variables	Frequency	Percent (%)
Knowledge on NSD		
Cleanup exercise on the first Saturday of every month to avoid disease	176	97.78
No idea	4	2.22
Total	180	100.0
Mode of organisation		
Community radio/FM	129	71.6
Personal contact/word of mouth by Assembly members	39	21.7
Self-initiative	12	6.7
Total	180	100.0

Commenting on the mode of organising respondents for the exercise, a member of the unit committee (MUC) mentioned that they either reach out to the participants in person, through the local radio station and/or at the mosque (Muslim's place of worship).

We reached out to the community members through personal contact by the organisers, through the local radio station in the community and the mosque. Basically, we reached out to them by word of mouth. [MUC, Aboabo No. 2]

3.2.2. Participation during NSD

As to whether respondents participated during the clean-up exercise, 65% confirmed their indulgence during the exercise whereas 35% do not participate. Regarding the rate of participation, 41% said they sometimes participate during the exercise, 32% participate in the exercise whenever it is organised, whereas 26% participate most of the time (see Table 3).

The following quote confirms the respondents claim on the frequency of participation.

Participation has been encouraging and increases from time to time. This was due to the presence of prominent personalities who visited our community during the clean-up exercises. [MUC, Aboabo No. 2]

The turnout was not what we expected or projected but we record increase in numbers when a prominent personality is coming to grace the activity. [Assemblyman, Aboabo No. 1]

As regard the activities undertaken during the exercise, 45% of the respondents swept the street and markets, 19% de-silted clogged gutters and 25% cleaned public places of convenience (see Table 4).

The following quote confirms the respondents' claim.

Mostly we cleaned the community by sweeping the streets and de-silt gutters. However, attention is paid to other activities such as cleaning the households and fumigating the gutters. [MUC, Aboabo No. 1]

Activities undertaken during the NSD ranged from de-silting choked gutters, sweeping the streets and fumigating gutters. [Assemblyman, Aboabo No. 2]

The gutters were de-silted, we swept the streets and fumigated the gutters during the clean-up exercises. [Senior Operations Officer, Zoomlion Ghana Limited]

The community members clean the streets, de-silt choked gutters, clean-up the market places and also fumigate the gutters [Principal Environmental Health and Sanitation Assistant of AMMA].

3.2.3. Challenges to participation

Various reasons were ascribed for non-participation during the NSD exercise. Insufficient equipment recorded (27%), busy work schedule (26%), poor and non-patriotic attitude towards the NSD (22%) and inadequate communication (18%). Most respondents are not able to partake in the exercise due to their work schedule which usually coincides with the NSD. This could be inferred from the

Table 3. Attitude towards the NSD		
	Frequency	**Percent (%)**
Participation		
Yes	117	65
No	63	35
Total	180	100.0
Frequency of participation		
Always	57	31.7
Most of the time	47	26.1
Sometimes	74	41.1
Never	2	1.1
Total	180	100.0

Table 4. Activities undertaken during the NSD exercise

Activities undertaken	Frequency	Percent (%)
Sweeping of streets and markets	82	45.5
De-silting of choked gutters	34	18.9
Weeding of streets, markets and other public places	19	10.6
Cleaning of public places of convenience	45	25
Total	180	100.0

Table 5. Reasons for non-participation in NSD

Challenges	Frequency	Percent (%)
Busy work schedule	32	26
Forgot and already preoccupied	27	21.9
Inadequate communication and information	22	17.9
Pressing economic needs	8	6.5
Insufficient equipment	34	27.6
Total	123	100

occupational background of the study participants where majority (31%) was petty traders whose businesses flourish during the weekends (see Table 5).

The following quotes throw more light on the reason for non-participation during the NSD.

The contributing factor to the problem of sanitation is apathy. You know that education is low in the communities, so community members do not really appreciate the importance of having a clean environment thus, the practice of disposing of refuse into open drains and practicing open defecation. [Senior Operations Officer of Zoomlion Ghana Limited]

Currently, we have twelve communal containers or skips in the community which are situated at vantage points. They are not enough considering the size of the two communities. However, we are working on adding more to the existing number. [Senior Operations Officer of Zoomlion Ghana Limited]

3.2.4. Punishment for non-participation
As regards whether respondents are punished for their non participation during the NSD, majority (90%) said that nothing is done to them. However, few of the study participants indicated that non-participants are made to pay fines (7%) and/or cautioned (2%). The majority's view was buttressed by the key informants.

Nothing was done to those who do not participate in clean-up exercises. Especially in the market, the best we could do was to ensure that their shops were locked till the activity was over. The surprising thing was, the shop owners would come to the market alright but would close their shops and wait for us to clean the market. They open the shops when we are done cleaning. [Senior Operation Officer of Zoomlion Ghana Limited]

Although there are bye-laws in Ghana which enjoins us to partake in sanitation activities, people do not partake during the exercise and nothing was done to them. We tried to educate those who are closer to us on the need to keep their environment clean because it is all about us. [MUC, Aboabo No. 1]

3.2.5. Effectiveness of the NSD exercise

In accordance with the study objective, the effectiveness of the NSD was assessed based on two distinct indicators: respondents' perception about the extent of environmental sanitation before the NSD exercise and their perception about the current environmental sanitation. Table 6 presents the result for the indicators. Whereas about 38% of the respondents' described the post-sanitary condition before the NSD exercise in a positive light (very clean/clean; 37.8%), some 54% described it as grubby (filthy/very filthy; 53.9%). Likewise, whereas 40% opined that the current sanitary condition of the community have been made better (highly improved/improved; 40%), some 48% posited that the situation have degenerated (worsened/significantly worsened; 48.3%). Relatively, majority of the respondents posited that the poor sanitary condition (filthy/very filthy) of the community have become worse (worsened/significantly worsened) though the NSD is still ongoing.

3.2.6. Predictors of respondents participation in NSD exercise

The results of the binary logistic regression analysis of the unadjusted predictors of community member's participation during NSD are presented in Table 7. Factors that predict individual's participation in the NSD were: gender, age, level of education and major occupation. Though females were less likely than males to participate during the NSD, the relationship was insignificant (OR = 0.911; 95% CI [0.432–1.923]; p = 0.807). As regards educational status, participants with primary education (OR = 0.101; 95% CI = [0.325–1.814]; p = 0.546) and first degrees (OR = 0.171; 95% CI = [0.033–0.880]; p = 0.035) had odds of engaging in the exercise than those with no formal education. The odds of participating in the NSD was about 24% among respondents who were artisans than those who were unemployed (OR = 0.239; 95% CI = [0.075–0.761]; p = 0.015).

Using the Nagelkerke R^2 from the summary result, it is concluded that though the type of occupation and respondents educational background significantly predict one's participation during the NSD exercise, only 21.1% explains the variation outcome. Since the pseudo R^2 value is not close to one, the strength of association between the dependent and independent variables are not strong enough. Hence, the model has a 0.211 measure of success of predicting the dependent variable from the independent variables (Nagelkerke, 1991). Moreover, the Hosmer and Lemeshow test value, p = 0.434 (<.05), buttresses the result of the Nagelkerke R^2 result (Table 8).

Table 6. Measuring the effectiveness of the NSD exercise

Indicators for effectiveness	N = 180 (%)
Perception about environmental sanitation before the NSD exercise	
Don't know	15 (8.3)
Very clean	18 (10)
Clean	50 (27.8)
Filthy	61 (33.9)
Very filthy	36 (20)
Perception about current environmental sanitation	
Highly Improved	14 (7.8)
Improved	58 (32.2)
Worsened	58 (32.2)
Significantly worsened	29 (16.1)
No change	21 (11.7)

Table 7. Binary logistic regression of predictors of community member's participation during National Sanitation Days [NSD]

| Covariate | OR | Participation during the National Sanitation Day in the last 12 months | p-Value |
		95.0% [C.I.]	
Gender			
Male	1		
Female	0.911	0.432–1.923	0.807
Age			
18–25	1		0.855
26–33	0.767	0.325–1.814	0.546
34–41	1.134	0.313–4.112	0.848
42–49	0.522	0.119–2.287	0.388
50+	0.631	0.064–6.196	0.693
Level of Education			
None	1		0.010
Primary	0.101	0.024–0.425	0.002*
Middle/J.S.S/J.H.S	0.567	0.167–1.925	0.363
Secondary/Technical/Vocational	0.312	0.080–1.216	0.093
Training College/University	0.171	0.033–0.880	0.035*
Major occupation			
Unemployed	1		0.040
Petty trader	0.387	0.135–1.109	0.077
Artisan/craftsmanship	0.239	0.075–0.761	0.015*
Civil servant/public servant	0.857	0.233–3.148	0.816
Student	0.714	0.246–2.072	0.536

Notes: OR = Odds ratio; CI = Confidence interval—unadjusted for other explanatory variables in the table
* < 0.05

Table 8. Model summary for binary logistic regression

Model summary	Test result
Hosmer and Lemeshow test	0.434
Nagelkerke R^2	0.211

4. Discussion

This is one of the first current surveys assessing the effectiveness of the NSD as a community participatory module toward improved environmental sanitation in Ghana. Studies have been carried out in developing countries to access the participation of community members towards improved environmental sanitation (Eawag, 2005; Kar, 2005; Lüthi & Tilley, 2008; Morel et al., 2008). To contribute to this debate in the Ghanaian context, the current study has detailed the contribution of community members towards improved environmental sanitation. The participants' description of the NSD as the first Saturday of every month set aside for the citizenry to clean their communities concurs with the declaration by the MLGRD. This high response rate is consistent with findings by Shrestha (2011) in Nepal where awareness on environmental sanitation by community members increased participation during cleanup exercise. The study found information sharing as an effective tool in mobilizing the study participants for the NSD. On the one hand, participants' mentioned the community radio and the use of the print media as means of gathering them for the exercise. This is in agreement other findings (Dhokhikah et al., 2015; Hotta & Aoki-Suzuki, 2014; Ramayah, Lee, & Lim,

2012; Salequzzaman & Stocker, 2001; Sukhor et al., 2011) where information about environmental sanitation disseminated through the media were effective in mobilizing the citizenry for cleanup exercises. On the other hand, respondents' were organised through personal contact. This could be attributed to the cordial relationship that exists among the Assemblymen, members of the unit committee and members of the study community. This stems from the fact that, prior to their election as Assemblymen, the communities have been their place of abode and as such, have developed a rapport with community members. Moreover, the Assemblymen and members of the unit committee always meet majority of the citizenry at their place of worship, the Mosque, due to the Islamic setting of the community.

Moreover, the study found among the general population in Aboabo a high indulgence in the NSD which was largely influenced by the presence of prominent personalities who grace the occasion. Though the high turn-out is good, this is a challenge that needs to be critically observed and addressed since participation during the exercise could drastically reduce in the absence of these personalities. NSD participation rate is comparable to findings of other studies in Asia, Central and Latin America and East and West Africa where majority of community members participated in keeping their surroundings clean (Eawag, 2005; Kar, 2005; Lüthi & Tilley, 2008; Minkler, 2005; Morel et al., 2008; Shrestha, 2011; Taylor et al., 2006). However, it is important to note that an increased participation helps communities remain economically and socially viable when community members are involved in community project (Taylor et al., 2006). This is due to the explicit link between participation and community benefit, that is, an improvement in environmental sanitation that would help promote a hygienic environment for the survival and growth of community members and the nation as a whole.

Respondents disposed of refuse indiscriminately and showed little concern to the sanitation of their environment as they openly threw rubbish into open drains as well as practiced "wrap and throw"(the act of defecating into polythene bags and dumping them in the environment). This attitude displayed by the respondents correspond to the theory of cognitive dissonance (Eagly & Chaiken, 2007; Festinger, 1957; Kassarjian & Cohen, 1965) which explains the inconsistency that a person perceives between one's behaviour and attitudes. That is, the theory examines why people continue to undertake activities which are injurious to their health and the environment despite their knowledge of it.

As part of their participation, sweeping of the street and markets recorded the highest percentages followed by cleaning of public places of convenience and de-silting of clogged gutters. These responses were consistent with the objective for the institution of the NSD by the MLGRD (2014). Meanwhile, busy work schedule, apathy, inadequate tools to work with, pressing economic needs, indiscipline and insufficient equipment for the clean-up exercise were factors that limited respondent's continuous participation in the exercise. Apathy, lack of sanitation equipment, inadequate communication and information as limiting factors for non-participation are consistent other studies (Grodzińska-Jurczak, Tarabuła, & Read, 2003; Robinson & Read, 2005; Shaw, Lyas, Maynard, & van Vugt, 2007; Singhirunnusorn, Donlakorn, & Kaewhanin, 2012; Water and Sanitation for the Urban Poor [WSUP], 2013) where apathy towards recycling and lack of public awareness deterred members of communities from participating in sanitation activities. These factors could be attributed to the lack of commitment of community members towards the NSD. As indicated earlier, some respondents' close their shops, wait till the exercise is over and then open the shop. Similarly, the fact that no punishment were meted out for non-participation in the exercise is enough incentive for respondent's to sit aloof whilst the exercise is being undertaken using the lack of sanitation equipment and inadequate communication and information as a front to cover for their lackadaisical attitude.

Moreover, busy work schedules and pressing economic needs were also articulated by the respondents as reasons for non-participation. Similar concerns have been raised elsewhere (McDonalds & Oates, 2003; Momoh & Oladebeye, 2010; Schultz, Oskamp, & Mainieri, 1995). These bottlenecks could be explained by the predominant economic activity, petty trading, engaged in by respondents in the municipality as well as the proximity of the community to the Central Business District (CBD). Petty traders in the community have their peak sale on Saturdays which coincide with the day set aside for the NSD. More importantly, the extent of apathy displayed by the citizenry to most public-spirited but populist interventions could be expressed in the disdainfully mocking statement "Yate abre", in the local Akan dialect, which is interpreted to mean "We've heard similar things before, nothing came out of them. (This one too will not work), so stop disturbing us". With such shrugs they signal the death knell to well-intentioned propositions, just because it infringes on their zones of comfort.

Albeit, no punishment was meted out for non-participation in the exercise. No policies and measures were put in place to deter respondents from non-participation. Shrestha (2011) observed in a regional study of Nepal that despite being aware of the importance of keeping proper sanitation, community members were reluctant to take initiatives themselves and no punishment was meted out to them. These responses agrees with Minkler (2005) who found that community participation was first and foremost about community benefit, arising spontaneously, and embedded in community narratives that supported it. Thus, forcing people to participate in developmental project downplays the essence of community participation. Hence, for community participation to be sustainable it must be spontaneous and not a coercion where non-participants are punished.

An overview of participants' responses based on the benchmark provided for measuring the effectiveness of the NSD exercise was quite interesting. It can be concluded that participants within the study prefecture viewed the exercise as being ineffective in achieving its core mandate. Perhaps, given the fact that increased participation in the exercise was mostly dependent on the occasional visitation of prominent personalities on the day of the exercise, inference can be made that whenever such personalities are absent, participation level plummets. The question the author(s) ask is, for how long can these prominent personalities keep honouring the community's invitation to participate in the exercise? This is a serious issue which requires urgent attention since the tenet of the NSD could be severely jeopardized when increased participation or otherwise is premised on the presence of distinguished personalities alone.

These challenges calls for sensitization of the respondents by the various stakeholders; Zoomlion Ghana Limited, Principal Environmental and Sanitation Assistant of the AMMA, Assemblymen and members of the Unit committee, on the need to keep their surroundings clean. Doing this, sensitization, will help the community members to identify with the sanitation problem and better contribute to keeping the environment clean. Similarly, the Municipality should make available enough sanitation equipment and tools during and after the clean-up exercises to house the waste generated. Doing this would help reduce the disposal of waste into drainage systems, on the street and may improve the health of the community members.

Different studies have been carried out to ascertain the influence of demographic variables on respondent's participation in environmental sanitation (Ilevbare, 2015; Momoh & Oladebeye, 2010; Schultz et al., 1995). Using gender, age, level of education and major occupation as predictive variables for respondents' participation in the NSD, the study discovered that one's educational status and type of occupation significantly influenced their involvement during the sanitation exercise. Particularly, respondents with primary education and first degree holders were likely to be involved in the NSD than those with no formal education. Though Scott and Willits (1994) posited that education significantly influence individuals participation in sanitation activities, Momoh & Oladebeye in their Nigerian reported that participants' educational level had no significant impact on their participation in environmental sanitation efforts. Similar findings were reported by Ilevbare (2015) and Ekong (2015). Interestingly, artisans significantly participated during the exercise as compared to those who were unemployed. This contradicts other studies where civil servants participated more in environmental sanitation efforts (Momoh & Oladebeye, 2010). Though Bell, Greene, Fisher, and Baum (2001) and Arcury and Christianson (1990) identified age to be the best predictors of environmentally concerned attitudes, this study discovered that age had no impact on participants involvement in the sanitation exercises. As regards gender as a predictive variable in participants' involvement in the NSD, the study discovered no relationship. This in line with study by Ilevbare (2015). However, it contradicts the study by Zeleeny and colleagues in which women showed more environmentally responsible behaviour than men (Zelezny, Chua, & Aldrich, 2000). Moreover, our finding evicts the study by Ekong (2015) who reported a significant association between gender and participation in environmental sanitation activities ($p < 0.05$), with males showing nine times more odds of participation than females (OR = 9.84, CI = 1.225–79.018).

5. Conclusion

The study provides an empirical evidence to suggest that the knowledge base of community members about the organisation of the NSD in Aboabo is high. The high response rate is largely explained by the presence of prominent personalities who frequently grace the exercise. Their presence serves as pull factors to the exercise. Media was identified as a vibrant route for the mobilization of participants and activities such as sweeping the streets and market places, de-silting choked gutters and cleaning of public places of convenience are undertaken. Non-participation was attributed mainly to busy work schedule, poor and non-patriotic attitude, insufficient equipment to work and pressing economic needs. The problem of attitude has been shown to be among the leading factors accounting for the wanton disregard for the environment and the continuous practice of indiscriminate dumping of refuse together with the equally reprehensible act of open defecation. Moreover, the programme was identified as being ineffective in the study prefecture.

However, given the importance of a clean environment on the socio-economic life and the health of community members and the nation in general, there is the need for recognizing community participation as an interaction, rather than a coercion, arising in a community of place and as a developmental process that must be undertaken willingly so as to reap its full potential. With developing countries identified as falling behind in sanitation goals, coupled with the recent cholera outbreak in Ghana in 2014, it is imperative for community members to willingly participate in the NSD so as to continually keep their surroundings clean from needless diseases and death (Uwaegbulum, 2004). Again, participants in clean-up exercises should be motivated by way of rewarding those who participate in the exercise to serve as an incentive for others to partake. Particularly, the youth should be encouraged by opinion leaders to be involved in the exercise as reported in previous studies (Arcury & Christianson, 1990; Bell et al., 2001; DeYoung, 1991). Moreover, if needs be that prominent people are invited to intermittently grace the NSD exercise, their invitations should be to empower the local people to esteem the importance of the programme by making it their own rather than depending on their presence to increase patronage. Given that the study made use of 180 respondents as a basis to ascertain the effectiveness of the NSD as community participatory approach, it is recommended that future studies should take into consideration a larger sample size and the spatial distribution of other communities to help validate the findings of this study.

Funding
The authors received no direct funding for this research.

Competing Interests
The authors declare no competing interest

Author details
Emmanuel Mawuli Abalo[1]
E-mail: eabalo92@gmail.com
Seth Agyemang[1]
E-mail: sagye123@yahoo.co.uk
Samuel Atio[1]

E-mail: atiosamel666@gmail.com
Derrick Ofosu-Bosompem[1]
E-mail: derrickbosompem28@gmail.com
Prince Peprah[1]
E-mail: princepeprah15@gmail.com
ORCID ID: http://orcid.org/0000-0002-3816-2713
Rita Ampomah-Sarpong[2]
E-mail: godrita.rs@gmail.com

[1] Department of Geography and Rural Development, PMB, Kwame Nkrumah University of Science and Technology (KNUST), Kumasi, Ghana.

[2] Department of Sociology and Social Work, PMB, Kwame Nkrumah University of Science and Technology (KNUST), Kumasi, Ghana.

References

Adubofour, K. (2010). *Sanitation survey of Aboabo and Asawase*. Kumasi: Master of Science (Environmental Science) Faculty of Bioscience, College of Science, Kwame Nkrumah University of Science and Technology.

Adubofour, K., Obiri-Danso, K., & Quansah, C. (2013). Sanitation survey of two urban slum Muslim communities in the Kumasi metropolis, Ghana. *Environment and Urbanization, 25*, 189–207. https://doi.org/10.1177/0956247812468255

Arcury, T. A., & Christianson, E. H. (1990). Environmental worldview in response to environmental problems. Kenturkey 1984 and 1988 compared. *Environment and Behaviour, 22*(3), 387–407. https://doi.org/10.1177/0013916590223004

Asokore Mampong Municipal Health Directorate Report. (2016). *Governemnt of Ghana*. Retrieved February 10, 2016, from.

Atkinson, A. (2007). Analysis of urban trends, culture, theory, policy, action. *City, 9* (3), 264–277.

Bell, P. A., Greene, T. C., Fisher, J. D., & Baum, A. (2001). *Environmental psychology* (5th ed.). Fort Worth, TX: Harcourt College Publishers.

Bill, B. (2007). Participatory planning approaches to community interventions. *The World Bank Participation Source book*. Retrieved October 18, 2017, from http://www/plannersweb.comorg/socialanalysissourcebook

Brudney, J. L., & England, R. E. (1983). Toward a definition of the co-production concept. *Public Administration Review, 43*(1), 59–65. https://doi.org/10.2307/975300

Bryman, A. (2001). *Social research methods*. London: Oxford University Press.

Business Advocay Network. (n.d.) *Determining sample size*. Retrieved October 18, 2017, from http://www.businessadvocacy.net/dloads/fsSampleSize.pdf

Chambers, R. (1983). *Rural development: Putting the last first*. London: Longman Publishers. Published by Routledge 21/11/1983; SBN 10: 0582644437/ ISBN 13: 9780582644434

Chambers, R. L., & Clark, R. G. (2012). *An introduction to model-based survey sampling with applications: Homogenous populations*. Print ISBN-13: 9780198566625. doi:10.1093/acprof:oso/9780198566625.003.0003

Chung, S. S., & Poon, C. S. (2001). A comparison of waste-reduction practices and new environmental paradigm of rural and urban Chinese citizens. *Journal of Environmental Management, 62*, 3–19. https://doi.org/10.1006/jema.2000.0408

Dakpallah, T. A. G. (2011). *Slum improvement in Ghana: The study of Aboabo and Asawase in Kumasi*. Kumasi: Master of Science; Development Planning and Management Department of Planning, College of Architecture and Planning.

DeYoung, R. (1991). Some psychological aspects of living lightly: Desired lifestyle partens and conservation behaviour. *Journal of Environmental Systems., 20*(1), 215–227.

Dhokhikah, Y., & Trihadiningrum, Y. (2012). Solid waste management in Asiandeveloping countries: Challenges and opportunities. *Journal of Applied Environmental and Biological Sciences, 2*(7), 329–335.

Dhokhikah, Y., Trihadiningrum, Y., & Sunaryo, S. (2015). Community participation in household solid waste reduction in Surabaya, Indonesia. *Resources, Conservation and Recycling, 102*, 153–162. doi:10.1016/j.resconrec.2015.06.013

Dukeshire, S., & Thurlow, J. (2002). Challenges and barriers to community participation in policy development. *Rural Communities Impacting Policy Project*. ISBN 0-9780913-2-9. Retrieved August 7, 2017, from https://pdfs.semanticscholar.org

Eagly, A. H., & Chaiken, S. (2007). The advantages of an inclusive definition of attitude. *Social Cognition, 25*(5), 582–602. https://doi.org/10.1521/soco.2007.25.5.582

Eawag: Swiss Federal Institute of Aquatic Science and Technology. (2005, June). *Household-centred environmental sanitation: Implementing the Bellagio principles in urban environmental sanitation*. ISBN 3-906484-35-1.

Ekong, I. E. (2015). An assessment of environmental sanitation in an urban community in Southern Nigeria. *African Journal of Environmental Science and Technology., 9*(7), 592–599. doi:10.5897/AJEST2015.1882

Evans, B., Colin, C., Jones, H., & Robinson, A. (2009, May). *Sustainability and equity aspects of total sanitation programmes: A study of recent wateraid-supported programmes in three countries* (Global synthesis report). Preprint prepared for the 34th WEDC Conference.

Festinger, L. A. (1957). *A theory of cognitive dissonance*. Stanford, CA: Stanford University Press.

Ghana Statistical Service. (2010). *Population and housing census*. Accra: National Analytical Report.

Grodzińska-Jurczak, M., Tarabuła, M., & Read, A. R. (2003). Increasing participation in rational municipal waste management—a case study analysis in Jaslo City (Poland). *Resources Conservation and Recycling, 38*, 67–88. https://doi.org/10.1016/S0921-3449(02)00124-6

Hamdi, N., & Goethert, R. (1997). *Action planning for cities – a guide to community practice*. New York, NY: Wiley & Sons.

Hotta, Y., & Aoki-Suzuki, C. (2014). Waste reduction and recycling initiatives in Japanese cities: Lessons from Yokohama and Kamakura. *Waste Management & Research, 32*(9), 857–866. https://doi.org/10.1177/0734242X14539721

Ilevbare, F. M. (2015). Socio-demographic characteristics associated with waste disposal behaviour among residents in selected communities of South-western, Nigeria. *Ife Research Publications in Geography, 13*(2015), 38–48.

Ishaku, H. T., & Majid, R. M. (2010). *Community participation: Alternative approach to water supply in Nigerian rural communities. The international conference on built environment in developing countries 2010 (ICBEDC 2010)*.

Isham, J., & Kahkonen, S. (1999). *What determines the effectiveness of community-based water projects? Evidence from Central Java, Indonesia on demand, responsiveness, service rules, and social capital. Social capital initiative* (Working Paper No. 14). Washington, DC: World Bank.

Kamara, A. J. (2006). *Household participation in domestic waste disposal and recycling in the Tshwane metropolitan area: An environmental education perspective*. Pretoria: Master of Education in the subject Environmental Education. University of South Africa.

Kar, K. (2005). *A practical guide to triggering community-led total sanitation project (CLTS)*. Brighton BN1 9RE: Independent Consultant and Visiting Fellow, Institute of Development Studies, University of Sussex.

Kar, K., & Chambers, R. (2008). *Handbook on community-led total sanitation*. Brighton: Institute of Development Studies at the University of Sussex and Plan.

Kassarjian, H. H., & Cohen, J. B. (1965). Cognitive dissonance and consumer behavior: Reactions to the surgeon general's report on smoking and health. *California Management Review, 8*(1), 55–64. https://doi.org/10.2307/41165660

Lawrence, A. (2006). No personal motive? Volunteers, biodiversity and the false dichotomies of participation. *Ethics, Place and Environment, 9*(3), 279–298. https://doi.org/10.1080/13668790600893319

Levine, C. H. (1984). Citizenship and service delivery: The promise of co-production. *Public Administration Review, 44*, 178–187. https://doi.org/10.2307/975559

Lüthi, C., McConville, J., & Kvarnström, E. (2010). Community-based approaches for addressing the urban sanitation challenges. *International Journal of Urban Sustainable Development, 1*(1–2), 49–63. doi:10.1080/19463131003654764

Lüthi, C., Schertenleib, R., & Tilley, E. (2007). HCES: A new approach to environmental sanitation planning. *Waterlines, 26*, 2–4. https://doi.org/10.3362/0262-8104.2007.044

Lüthi, C., & Tilley, E. (2008). *Access to sanitation and safe water: Global partnership and local actions, 33rd WEDC international conference*, Accra, Ghana.

McDonalds, S., & Oates, C. (2003). Reasons for non-participation in a kerbside recycling scheme. *Resources, Conservation and Recycling, 39*(4), 369–385. https://doi.org/10.1016/S0921-3449(03)00020-X

MDG Report. (2015). *Ghana gets tagged as 7th dirtiest country?* Retrieved August 4, 2017, from https://www.newsghana.com.gh/ghana-gets-tagged-as-7th-dirtiest-country-mdg-report/

Ministry of Local Government and Rural Development, Report. (2014). *Government of Ghana, Accra.* Retrieved October 18, 2017, from http://www.mofep.gov.gh/sites/default/files/pbb_/2014/Local.pdf

Minkler, M. (2005). *Community organizing and community building for health* (2nd ed.). NewBrunswick, NJ: Rutgers University Press.

MoLG. (2005). *National sanitation strategy Bangladesh.* Bangladesh: Ministry of Local Government, Rural Government and Cooperatives. Retrieved October 18, 2017, from http://www.watersanitationhygiene.org/References/EH_KEY_REFERENCES/SANITATION/General%20Sanitation%20References/Bangladesh%20National%20Sanitation%20Strategy%20(MRDC).pdf

Momoh, J. J., & Oladebeye, D. H. (2010). Assessment of awareness, attitude and willingness of people to participate in household solid waste recycling programme in Ado-Ekiti, Nigeria. *Journal of Applied Sciences in Environmental Sanitation., 5*(1), 93–105.

Mongkolnchaiarunya, J. (2005). Promoting a community-based solid wastemanagement initiative in local government: Yala municipality, Thailand. *Habitat International, 29*, 27–40. https://doi.org/10.1016/S0197-3975(03)00060-2

Morel, A., Luethi, C., & Schertenleib, R. (2008). *Integrate at the top, involve at the bottom – The household centred approach to environmental sanitation.* Dübendorf: Eawag - Swiss Federal Institute of Aquatic Science and Technology.

Nagelkerke, N. J. D. (1991). A note on a general definition of the coefficient of determination. *Biometrika, 78*(3), 691–692. https://doi.org/10.1093/biomet/78.3.691

Nance, E. B. (2004). *Putting participation in context: An evaluation of urban sanitation in Brazil* (PhD diss.). Stanford University, Stanford, CA.

Nance, E., & Ortolano, L. (2007). Community participation in urban sanitation experiences in Northeastern Brazil. *Journal of Planning Education and Research, 26*, 284–300. doi:10.1177/0739456X06295028. © 2007 Association of Collegiate Schools of Planning

Narayan, D. (1995). *The contribution of people's participation: Evidence from 121 rural water supply projects. Environmentally sustainable development occasional* (Paper Series No. 1). Washington, DC: World Bank.

Nour, A. M. (2011). Challenges and advantages of community participation as an approach for sustainable urban development in Egypt. *Journal of Sustainable Development, 4*(1), 86.

Ostrom, E. (1992). *Crafting institutions for self-governing irrigation systems.* San Francisco, CA: ICS Press.

Percy, S. L. (1984). Citizen participation in the co-production of urban services. *Urban Affairs Review, 19*(4), 431–446. https://doi.org/10.1177/004208168401900403

Prokopy, L. S. (2002). *The relationship between participation and project outcomes: A study of rural drinking water projects in India* (PhD diss.). University of North Carolina, Chapel Hill, NC.

Prokopy, L. S. (2005). The relationship between participation and project outcomes: Evidence from rural water supply projects in India. *World Development, 33*(11), 1801–1819. https://doi.org/10.1016/j.worlddev.2005.07.002

Ramayah, T., Lee, J. W. C., & Lim, S. (2012). Sustaining the environment through recycling: An empirical study. *Journal of Environmental Management, 102*, 141–147. https://doi.org/10.1016/j.jenvman.2012.02.025

Repetto, R., Dower, R. C., Jenkins, R., & Geoghegan, J. (1992). *Green fees: How a tax shift can work for the environment and the economy.* World Resources Institute. Retrieved August 8, 2017, fromhttp://pdf.wri.org/greenfees_bw.pdf

Robinson, G. M., & Read, A. D. (2005). Recycling behaviour in a London Borough: Results from large-scale household surveys. *Resources, Conservation and Recycling, 45*, 70–83.

Rosemarin, A., Ekane, N., Caldwell, I., Kvarnström, E., McConville, J., Ruben, C., & Fogde, M. (2008). *Pathways for sustainable sanitation: Achieving the millennium development goals.* Stockholm: IWA Publishing, EcoSanRes Programme, Stockholm Environment Institute.

Salequzzaman, M., & Stocker, L. (2001). The context and prospects for environmental education and environmental career in Bangladesh. *International Journal of Sustainability in Higher Education, 2*(2), 104–127. https://doi.org/10.1108/14676370110388309

Sara, J., & Katz, T. (1997). *Making rural water supply sustainable: Report on the impact of project rules.* Washington, DC: UNDP-World Bank Water and Sanitation Program.

Schultz, P. W., Oskamp, S., & Mainieri, T. (1995). Who recycles and when? A review of personal and situation actors. *Journal of Environmental Psychology., 15*(2), 105–121. https://doi.org/10.1016/0272-4944(95)90019-5

Schumacher, F. (1976). *Small is beautiful – Economics as if people mattered.* London: Abacus.

Scott, D., & Willits, F. K. (1994). Environmental attitude and behaviour. A Pennislavia survey.. *Environmental and Behaviour., 26*(2), 239–260. https://doi.org/10.1177/001391659402600206

Stockholm Environment Institute. (2008, August 25–26). *Proceedings from SEI/EcoSanRes2 workshop: Planning and implementation of sustainable sanitation in peri/semi-urban settings – a need for development of existing tools?,* Stockholm.

Shaw, P. J., Lyas, J. K., Maynard, S. J., & van Vugt, M. (2007). On the relationship between set-out rates and participation ratios as a tool for enhancement of kerbside household waste recycling. *Journal of Environmental Management, 83*, 34–43. https://doi.org/10.1016/j.jenvman.2006.01.012

Shrestha, R. L. (2011). *WaterAid in Nepal report - People's perception on sanitation: Findings from Nepal.* Gulmi: A Water Aid in Nepal Publication.

Singhirunnusorn, W., Donlakorn, K., & Kaewhanin, W. (2012). Contextual factors influencing household recycling behaviours: A case of waste bank project in Mahasarakham municipality. *Procedia-Social and Behavioural Sciences, 36*, 688–697. https://doi.org/10.1016/j.sbspro.2012.03.075

Skumatz, L. A. (1996). *Nationwide diversion rate study: Quantitative effects of program choices on recycling and green waste diversion–beyond case studies.* Superior, CO: Report Prepared by Skumatz Economic Research Associates, Inc.

Smith, M. K. (2006). *'Community participation', the encyclopaedia of informal education.* Retrieved October 18, 2017, from www.infed.org/community/b-compar.htm

Sukhor, F. S. A., Mohammed, A. H., Sani, S. I. A., & Awang, M. (2011). *A review on the success factors for community* participation in solid waste management. *International Conference on Management (ICM) Proceeding.*

Sundeen, R. A. (1985). Co-production and communities: Implications for local administrators. *Administration and Society, 16*(4), 387–402. https://doi.org/10.1177/009539978501600401

Taylor, J., Wilkinson, D., & Cheers, B. (2006). Community participation in organising rural general practice: Is it sustainable? *Australian Journal of Rural Health, 14*, 144–147. doi:10.1111/j.1440-1584.2006.00790.x

Tilley, E., Atwater, J., & Mavinic, D. (2008). Recovery of struvite from stored human urine. *Environmental Technology, 29*(7), 807–816. https://doi.org/10.1080/09593330801987145

Ton, S., & Patrick, M. (2003). *Community water management. From system to service in rural areas* (pp. 103–105). Southampton Row, London: ITDG publishing.

Tukahirwa, J. T., Mol, A. P. J., & Oosterveer, P. (2010). Civil society participation in urban sanitation and solid waste management in Uganda. *Local Environment, 15*(1), 1–14. doi:10.1080/13549830903406032

UN JMP. (2008). *Progress on drinking water and sanitation – special focus on sanitation.* Geneva: Joint Monitoring Programme for Water Supply and Sanitation. Retrieved October 18, 2017, from http://www.who.int/water_sanitation_health/monitoring/jmp_report_7_10_lores.pdf

United Nations Population Fund. (2007). *State of the world population 2007.* New York, NY: Author. Retrieved October 18, 2017, from https://www.unfpa.org/sites/default/files/pub-pdf/695_filename_sowp2007_eng.pdf

Uwaegbulum, C. (2004). World is meeting goals of safe drinking water but failing behind sanitation, says UN. *The Guardian Newspaper*, p. 50.

Wade, R. (1988). The management of irrigation systems: How to evoke trust and avoid the prisoners' dilemma. *World Development, 16*(4), 489–500. https://doi.org/10.1016/0305-750X(88)90199-4

Wahabu, A., Oduro-Kwarteng, S., Monney, I., & Kotoka, P. (2014). Characteristics of diverted solid waste in Kumasi: A Ghanaian city. *American Journal of Environmental Protection, 3*(5), 225–231. doi:10.11648/j.ajep.20140305.13

Water and Sanitation for the Urban Poor. (2013). *Getting communities engaged in water and sanitation projects: Participatory design and consumer feedback, USAID.* Retreived October 18, 2017, from http://www.wsup.com/resource/getting-communities-engaged-in-water-and-sanitation-projects-participatory-design-and-consumer-feedback/

Water and Sanitation Program. (2012). *Economic impacts of poor sanitation in Africa.* Retrieved June 1, 2016, from http://siteresources.worldbank.org/INTAFRICA/Resources/economic-impacts-of-poor-sanitation-in-africa-factsheet.pdf

WHO & UNICEF. (2015). *Progress on sanitation and drinking water. 2015 Update and MDG assessment.* Retrieved August 7, 2017, from http://files.unicef.org/publications/files/Progress_on_Sanitation_and_Drinking_Water_2015_Update_.pdf

World Bank. (2004). *World development report 2004: Making services work for poor people.* New York, NY: Oxford University Press.

Water & Sanitation Programme. (2007, January). *From burden to communal responsibility. A success story from southern region in Ethiopia.* Nairobi: Author Field Note.

Zelezny, L. C., Chua, P., & Aldrich, C. A. (2000). New ways of thinking about environmentalism: Elaborating on gender differences environmentalism. *Journal of Social Issues, 56*(3), 443–457. https://doi.org/10.1111/0022-4537.00177

Zhu, D., Asnani, P. H., Zurbrügg, C., Anapolsky, S., & Mani, S. (2008). *Improving municipal solid waste management in India, A source book for policy makers and practictioners.* Washington, DC: World Bank.

Zurbrügg, C. (2002). Urban solid waste management in low-income countries of Asia: How to cope with the garbage crisis. *Proceeding of the scientific committee on problems of the environment (SCOPE), urban solid waste management review session,* Durban, South Africa (p. 1).

Zurbrügg, C., Drescher, S., Patel, A., & Sharatchandra, H. C. (2004). Decentralisedcomposting of urban waste–an overview of community and private initiativesin Indian cities. *Waste Management, 24*, 655–662. https://doi.org/10.1016/j.wasman.2004.01.003

7

A multipollutant evaluation of APEX using microenvironmental ozone, carbon monoxide, and particulate matter (PM$_{2.5}$) concentrations measured in Los Angeles by the exposure classification project

Ted R. Johnson[1], John E. Langstaff[2]*, Stephen Graham[2], Eric M. Fujita[3] and David E. Campbell[3]

*Corresponding author: John E. Langstaff, U.S. Environmental Protection Agency, 109 TW Alexander Drive, Research Triangle Park, NC 27711, USA

E-mail: Langstaff.John@epa.gov

Reviewing editor: Alberto Bezama, Helmholtz Center for Environmental Research, Germany

Abstract: This paper describes an operational evaluation of the US Environmental Protection Agency's (EPA) Air Pollution Exposure Model (APEX). APEX simulations for a multipollutant ambient air mixture, i.e. ozone (O$_3$), carbon monoxide (CO), and particulate matter 2.5 microns in diameter or less (PM$_{2.5}$), were performed for two seasons in three study areas in central Los Angeles. APEX predicted microenvironmental concentrations were compared with concentrations of these three pollutants monitored in the Exposure Classification Project (ECP) study during the same periods. The ECP was designed expressly for evaluating exposure models and measured concentrations inside and outside 40 microenvironments. This evaluation study identifies important uncertainties in APEX inputs and model predictions useful for guiding further exposure model input data and algorithm development efforts. This paper also presents summaries of the concentrations in the different microenvironments.

ABOUT THE AUTHORS

Ted R. Johnson is the President and Research Director of TRJ Environmental, Inc., a consulting company specializing in analysis of air quality data and estimation of population exposure. He has designed exposure models used by the US EPA to simulate the exposure of urban populations to air pollution.

John E. Langstaff and Dr Stephen Graham are exposure modelers at the US EPA. Their research encompasses the development and application of exposure modeling techniques.

Dr Eric M. Fujita, emeritus research professor at the Desert Research Institute (DRI), was the Principal Investigator for the ECP sampling and analysis project. His research interests included chemical characterization of emission sources and measurement and characterization of exposures to toxic air contaminants.

Dave E. Campbell is an associate research scientist at DRI, whose research interests include characterization and apportionment of gaseous and aerosol pollutants from mobile sources, and the influence of mobile source contributions on photochemical processes.

PUBLIC INTEREST STATEMENT

Decades of research and numerous studies have consistently indicated air pollution contributes to sickness, disease development, and premature death. To best understand the relationship between air pollution and the negative impacts to human health, it is crucial to account for how people might come in contact with pollutants and experience the important features of exposure, such as the magnitude, duration, frequency, and pattern of pollutant concentrations that occur in their immediate surroundings. In this study, we use a novel multipollutant exposure modeling approach that combines the complexities of human behavior with air pollutant concentrations that vary across an urban area considering movement across space, time, and interaction within built-environments. While reasonable agreement was observed between model estimations and measured concentrations, the most significant uncertainties are identified to further enhance the benefits associated with using a model based approach to estimate multipollutant exposures.

Subjects: Environmental Sciences; Environment & Health; Pollution

Keywords: multipollutant exposure model; microenvironment measurements; ozone; particulate matter; carbon monoxide

1. Introduction

The EPA has used APEX to estimate human exposure to ozone (O_3), carbon monoxide (CO), sulfur dioxide (SO_2), nitrogen dioxide (NO_2), and other air pollutants (US EPA, 2008, 2009a, 2010a, 2014), in support of reviews of the National Ambient Air Quality Standards (NAAQS). APEX is used because of its flexible, user-defined physical-based microenvironmental approach and, when combined with extensive activity pattern and population databases, temporally resolved ambient concentrations, and well-parameterized distributions that capture variability in other model inputs, can probabilistically and more realistically estimate population-based human exposures to air pollutants. In each of these past NAAQS exposure assessments, the agency has applied APEX to a single pollutant. Because of growing interest in health effects associated with multiple chemical stressors, APEX has also been recently developed to estimate simultaneous exposures to ambient air pollutants.

This paper describes an evaluation of a multipollutant application of APEX by comparing microenvironmental concentrations of O_3, CO, and fine particulates ($PM_{2.5}$) estimated by APEX in each of three study areas within central Los Angeles with corresponding microenvironmental concentrations measured by the Exposure Classification Project (ECP) in the same three study areas. There are a number of other air pollution exposure models, for example, INDAIR-2/EXPAIR (Dimitroulopoulou, Ashmore, & Terry, 2017), LHEM (Smith et al., 2016), EMI (Breen et al., 2015), HAPEM (US EPA, 2015), EXPAND (Soares et al., 2014), MENTOR (Georgopolis and Lioy, 2006), EXPOLIS (Kruize, Hanninen, Breugelmans, Lebret, & Jantunen, 2003), and SHEDS (Burke, Zufall, & Ozkaynak, 2001). APEX was selected for this evaluation since it is used in regulatory applications and it can model short-term exposures to multiple pollutants in multiple microenvironments.

The three pollutants were selected in this evaluation due to their having been the subject of NAAQS-related exposure and/or risk assessments (US EPA, 2010a, 2010c, 2014) and, as such, are widespread air pollutants having multiple sources and are reasonably expected to endanger public health. More specifically, O_3 is formed photo-chemically via sunlight and precursor chemical emissions from anthropogenic (largely combustion-related) and natural sources. The strongest evidence is for adverse health effects that are respiratory-related and result from short-term (hours to weeks) O_3 exposures, as O_3 has been determined to cause clinically significant lung function impairment and is associated with increased hospital admissions and emergency department visits (US EPA, 2013). Although there are a limited number of studies, short-term O_3 exposure has also been linked with cardiovascular-related morbidity (US EPA, 2013). Regarding CO, a pollutant largely emitted from internal combustion engines (e.g. gasoline powered automobiles), clinical studies among individuals with coronary artery disease showed consistent decreases in the time to onset of exercise-induced angina and ventricular repolarization (or ST-segment) changes following short-term (one hour to a few hours) CO exposures (US EPA, 2010b). Combustion sources are also largely responsible for ambient $PM_{2.5}$, and epidemiological studies show consistent, significant associations of both short-term and long-term $PM_{2.5}$ exposures with a variety of cardiovascular- and respiratory-related health effects, including mortality (US EPA, 2009b). Therefore, given a general correspondence of select exposure- and health-related attributes for each individual chemical (e.g. short-term exposure and cardiovascular effects), realistically quantifying the simultaneous time series exposure profiles for each of these pollutants by understanding when the highest multi-chemical exposures occur could be very important in better understanding their potential cumulative health effects.

2. The air pollution exposure model (APEX)

The Air Pollution Exposure Model (APEX; US EPA, 2017), has its origins in the NAAQS Exposure Model (NEM) initially developed in the early 1980s (Biller et al., 1981; McCurdy, 1994, 1995). APEX simulates the movement of individuals through time and space and their exposure to a given pollutant in indoor, outdoor, and in-vehicle microenvironments.

The model stochastically generates simulated individual characteristics and behaviors using Census-derived probability distributions for their demographics. Any number of simulated individuals can be modeled; by design, they can represent a random sample of the study area population. Survey-derived time activity data, or diaries, are used to construct a sequence of activity events (locations visited and activities performed) for each simulated individual for a day or longer up to a year. The selected diary data are consistent with the individual's demographic characteristics (e.g. age, gender) and account for the influential effects of day-type (e.g. weekday, weekend) and outdoor temperature on daily activities. APEX calculates the concentration in the microenvironment associated with each event in an individual's activity pattern and time-averages the event-specific exposures within each user-specified timestep (typically 1 h). It then uses this information to obtain a continuous time series of exposures spanning the duration of interest. From these exposure estimates, APEX calculates exposures for averaging times greater than the timestep—8 h and 24 h averages from 1 h timesteps; 1 h averages from 5 min timesteps (US EPA, 2014).

APEX employs a flexible approach for simulating microenvironmental concentrations, where the user can define any number of microenvironments to be modeled and their associated characteristics. The concentrations in each microenvironment can be calculated using either a factors or mass-balance approach, depending upon data availability. Probability distributions are used to represent the variability (rather than the uncertainty) of the input data that enter into the calculations (e.g. indoor–outdoor air exchange rates). The parameters of the distributions can vary temporally and spatially and can be set-up to depend on the values of other model input variables. For example, the distribution of air exchange rates in a home may depend on average outdoor temperature and whether air conditioning is present. The value of a stochastic variable can be kept constant for an individual for the entire simulation (e.g. house volume), or a new value can be drawn hourly, daily, or seasonally from specified distributions. APEX also allows the specification of diurnal, weekly, and seasonal patterns for microenvironmental variables (US EPA, 2014).

3. The exposure classification project (ECP)

Field studies that measure pollutant concentrations in microenvironments can be useful in developing and validating models for specific microenvironments. The Exposure Classification Project (ECP) was designed specifically to provide such data. The Desert Research Institute (DRI) measured personal breathing-zone concentrations within several microenvironments including in-vehicle, near-road, and various public indoor and outdoor microenvironments. Air pollutants measured included O_3, $PM_{2.5}$ mass, ultrafine particles, black carbon, volatile air toxics (benzene, toluene, ethylbenzene, xylenes, 1,3-butadiene), CO, carbon dioxide (CO_2), additional volatile organic compounds (VOCs), and nitrogen oxides (NO_x). The microenvironmental measurements discussed in this paper were made during two field campaigns: Season 1 (12 days between 9/13/08 and 10/8/08) and Season 2 (11 days between 3/2/09 and 3/19/09) in three study areas in the Los Angeles metropolitan area: Carson/Long Beach, downtown Los Angeles, and Alhambra/Monterey Park (Figure 1). Corresponding outdoor measurements were made immediately following or preceding many of the indoor microenvironmental measurements. In-vehicle measurements were made throughout Los Angeles County for various types of roads and traffic conditions. The lower detection limits were ~1.5 ppb for O_3, ~1 ppm for CO, and ~1 µg/ m^3 for $PM_{2.5}$. The experimental protocols are described by Fujita, Campbell, Arnott, Johnson, and Ollison (2014) and in two interim reports prepared by Fujita, Campbell, and Zielinska (2009a, 2009b).

The ECP microenvironmental measurements were made during a series of tests that followed a prepared script. During each test, the technician measured concentrations of various air pollutants in a single microenvironment for a time period that varied from 1 min to 60 min (median = 12 min), during which local temperature, relative humidity, and wind speed were also recorded. In addition, the technician provided a written description of the location and microenvironment. Each multipollutant measurement in the ECP database was assigned a unique test number and was labeled as to date,

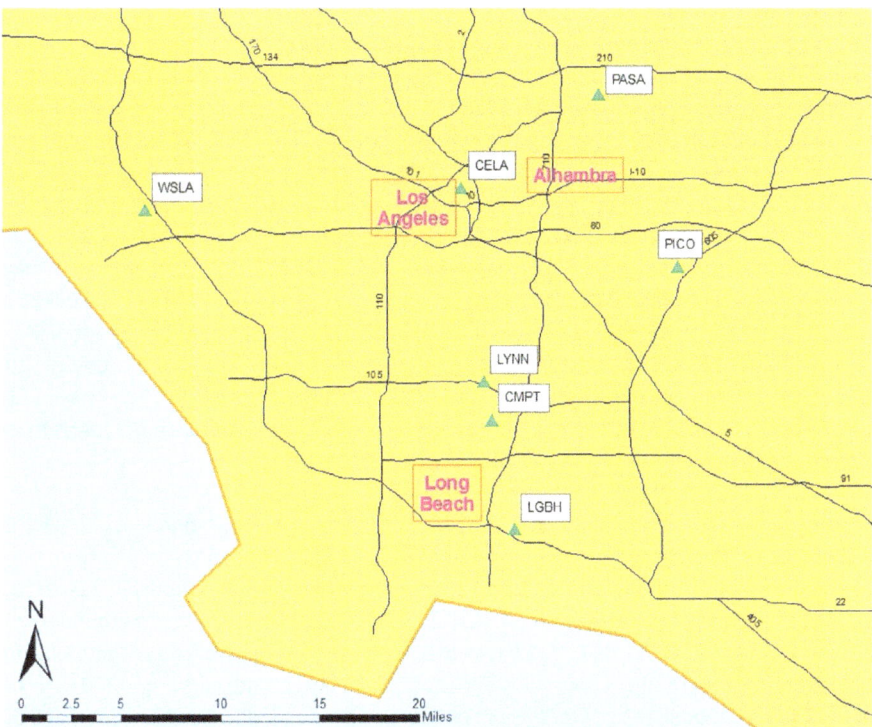

Figure 1. Microenvironmental measurements were made in three study areas, downtown Los Angeles, Alhambra, and Long Beach.

Note: Triangles indicate locations of nearby air monitoring stations of the South Coast Air Quality Management District (SCAQMD).

time, geographic location, and microenvironment. Each test belongs to a group of tests that were made at the same geographic location on the same date within a period of two hours or less. In total there were 87 of these "test groups" having pollutant concentrations measured in one or more microenvironments; 39 occurred during the 2008 measurement campaign while 48 occurred in 2009. These test groups provide a means of analyzing the relationships (e.g. ratios, differences, correlations) between the air pollutant concentrations measured in two different microenvironments at the same geographic location and occurring at approximately the same time. The final microenvironmental measurements totaled 419 tests. Technicians attempted to measure all of the target pollutants in 331 of the 419 tests; the target pollutants were limited to $PM_{2.5}$, VOC, and CO in the remaining 88 tests.

Ideally in this study, corresponding pollutant concentrations estimated by APEX for the same three Los Angeles locations under similar conditions would have statistical properties similar to those of the ECP data. In particular, we would expect a similarity in the ratios of pollutant concentrations measured simultaneously in specific pairs of related microenvironments, such as inside a school and immediately outside the same school. We calculated three sets of ratios for each the APEX generated and ECP measurement concentrations: (1) microenvironmental concentrations to outside concentrations, (2) microenvironmental concentrations to ambient monitor concentrations and, (3) microenvironmental concentrations to near-road concentrations (ECP measurement data only). Descriptive statistics (minimum, median, and maximum) were used to compare and contrast the overall distribution of each the corresponding APEX and ECP datasets.

The Supplemental Material presents statistical summaries of the ECP data. Descriptive statistics are presented for the O_3, CO, and $PM_{2.5}$ concentrations in each of the 40 microenvironments assigned to the ECP data. Percentiles of the ratios of mean O_3, CO, and $PM_{2.5}$ concentrations for specified pairs of microenvironments based on measurements made during the same test group are also given for 19 pairings of microenvironments.

4. Similar measurement studies

Several studies have made measurements of pollutants in microenvironments. Some studies report indoor/outdoor (I/O) ratios where the outdoor measurements are just outside the microenvironment, others report indoor/ambient ratios, where "ambient" is at fixed site central monitor(s). In this paper, we distinguish between these. The majority of these studies are for $PM_{2.5}$ in residential microenvironments (Mohammed et al., 2015). Breen et al. (2015), in an evaluation of the Exposure Model for Individuals (EMI), measured a mean $PM_{2.5}$ I/O ratio of 0.58 (5th to 95th percentiles 0.32 to 0.84) in residences. Allen et al. (2012) in a study of 353 homes, found a mean I/O ratio of 0.62 (standard deviation 0.21), ranging from 0.47 to 0.82 across seven communities. Long, Suh, and Koutrakis (2000) measured $PM_{2.5}$ at 9 residences and report the mean I/O ratios as 2.4 ± 14 (range 0.03–257) for daytime and 0.74 ± 0.41 (range 0.03–3.7) for nighttime. Lai et al. (2004) found a residential $PM_{2.5}$ I/O ratio of means to be 1.9. Meng, Spector, Colome, and Turpin (2009), in a 3-city study of 114 homes found a mean I/O ratio of 0.69 (standard deviation 0.23), ranging from 0.1 to 1.3. The PTEAM tudy (Pellizzari, Thomas, Clayton, Whitmore, & Shores, 1992) found the median 24 h I/O ratios to be 0.94, with 10th and 90th percentiles 0.40 and 2.05, for 178 residences.

There are some studies of non-residential microenvironments and of O_3, CO, and $PM_{2.5}$. Challoner and Gill (2014) looked at 10 commercial buildings and measured $PM_{2.5}$ I/O ratios ranging from 0.47 to 4.68 with a median of 1.26. Zhang and Zhu (2012) measured $PM_{2.5}$ at five schools in Texas and found I/O ratios at the schools to be 0.34, 0.30, 0.38, 0.27, and 0.53. Crist, Liu, Kim, Deshpande, and John (2008) found I/O ratios for $PM_{2.5}$ of 2.61, 1.71, and 2.98 on school days at three schools in Ohio. Ratios on non-school days were much lower. Hanninen, Kruize, Lebret, and Jantunen (2003), in an evaluation of the EXPOLIS model, reports for $PM_{2.5}$ in residences a mean indoor/ambient ratio of 0.93 and a work indoors/ambient ratio of 1.01. Chaloulakou, Mavroidis, and Duci (2003) measured CO at an office and a school. Mean daily I/O ratios ranged from 0.74 to 1.00 at the office and 0.53 to 0.89 at the school. Weschler (2000) summarized the I/O ratios for O_3 from 55 sets of measurements in homes and several other microenvironments, ranging from < 0.1 to 1.0. These previous measurements are generally consistent with the ECP measurements in this study, except for the high values recorded by Long et al. (2000), which were due to indoor sources. Our study differs from previous studies in that ECP sampled several pollutants in 40 microenvironments, both inside and outside the microenvironments.

Branco, Alvim-Ferraz, Martins, and Sousa (2014) point out the need for evaluation of the microenvironment modeling approach and Milner, Vardoulakis, Chalabi, and Wilkinson (2011) state "There is a need for further measurement studies on indoor air exposure to provide independent monitoring data-sets for testing of exposure models. Large-scale studies should, where possible, make simultaneous observations on indoor and outdoor concentrations …" Kruize et al. (2003) noted "it would be very helpful if more databases on (indoor) concentration data … were published and made available." This study makes a significant addition to the literature reporting on indoor and outdoor microenvironmental concentrations.

5. APEX inputs for the multipollutant evaluation

As described above, based on an evaluation of ECP and fixed-site monitoring data availability, we focused our multipollutant evaluation on three pollutants: O_3, CO, and $PM_{2.5}$. Therefore, we set up a multipollutant APEX simulation which modeled population exposures to all three pollutants using a single set of simulated individuals. In this way, the daily activity patterns used to estimate the exposures of each individual are the same for all three pollutants (i.e. the same locations visited, activities performed, and microenvironmental settings, all occurring at the same times of the day). One thousand individuals were simulated for each season and each of the three communities, for a total of six APEX simulations. Preliminary simulations showed that modeling this number of individuals provided sufficient numbers of values (at least 300 in each microenviroment) for the statistical analyses conducted. Note that we are not comparing exposures, but concentrations in microenviroments, so the population demographics are not relevant.

In setting up an APEX simulation that includes CO and $PM_{2.5}$ exposures, the user can enter parameter values that account for specific indoor sources of CO (e.g. smoking and gas stoves) and $PM_{2.5}$ (numerous sources). We determined that modeling these sources within the multipollutant APEX run would unnecessarily complicate the comparison of APEX exposure estimates and ECP pollutant measurements, since we can theoretically identify the APEX events affected by indoor sources but not necessarily the ECP measurements. Therefore, the contribution of CO and $PM_{2.5}$ from indoor sources was not modeled by APEX.

APEX has several options and types of input data that allow an application to be tailored to a specific area and scenario. In order to conduct an unbiased evaluation of a typical model application, we used the model options and inputs that were used in a recent application of APEX for O_3 to the Los Angeles area (US EPA, 2014), except adjusting the time periods modeled to coincide with the ECP sample collection dates. The microenvironmental definitions were revised to match them with the microenvironments in the ECP database, but the microenvironmental parameter settings were not changed, except for the proximity factors, which were updated to account for the spatial variability of concentrations between monitors. Proximity factors are stochastic ratios between monitors and other outdoor locations.

The APEX input data fall into the following general categories: human activity data; population, employment, and commuting data; air quality data; temperature data; physiological data; and microenvironmental data or variables. The human activity data are from the Consolidated Human Activity Database (CHAD) (McCurdy, Glen, Smith, & Lakkadi, 2000; US EPA, 2002), which contains over 50,000 daily activity diaries. The population demographics were obtained from the 2000 US Census data at the tract level, and a national commuting database based on 2000 Census data provides home-to-work commuting flows between tracts. Hourly surface temperature measurements were obtained from the National Weather Service data files (http://www.ncdc.noaa.gov/oa/ climate/surfaceinventories.html). APEX assigns the data from the closest weather station to each Census tract. Temperatures are used by APEX both in selecting activity diaries used to simulate the exposed individual and in estimating air exchange rates (AERs) for indoor microenvironments. The default APEX physiological data were used (e.g. distributions of body mass by age and gender, see Isaacs & Smith, 2005). Microenvironmental variables include AERs, decay rates, penetration rates, and proximity factors. The $PM_{2.5}$ decay (deposition) rates are taken from Bouilly, Karim, Claudine, and Allard (2005). The data underlying the other microenvironment variables are described in US EPA (2014).

The parameters describing the distributions used in estimating microenvironmental concentrations are listed in Table 1. Because air exchange is a physical characteristic of the indoor microenvironment, the AERs appropriately apply to all pollutants. The mass balance model is used for the indoor microenvironments and the regression factors model is used for outdoor and in-vehicle microenvironments. 100% pollutant indoor penetration rates are assumed in lieu of data; as penetration rate measurements become available, the distributions input to APEX can be updated. These models and the various inputs are described in greater detail in the APEX user's guides (EPA, 2017).

Air quality data reported by seven fixed-site monitors within and around the ECP sampling locations were used for input to APEX. APEX uses the ambient concentration data from the closest air quality monitoring site to each Census tract. Figure 1 shows the locations of the seven monitoring sites: Central Los Angeles (CELA), Compton (CMPT), Long Beach (LGBH), Lynwood (LYNN), Pasadena (PASA), Pico Rivera (PICO), West Los Angeles (WSLA). Hourly measurements of ambient O_3, $PM_{2.5}$, and CO and daily measurements of $PM_{2.5}$ were compiled from EPA's Air Quality System (https://www.epa.gov/aqs). Adequate 1 h monitoring data were available surrounding the three study areas for CO and O_3. However, the availability of 1 h $PM_{2.5}$ data were limited. There are two sites (CELA and Glendora) within 20 miles of the study areas that have 1 h $PM_{2.5}$ data for 2008. There are six monitor sites (CELA, LGBH, Anaheim, Burbank, Glendora, and Reseda) within that radius that have 1 h $PM_{2.5}$ data for 2009. Monitored data for the remaining ECP pollutants were insufficient to perform comparisons with APEX.

Table 1. APEX microenvironment parameters

Microenvironment	Parameter	Conditions	Distribution
Indoors—residence	AER	Temp < 68; A/C: central	LogN(0.577, 1.897, 0.1, 10)
		Temp 68–76; A/C: central	LogN(1.084, 2.336, 0.1, 10)
		Temp 77–85; A/C: central	LogN(0.861, 2.415, 0.1, 10)
		Temp > 85; A/C: central	LogN(0.861, 2.344, 0.1, 10)
		Temp < 68; A/C: room	LogN(0.672, 1.863, 0.1, 10)
		Temp 68–76; A/C: room	LogN(1.674, 2.223, 0.1, 10)
		Temp > 76; A/C: room	LogN(0.949, 1.644, 0.1, 10)
		Temp < 50; A/C: none	LogN(0.543, 3.087, 0.1, 10)
		Temp 50–67; A/C: none	LogN(0.747, 2.085, 0.1, 10)
		Temp 68–76; A/C: none	LogN(1.372, 2.283, 0.1, 10)
		Temp > 76; A/C: none	LogN(0.988, 1.967, 0.1, 10)
Indoors–school	AER	All	Discrete (range 0.1 to 3.0)
Indoors–restaurant, bar, night club, café	AER	All	LogN(3.712, 1.855, 1.46, 9.07)
Indoors–other	AER	All	LogN(1.109, 3.015, 0.07, 13.8)
Indoors–All	O_3 Decay rate	All	LogN(2.51, 1.53, 0.95, 8.05)
	PM_{25} Decay rate	All	Uniform(0.1,1.1)
	CO Decay rate	All	No decay
	O_3 Proximity	All	Normal(1.0, 0.07, 0.9, 1.1)
All MEs	PM_{25} Proximity	All	Normal(1.0, 0.07, 0.9, 1.1)
All MEs	CO Proximity	All	Normal(1.0, 0.15, 0.8, 1.2)
Outdoors–near road	O_3 Proximity	All	Normal(0.755, 0.203, 0.422, 1.0)
Outdoors–other	O_3 Proximity	All	Normal(1.0, 0.07, 0.9, 1.1)
In-vehicle	O_3 Proximity	Local roads (6%)	Normal(0.755, 0.203, 0.422, 1.0)
		Urban roads (65%)	Normal(0.754, 0.243, 0.355, 1.0)
		Interstates (29%)	Normal(0.364, 0.165, 0.093, 1.0)
	O_3 Penetration	All	Normal(0.300, 0.232, 0.1, 1.0)

Notes: Temp is daily average temperature in degrees Fahrenheit. A/C indicates the type of air conditioning. LogN is lognormal(geometric mean, geometric standard deviation, minimum, maximum). Normal is Gaussian(mean, standard deviation, minimum, maximum). When a sampled value is below the minimum or above the maximum, it is resampled by APEX (Gaussian and lognormal distributions).

The APEX results presented in this paper were calculated from the APEX events output files (i.e. the complete time-series of concentrations in each microenviroment, on a minute-by-minute basis) and were subset to times between 7 am and 8 pm, since the ECP data were collected during these hours of the day.

6. Preparation of the ECP database

To facilitate future statistical analyses, we created a set of 40 ECP codes for classifying the microenvironments where measurements were collected, based on the ECP descriptions of the sampled microenvironments. The codes include 14 indoor microenvironments, 17 outdoor microenvironments, and 9 in-vehicle microenvironments. This code set was considered to provide adequate specificity while also increasing sample sizes for most microenvironments. In addition, the code set provided several microenvironmental classifications that were specific to certain locations within the study area, such as the Metro transit center.

To further increase the number of measurements representing each microenvironment while maintaining sufficient specificity, we defined a condensed set of 24 microenvironments which we used in the APEX modeling, such that each of the 40 ECP microenvironments could be mapped to one of the 24 APEX microenvironments (Table 2). CO and $PM_{2.5}$ were measured in all 24 microenvironments, while for O_3, there are measurements in all of the APEX microenvironments except

Table 2. Mapping of 40 ECP microenvironments to condensed set of 24 APEX microenvironments	
APEX microenvironment	**ECP microenvironment**
Indoors	
01: indoors–residence	11: indoors–apartment
02: indoors–community center or auditorium	12: indoors–community center or auditorium
03: indoors–restaurant	13: indoors–food court
	17: indoors–restaurant-fast food
04: indoors–hotel/motel	14: indoors–hotel
05: indoors–office building, bank, post office	15: indoors–office/office building
	16: indoors–public building (post office, etc.)
06: indoors–bar, night club, café	18: indoors–restaurant-café or other
07: indoors–school	19: indoors–school
08: indoors–shopping mall/non-grocery store	20: indoors–shopping mall or enclosed courtyard
	24: indoors–non-food store (department, pharmacy, etc.)
09: indoors–grocery store/convenience store	21: indoors–food store (bakery, supermarket, etc.)
10: indoors–metro-subway-train station	22: indoors–subway station or train station
11: indoors–hospital/medical care facility	23: indoors–hospital
Outdoors-other	
12: outdoors–residential	35: outdoors–neighborhood background or residential grounds
13: outdoors–general non-residential	32: outdoors–community or retirement center
	34: outdoors–mall, market, patio, or plaza
	41: outdoors–public building
14: outdoors–park or golf course	36: outdoors–park or golf course
15: outdoors–restaurant or café	42: outdoors–restaurant/picnic
16: outdoors–school grounds	43: outdoors–school grounds
Outdoors-near road	
17: outdoors–metro-subway-train stop	31: outdoors–bus stop
	46: outdoors–metro station platform
	47: outdoors–bus transit center
18: outdoors–within 10 yards of street	33: outdoors–freeway edge or gradient, pedestrian overpass
	40: outdoors–pedestrian walk
	45: outdoors–street–residential
19: outdoors–garage (covered or below ground)	37: outdoors–parking garage or covered parking
	38: outdoors–parking garage–below ground
20: outdoors–parking lot (open), street parking	39: parking lot (open), street parking, window shopping
21: outdoors–service station	44: outdoors–service station
In-Vehicle	
22: vehicle–car	51: vehicle–auto-commercial strip, surface street
	52: vehicle–auto-freeway
	53: vehicle–auto-garage-underground
	54: vehicle–auto-refueling
	55: vehicle–auto-residential street
	56: vehicle–auto-restaurant drive-through
	57: vehicle–auto-urban canyon or urban streets
23: vehicle–bus	58: vehicle–bus
24: vehicle–train or subway	59: vehicle–train

microenvironments 6, 11, 23, and 24. Table 2 also shows how the microenvironments are grouped within four general microenvironments ("Indoors," "Outdoors-near road," "Outdoors-other," and "in-Vehicle"). These general microenvironments are used in statistical analyses to increase sample size and overall strength of inference.

7. Fixed-site monitors selected for analysis
We identified fixed-site monitors that were relatively close to each of the three ECP study areas and that provided relatively complete data for each of the two ECP study periods. We assigned the central Los Angeles site (CELA) to the Alhambra and downtown Los Angeles study areas, and the Long Beach site (LGBH) to the Carson/Long Beach study area. These monitors were used for all analyses that paired ECP measurements with concurrent fixed-site measurements. In the case of $PM_{2.5}$, fixed-site monitoring data were only available for the LGBH site for 2009. Consequently, the $PM_{2.5}$ analyses that pair ECP data to LGBH data are limited to 2009 data.

8. Ratios of microenvironmental concentrations to outdoor concentrations
As discussed above, each multipollutant measurement (referred to as a "test") is labeled with the corresponding date, time, geographic location, and microenvironment. Each test belongs to a group of tests that were made at the same geographic location on the same date within a period of two hours or less. There are 87 of these "test groups" in which DRI measured pollutant concentrations in one or more microenvironments; 39 occurred in 2008 and 48 in 2009. The use of test groups provides a means of analyzing the relationships (ratios, differences, correlations, etc.) between pollutant concentrations measured in two different microenvironments at the same geographic location at approximately the same time. The APEX distributions in microenvironments used in these analyses are the distributions of the daytime hourly values predicted by APEX in the microenvironments. We have not attempted to match the APEX and ECP concentrations by day or time-of-day. In both cases we are consolidating concentrations for the same period of time. To the extent that the concentrations vary by time of day there will be some mismatch in time. However, the indoor-outdoor ratios modeled in APEX do not directly depend on concentration levels, so this potential mismatch will not affect comparisons of indoor-outdoor ratios, which is why we focus the comparisons on these ratios.

We have a particular interest in the relationships between indoor and nearby outdoor concentrations, since this is one of the key relationships being modeled by APEX. Table 3 presents the minimum, median, and maximum indoor/outdoor ratios for ECP O_3 concentrations for all indoor microenvironments and for cases when the outdoor microenvironment is either "Outdoors-other" or "Outdoors-near road." We expect indoor/outdoor ratios to be less than one due to decay of O_3 indoors, absent substantial positive O_3 monitor interferences. The median ratio for all indoors to outdoors-other (0.30) is lower than the median ratio for all indoors to outdoors-near road (0.37). This result is likely because the titration of O_3 by on-road NO emissions tends to cause a reduction in near-road concentrations of O_3 compared with other outdoor locations.

Table 3 also lists the median indoor/outdoor ratios determined by applying APEX to the ECP study area. These are limited to the "Outdoors-other" category only because APEX does not continuously estimate the concentrations at the near-road location closest to a given (non-near-road) microenvironment. The median of the indoor/outdoor ratios for each indoor microenvironment range from 0.16 to 0.58 for ECP and from 0.22 to 0.60 for APEX. The APEX median ratio for all indoors to outdoors-other is 0.24, lower than the corresponding ECP median ratio (0.30).

Table 4 lists statistics for CO ratios. In general, we expect the indoor/outdoor ratios to be about one due to the limited reactivity of CO indoors. The APEX indoor/outdoor median ratios are all between 1.0 and 1.08, while the ECP ratios exhibit wider variation. The APEX median ratio for all indoors

Table 3. Comparisons of APEX estimated ratios of indoor microenvironmental O_3 concentration to simultaneous nearby outdoor microenvironmental O_3 concentration obtained from ECP measurements

Indoor microenvironment		Outdoor microenvironment	Indoor-to-Outdoor O_3 ratio						
			PEC measurements				APEX		
	Description		N^a	Min	Med	Max	Min	Med	Max
1	Residence	Other	3	0.11	0.24	1.41	0.01	0.23	2.71
		Near-road	1	–	1.02	–	–	–	–
2	Community Center or auditorium	Other	4	0.07	0.18	0.28	0.03	0.30	0.86
		Near-road	7	0.07	0.13	0.46	–	–	–
3	Restaurant	Other	11	0.07	0.30	1.00	0.16	0.59	1.79
		Near-road	10	0.16	0.48	14.3	–	–	–
4	Hotel/motel	Other	1	–	0.54	–	0.01	0.30	1.85
		Near-road	3	0.56	0.65	0.77	–	–	–
5	Office building, bank, post office	Other	6	0.16	0.58	1.94	0.01	0.29	2.04
		Near-road	11	0.14	0.42	1.38	–	–	–
7	School	Other	5	0.34	0.37	0.65	0.01	0.22	1.98
		Near-road	2	0.37	0.63	0.89	–	–	–
8	Shopping mall, non-grocery store	Other	5	0.10	0.31	0.40	0.01	0.29	1.43
		Near-road	4	0.10	0.25	0.42	–	–	–
9	Grocery store, convenience store	Other	3	0.14	0.16	0.90	0.01	0.30	1.79
		Near-road	6	0.14	0.29	0.66	–	–	–
All	All indoors[b]	Other	25	0.07	0.30	1.94	0.01	0.38	2.71
		Near-road	29	0.09	0.37	14.3	–	–	–

Notes: The "all indoors ME" value for a particular test group may combine (average) pollutant values from two or more indoor MEs that were measured at about the same time. So the "all" N value may be less than the sum of the N values for individual MEs.

[a]N is the number of ratios of measured values. The APEX statistics are based on >1000 simulated values.

[b]The values in the "all indoors" rows (e.g. the 29 and 0.09 in the last row) are not expected to be consistent with the numbers above.

to outdoors-other is 1.01, slightly lower than the corresponding ECP median ratio (1.10). Very high APEX indoor/outdoor ratios occur when the outdoor concentration drops rapidly to close to zero, while the indoor concentration takes longer to decrease if all windows and doors are closed. There are no measured ratios as high as these since measurements were not made under these conditions.

Table 5 presents the corresponding statistics for $PM_{2.5}$ ratios. In general, we expect indoor/outdoor ratios to be less than one due to $PM_{2.5}$ removal processes indoors, though indoor sources (e.g. frying) and other human activities could (e.g. vacuuming) could increase this ratio above unity. The ECP median ratios for all indoors to outdoors-other and indoors to outdoors-near road are very close (0.55 and 0.52). The median indoor-to-outdoors-other ratio from APEX is 0.76, somewhat higher than the ECP ratio (0.55).

Table 4. Comparisons of APEX estimated ratios of indoor microenvironmental CO concentration to simultaneous nearby outdoor microenvironmental CO concentration obtained from ECP measurements

Indoor microenvironment		Outdoor ME	Indoor-to-Outdoor CO ratio						
			ECP measurements				APEX		
	Description		N^a	Min	Med	Max	Min	Med	Max
1	Residence	Other	2	0.71	0.85	1.00	0.07	1.00	336
		Near-road	0	–	–	–	–	–	–
2	Community Center or auditorium	Other	3	0.48	1.28	2.00	0.39	1.04	380
		Near-road	7	0.73	1.13	2.88	–	–	–
3	Restaurant	Other	12	0.63	1.62	19.0	0.48	1.00	95
		Near-road	10	0.64	1.06	7.00	–	–	–
4	Hotel/motel	Other	1	–	0.39	–	0.16	1.00	315
		Near-road	3	0.88	0.93	1.20	–	–	–
5	Office building, bank, post office	Other	7	0.78	1.00	1.60	0.05	1.05	485
		Near-road	11	0.50	1.00	1.80	–	–	–
7	School	Other	5	0.09	0.69	2.00	0.10	1.08	416
		Near-road	2	1.00	2.00	3.00	–	–	–
8	Shopping mall, non-grocery store	Other	5	0.55	1.33	1.80	0.06	1.03	371
		Near-road	5	0.69	1.00	1.92	–	–	–
9	Grocery store, convenience store	Other	3	1.26	1.80	17.6	0.08	1.02	507
		Near-road	7	0.52	1.00	1.93	–	–	–
All	All indoors[b]	Other	25	0.55	1.10	19.0	0.05	1.01	507
		Near-road	29	0.64	1.00	3.33	–	–	–

Notes: The "all indoors ME" value for a particular test group may combine (average) pollutant values from two or more indoor MEs that were measured at about the same time. So the "all" N value may be less than the sum of the N values for individual MEs.

[a]N is the number of ratios of measured values. The APEX statistics are based on >1000 simulated values.

[b]The values in the "all indoors" rows are not expected to be consistent with the numbers above.

Table 5. Comparisons of APEX estimated ratios of indoor microenvironmental $PM_{2.5}$ concentration to simultaneous nearby outdoor microenvironmental $PM_{2.5}$ concentration obtained from ECP measurements

Indoor microenvironment		Outdoor ME	Indoor-to-Outdoor $PM_{2.5}$ ratio						
			ECP measurements				APEX		
	Description		N^a	Min	Med	Max	Min	Med	Max
1	Residence	Other	3	0.29	0.33	0.89	0.04	0.60	7.37
		Near-road	1	–	0.37	–	–	–	–
2	Community Center or auditorium	Other	4	0.16	0.37	0.61	0.07	0.69	4.57
		Near-road	8	0.13	0.36	0.67	–	–	–
3	Restaurant	Other	12	0.30	0.98	34.25	0.39	0.87	6.97
		Near-road	10	0.33	1.00	15.36	–	–	–
4	Hotel/motel	Other	1	–	0.80	–	0.04	0.68	4.71
		Near-road	3	0.16	0.70	1.78	–	–	–
5	Office building, bank, post office	Other	7	0.10	0.28	0.50	0.04	0.68	7.34
		Near-road	11	0.13	0.24	0.78	–	–	–
7	School	Other	5	0.13	0.41	0.53	0.04	0.58	7.41
		Near-road	2	0.10	3.01	5.91	–	–	–
8	Shopping mall, non-grocery store	Other	5	0.08	0.55	2.00	0.03	0.68	6.59
		Near-road	5	0.10	0.34	0.57	–	–	–
9	Grocery store, convenience store	Other	3	0.51	0.56	1.31	0.04	0.69	6.84
		Near-road	7	0.48	2.11	12.43	–	–	–
All	All indoors[b]	Other	26	0.08	0.55	24.14	0.03	0.76	7.41
		Near-road	30	0.10	0.52	15.36	–	–	–

Notes: The "all indoors ME" value for a particular test group may combine (average) pollutant values from two or more indoor MEs that were the same time. So the "all" N value may be less than the sum of the N values for individual MEs.

[a]N is the number of ratios of measured values. The APEX statistics are based on >1000 simulated values.

[b]The values in the "all indoors" rows are not expected to be consistent with the numbers above.

9. Ratios of microenvironmental concentrations to fixed-site concentrations

9.1. Ozone
Table 6 provides minimum, median, and maximum microenvironment-to-monitor ratios based on ECP O_3 data and estimates obtained from the APEX runs for each of the 24 APEX microenvironments.

General patterns in the O_3 results can be identified by examining the median ratios for the four general microenvironments for the two study periods combined (see the code = "All" listings in Table 6). In the indoor microenvironments, the median ECP ratio (0.29) is 20 percent smaller than the APEX ratio (0.36), although 5 of the 9 indoor microenvironments have ECP median ratios higher than the APEX median ratios. The ECP and APEX median ratios for the other outdoor microenvironments are quite close, however, the ECP measurements have a much wider range than APEX. ECP ratios also have a wider range than APEX for the near-road outdoor microenvironments, while the median ECP ratio (0.84) is 14 percent higher than the APEX ratio (0.74). The ECP and APEX median ratios differ the most for the in-vehicle microenvironment. In this case, the ECP median ratio (0.46) is approximately 30 percent lower than the APEX median ratio (0.63). The direction of this difference is unexpected due to the fact that the ECP vehicles were operated under high ventilation conditions during the tests, whereas APEX models a variety of ventilation conditions.

For three of the general microenvironments (Outdoors-near road, Outdoors-other, and in-Vehicle), there is little or no variation in the median APEX ratios listed for the individual microenvironments (Table 6). This finding is the result of the method currently employed by APEX to estimate O_3 concentrations in these microenvironments. In each case, the microenvironmental O_3 concentration is estimated as a linear function (with variability) of the O_3 concentration at a nearby fixed-site monitor, and these functions do not vary substantially among the individual microenvironments.

Additional differences between the ECP and APEX ratios can be observed in Table 7. This table provides descriptive statistics that characterize the distributions of the ECP and APEX ME-to-monitor ratios. To provide adequate sample size for calculating the ECP statistics, the tables are limited to the four general microenvironments used in previous analyses (Indoors, Outdoors-near road, Outdoors-other, and in-Vehicle).

General patterns observed in the ECP and APEX ratios in Table 7 and the corresponding Figure 2 for each of the four general microenvironments are as follows:

(1) *Indoors*: The ECP values tend to be roughly equal to the APEX values for the 5th through 75th percentiles. The 90th and 95th percentile ECP values are significantly larger than the corresponding APEX values. However, the ECP values for these high percentiles are based on relatively small sample sizes.

(2) *Outdoors-near road*: The median (50th percentile) ECP ratio (0.84) is larger than the median APEX value (0.74). The ECP values also exhibit a larger variance than the APEX values. ECP values for percentiles below the median are equal to or less than the corresponding APEX values. ECP values for percentiles above the median are larger than the corresponding APEX values.

(3) *Outdoors-other*: The APEX values vary over a narrow range from 0.90 to 1.10. This results from the distribution of proximity factors, which ranges from 0.90 to 1.10. The median APEX value (1.00) is comparable to the median percentile value listed for the ECP values (0.98). However, the ECP ratio values exhibit a much wider range than the APEX ratios, with a 10th percentile value of 0.60 and a 90th percentile value of 1.60.

(4) *in-Vehicle*: The median ECP ratio is 0.46; the median APEX ratio is 0.63. Up to the 75th percentile, ECP values are always lower than the corresponding APEX values. ECP ratios values exceed the corresponding APEX ratio values for higher percentiles.

Microenvironment			Microenvironment-to-monitor O_3 ratios					
			ECP measurements			APEX estimates		
Code	Description	N^a	Min	Med	Max	Min	Med	Max
Indoors								
1	Residence	3	0.08	0.26	1.55	0.01	0.23	2.74
2	Community Center or auditorium	6	0.12	0.19	0.57	0.02	0.29	0.90
3	Restaurant	15	0.05	0.26	10.00	0.16	0.58	1.81
4	Hotel/motel	2	0.24	0.28	0.33	0.01	0.30	1.80
5	Office building, bank, post office	8	0.15	0.55	1.65	0.01	0.29	1.88
7	School	8	0.13	0.31	0.84	0.01	0.22	1.85
8	Shopping mall, non-grocery store	5	0.07	0.20	0.37	0.01	0.29	1.41
9	Grocery store, convenience store	7	0.14	0.33	0.51	0.01	0.30	1.91
10	Metro–sub-way–train	2	0.86	0.90	0.93	0.03	0.27	0.92
All	All indoors	56	0.05	0.29	10.00	0.01	0.36	2.74
Outdoors-Other								
12	Residential grounds	4	0.85	0.97	1.10	0.90	1.00	1.10
13	General–non-residential	16	0.36	0.91	2.06	0.90	1.00	1.10
14	Park or golf course	17	0.47	0.98	1.53	0.90	1.00	1.10
15	Restaurant or café	15	0.55	1.00	3.40	0.90	1.00	1.10
16	School grounds	11	0.50	1.02	2.82	0.90	1.00	1.10
All	All outdoors—other	63	0.36	0.98	3.40	0.90	1.00	1.10
Outdoors-Near road								
17	Metro–sub-way–train stop	15	0.33	0.81	1.12	0.42	0.73	1.00
18	Within 10 yards of street	28	0.14	0.73	6.40	0.42	0.74	1.00
19	Parking garage	10	0.18	0.58	1.14	0.42	0.72	1.00
20	Parking lot (open), street parking	36	0.20	0.92	6.10	0.42	0.74	1.00
21	Service station	7	0.55	0.98	2.03	0.42	0.73	1.00
All	All outdoors-near road	96	0.14	0.84	6.40	0.42	0.74	1.00
Vehicle								
22	Car	57	0.11	0.46	1.90	0.09	0.63	1.00
All	All vehicle	57	0.11	0.46	1.90	0.09	0.63	1.00

Table 6. Comparison of APEX estimated ratios of microenvironmental O_3 concentrations to simultaneous fixed-site O_3 concentrations obtained from ECP measurements

^{a}Number of ratios.

9.2. Carbon monoxide

Tables 8 and 9 provide statistics for microenvironment-to-monitor CO ratios that follow the table formats used by Tables 6 and 7. Table 8 provides minimum, median, and maximum microenvironment-to-monitor ratios based on ECP CO data and on estimates obtained from the APEX CO runs for each of the 24 APEX microenvironments. General patterns in the CO results can be identified by examining the median ratios for the four general microenvironments for the two study periods combined (see Table 8). The ECP medians are consistently higher than the corresponding APEX median ratios (which tend to be near 1.0).

The APEX medians listed for individual microenvironments in Table 8 for the general indoor microenvironment are all slightly larger than 1.0. This pattern is due to the frequency with which ambient CO concentrations rapidly decrease at certain times of the day. If this happens when the indoor ventilation rate is low (e.g. when windows are closed), the outdoor concentration falls more rapidly than the indoor concentration, resulting in an indoor/outdoor concentration ratio less than 1. When the ambient CO concentrations fall to very small values, this ratio can be relatively large, even exceeding 500 (see Table 9).

If we focus on the median ECP ratios in Table 8, we see values above 1.7 for individual microenvironments #3: restaurant (ratio = 2.50), #8: shopping mall or non-grocery store (1.79), #9: grocery store or convenience store (1.88), #16: school grounds (2.00), #18: within 10 yards of street (2.10), #22: car (1.91), and #23: bus (2.18). Some of these microenvironments would be expected to contain CO sources associated with gas stoves (e.g. restaurant) or motor vehicles (e.g. within 10 yards of street, car and bus).

Median ratio values below 1.2 are associated with #1: residence (1.09), #2: community center or auditorium (1.08), #5: office building, bank, or post office (1.00), #7: school (1.06), #12: residential grounds (0.28), #14: park or golf course (1.14), #15: outside restaurant or café (1.14), and #24: train (0.94). Note that some of these ratios are based on a small sample size (n < 5) (microenvironment #s 1, 12, and 23).

Table 7. Comparisons of descriptive statistics for ratios of microenvironmental O_3 concentrations to simultaneous fixed-site O_3 concentrations obtained from ECP measurements and APEX estimates—grouped microenvironments

Statistic	Microenvironment-to-monitor O_3 ratio							
	Indoors		Outdoors-Other		Outdoors-Near road		Vehicle	
	ECP	APEX	ECP	APEX	ECP	APEX	ECP	APEX
Number of values	56	>1000	63	>1000	96	>1000	57	>1000
Mean	0.58	0.39	1.09	1.00	1.05	0.73	0.57	0.62
SD	1.33	0.24	0.50	0.05	0.99	0.15	0.39	0.22
Minimum	0.05	0.01	0.36	0.90	0.14	0.42	0.11	0.09
5th pct	0.08	0.07	0.50	0.92	0.26	0.48	0.13	0.24
10th pct	0.12	0.10	0.60	0.93	0.34	0.52	0.20	0.32
25th pct	0.19	0.19	0.83	0.96	0.62	0.62	0.32	0.46
50th pct	0.29	0.36	0.98	1.00	0.84	0.74	0.46	0.63
75th pct	0.56	0.56	1.30	1.04	1.10	0.85	0.64	0.80
90th pct	0.97	0.69	1.59	1.07	1.86	0.93	1.17	0.91
95th pct	1.56	0.76	1.99	1.08	2.67	0.96	1.47	0.95
Maximum	10.00	2.74	3.40	1.10	6.40	1.00	1.90	1.00

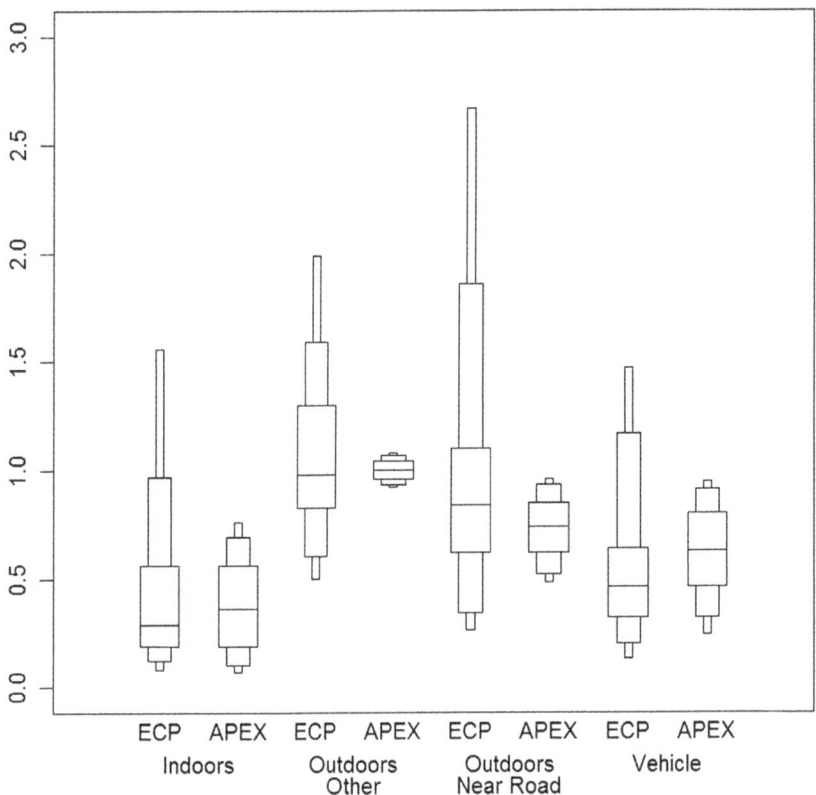

Thickest bars indicate the 25th, 50th, and 75th percentiles. Secondary bars extend
to the 10th and 90th percentiles. Tertiary bars extend to the 5th and 95th percentiles.

Figure 2. Microenvironment-to-monitor O$_3$ ratios.

Table 8. Comparison of ratios of microenvironmental CO concentrations to simultaneous fixed-site CO concentrations obtained from ECP measurements and APEX

Microenvironment		Microenvironment-to-monitor CO ratios					
		ECP measurements			APEX estimates		
Code	Description	Nº	Min	Med	Max	Min	Med	Max
Indoors								
1	Residence	2	0.92	1.09	1.25	0.07	1.01	389
2	Community Center or auditorium	5	0.67	1.08	1.38	0.36	1.08	339
3	Restaurant	16	0.63	2.50	35.7	0.39	1.02	97
4	Hotel/motel	6	0.71	1.50	2.18	0.16	0.98	329
5	Office building, bank, post office	9	0.33	1.00	2.33	0.06	1.07	511
6	Bar, night club, cafe	8	0.44	1.24	4.78	0.53	1.00	53
7	School	10	0.44	1.06	1.63	0.09	1.10	349
8	Shopping mall, non-grocery store	20	0.57	1.79	5.00	0.06	1.07	419

(Continued)

Table 8. (Continued)

Microenvironment		Microenvironment-to-monitor CO ratios					
		ECP measurements			APEX estimates		
Code	Description	N[a]	Min	Med	Max	Min	Med	Max
9	Grocery store, convenience store	9	0.33	1.88	6.29	0.08	1.06	414
10	Metro–sub-way–train	16	0.29	1.46	24.2	0.33	1.04	72
11	Hospital, medical care facility	6	0.83	1.32	4.75	0.07	1.07	331
All	All indoors	107	0.29	1.36	35.7	0.06	1.04	511
Outdoors-Other								
12	Residential grounds	4	0.00	0.28	1.32	0.80	1.00	1.20
13	General–non-residential	15	0.43	1.40	2.50	0.80	1.00	1.20
14	Park or golf course	19	0.20	1.14	12.0	0.80	1.00	1.20
15	Restaurant or café	14	0.22	1.14	5.67	0.80	1.00	1.20
16	School grounds	11	0.25	2.00	18.1	0.80	1.00	1.20
All	All outdoors-other	63	0.00	1.33	18.1	0.80	1.00	1.20
Outdoors-Near road								
17	Metro–sub-way–train stop	23	0.35	1.27	4.75	0.80	1.00	1.20
18	Within 10 yards of street	28	0.29	2.10	5.60	0.80	1.00	1.20
19	Parking garage	10	0.00	1.52	37.4	0.80	1.00	1.20
20	Parking lot (open), street parking	36	0.25	1.40	6.00	0.80	1.00	1.20
21	Service station	6	0.86	1.60	4.00	0.80	1.00	1.20
All	All outdoors-near road	103	0.00	1.60	37.4	0.80	1.00	1.20
Vehicle								
22	Car	57	0.17	1.91	6.67	0.80	1.00	1.20
23	Bus	1	–	2.18	–	0.80	1.00	1.20
24	Train	6	0.71	0.94	5.18	0.80	1.00	1.20
All	All vehicle	64	0.17	1.79	6.67	0.80	1.00	1.20

[a]Number of ratios.

Table 9 and the corresponding Figure 3 provide descriptive statistics that characterize the distributions of the ECP and APEX microenvironment-to-monitor ratios by general microenvironment. For the indoors microenvironment, the ECP value is higher than the APEX values for the 50th through 95th percentiles. The high APEX standard deviation is the result of a few very high values (explained above). Approximately 0.9% of the APEX indoor/monitor ratios are > 35. For the outdoors-other and outdoors-near road microenvironments, the ECP value exceeds the APEX value for the 50th through

| Statistic | Microenvironment-to-monitor CO ratio | | | | | | | |
| | Indoors | | Outdoors-Other | | Outdoors-Near road | | Vehicle | |
	ECP	APEX	ECP	APEX	ECP	APEX	ECP	APEX
Number of values	107	>1000	63	>1000	103	>1000	64	>1000
Mean	2.54	3.14	1.90	1.00	2.21	1.00	2.05	1.00
SD	4.29	23.28	2.70	0.10	3.73	0.10	1.38	0.10
Minimum	0.29	0.06	0.00	0.80	0.00	0.80	0.17	0.80
5th pct	0.46	0.74	0.23	0.83	0.37	0.83	0.39	0.83
10th pct	0.63	0.81	0.28	0.86	0.52	0.86	0.60	0.86
25th pct	0.91	0.91	0.57	0.92	0.91	0.92	0.94	0.92
50th pct	1.36	1.04	1.33	1.00	1.60	1.00	1.79	1.00
75th pct	2.70	1.20	2.00	1.08	2.50	1.08	2.93	1.08
90th pct	4.36	1.58	3.43	1.14	3.90	1.14	4.00	1.14
95th pct	7.51	2.13	5.33	1.17	5.35	1.17	4.98	1.17
Maximum	35.7	510	18.13	1.20	37.40	1.20	6.67	1.20

Table 9. Comparisons of descriptive statistics for ratios of microenvironmental CO concentrations to simultaneous fixed-site CO concentrations obtained from ECP measurements and APEX estimates—grouped microenvironments

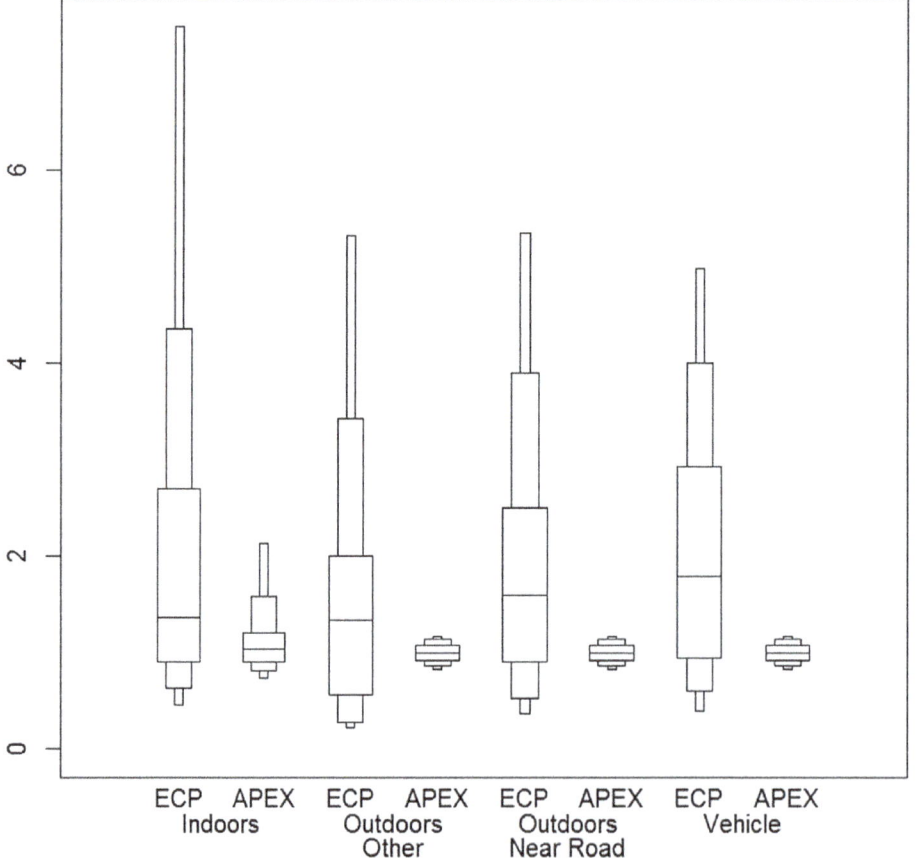

Thickest bars indicate the 25th, 50th, and 75th percentiles. Secondary bars extend to the 10th and 90th percentiles. Tertiary bars extend to the 5th and 95th percentiles.

Figure 3. Microenvironment-to-monitor CO ratios.

95th percentiles. For the in-vehicle general microenvironment, the ECP value is larger than the APEX value for the 25th through 95th percentiles. The APEX distributions for the outdoors-other, outdoors-near road, and vehicle microenvironments are the same as a result of having the same distributions of proximity factors.

9.3. $PM_{2.5}$

Tables 10 and 11 provide statistics for microenvironment-to-monitor $PM_{2.5}$ that follow the table formats used by Tables 6 and 7. These values are consistent with expectations; they show generally lower ratios for indoor microenvironments and higher ratios for the vehicle and outdoors-near road microenvironments.

Table 10 provides minimum, median, and maximum microenvironment-to-monitor ratios based on ECP $PM_{2.5}$ data and estimates obtained from the APEX $PM_{2.5}$ runs for each of the 24 APEX microenvironments. Reviewing the results for the four general microenvironments, we see that the ECP median ratio of 1.19 for the general indoor microenvironment is almost double the median APEX ratio for this microenvironment (0.73). A value less than 1 would be expected for the APEX median ratio since APEX currently accounts for indoor particle deposition (which lowers the ratio) but not for indoor $PM_{2.5}$ sources (which increases the ratio). The median APEX ratio is 1.0 for each of the individual microenvironments in the outdoors-other, outdoors-near road, and vehicle general microenvironments. In each of these cases, APEX currently models the microenvironment using a proximity factor with mean = 1.

Focusing on the median ECP ratios in Table 10, we see values above 1.50 for individual microenvironments #3: restaurant (1.93), #6: bar, night club, or café (1.52), #9: grocery store or convenience store (2.33), #10: inside metro, subway, or train (1.58), #14: park or golf course (1.80), #15: outdoor restaurant or café (1.95), #16: school grounds (2.10), #17: outdoor metro, subway, or train stop (1.61), #18: within 10 yards of street (1.71), #20: open parking lot or street parking (1.68), #22: car (1.96), and #23: bus (1.62). Median ratio values below 1.00 are associated with #1: residence (0.51), #2: community center or auditorium (0.70), #5: office building, bank, or post office (0.33), #7: school (0.70), #8: shopping mall or non-grocery store (0.91), and #11: hospital or medical care facility (0.84). Note again that some of these ratios (microenvironment #s 1, 12, and 23) are based on small samples ($n < 5$).

Table 11 and the corresponding Figure 4 provide descriptive statistics that characterize the distributions of the ECP and APEX microenvironment-to-monitor ratios by general microenvironment. Listed below are general patterns that can be observed for each of the four general microenvironments. The APEX distributions for the outdoors-other, outdoors-near road, and vehicle microenvironments are the same as a result of having the same distributions of proximity factors.

(1) *Indoors*: The ECP values tend to be larger than the corresponding APEX values with the difference increasing toward the higher percentiles. The median ratios are 1.19 for ECP and 0.73 for APEX.

(2) *Outdoors-other*: The median ECP ratio (1.500) is larger than the median APEX value (1.00). The APEX ratios vary over a narrow range from 0.90 to 1.10. The ECP ratios exhibit a much larger variability, with a 10th percentile value of 0.75 and a 90th percentile value of 3.64.

(3) *Outdoors-near road*: The median ECP ratio (1.61) is larger than the APEX ratio (1.00). The APEX ratios values vary over the same narrow range as the outdoors-other ratios with the same percentile values. The ECP ratios exhibit a much larger variability, with a 10th percentile value of 0.88 and a 90th percentile value of 3.67.

Table 10. Comparison of ratios of microenvironmental PM$_{2.5}$ concentrations to simultaneous fixed-site PM$_{2.5}$ concentrations obtained from ECP measurements and APEX estimates

Microenvironment		Microenvironment-to-monitor PM$_{2.5}$ ratios						
		ECP measurements			APEX estimates			
Code	Description	N^a	Min	Med	Max	Min	Med	Max
Indoors								
1	Residence	2	0.43	0.51	0.58	0.04	0.60	7.74
2	Community Center or auditorium	5	0.33	0.70	0.84	0.08	0.68	4.81
3	Restaurant	14	0.97	1.93	31.17	0.38	0.87	7.03
4	Hotel/motel	6	0.33	1.08	3.92	0.04	0.68	5.03
5	Office building, bank, post office	9	0.27	0.33	1.12	0.03	0.68	7.89
6	Bar, night club, cafe	7	1.11	1.52	2.92	0.45	0.86	2.33
7	School	10	0.09	0.70	12.36	0.04	0.58	7.96
8	Shopping mall, non-grocery store	17	0.13	0.91	2.22	0.03	0.68	6.90
9	Grocery store, convenience store	9	0.83	2.33	6.00	0.04	0.68	7.16
10	Metro–sub-way–train	14	0.70	1.58	4.40	0.08	0.62	2.81
11	Hospital, medical care facility	6	0.44	0.84	1.34	0.04	0.67	7.34
All	All indoors	99	0.09	1.19	31.17	0.03	0.73	7.96
Outdoors-Other								
12	Residential grounds	3	1.14	1.28	1.61	0.90	1.00	1.10
13	General–non-residential	15	0.55	1.32	2.04	0.90	1.00	1.10
14	Park or golf course	18	0.46	1.80	8.00	0.90	1.00	1.10
15	Restaurant or café	12	0.96	1.95	18.00	0.90	1.00	1.10
16	School grounds	9	0.58	2.10	4.23	0.90	1.00	1.10
All	All outdoors-other	57	0.46	1.50	18.00	0.90	1.00	1.10
Outdoors-Near road								
17	Metro–sub-way–train stop	22	0.41	1.61	5.40	0.90	1.00	1.10
18	Within 10 yards of street	24	0.41	1.71	4.60	0.90	1.00	1.10
19	Parking garage	10	0.80	1.11	4.00	0.90	1.00	1.10
20	Parking lot (open), street parking	30	0.34	1.68	6.67	0.90	1.00	1.10

(*Continued*)

Table 10. (Continued)

| Microenvironment | | Microenvironment-to-monitor PM$_{2.5}$ ratios | | | | | | |
| | | ECP measurements | | | APEX estimates | | | |
Code	Description	N[a]	Min	Med	Max	Min	Med	Max
21	Service station	6	0.73	1.21	1.60	0.90	1.00	1.10
All	All outdoors–near road	92	0.34	1.61	6.67	0.90	1.00	1.10
Vehicle								
22	Car	45	0.46	1.96	13.10	0.90	1.00	1.10
23	Bus	1	–	1.62	–	0.90	1.00	1.10
24	Train	6	0.70	1.17	1.40	0.90	1.00	1.10
All	All vehicle	52	0.46	1.64	13.10	0.90	1.00	1.10

[a]Number of ratios.

Table 11. Comparisons of descriptive statistics for ratios of microenvironmental PM$_{2.5}$ concentrations to simultaneous fixed-site PM$_{2.5}$ concentrations obtained from ECP measurements and APEX estimates—grouped microenvironments

| Statistic | Microenvironment-to-monitor PM$_{2.5}$ ratio | | | | | | | |
| | Indoors | | Outdoors-Other | | Outdoors-Near road | | Vehicle | |
	ECP	APEX	ECP	APEX	ECP	APEX	ECP	APEX
Number of values	99	>1000	57	>1000	92	>1000	52	>1000
Mean	2.23	0.73	2.16	1.00	1.98	1.00	2.61	1.00
SD	4.27	0.50	2.47	0.05	1.23	0.05	2.66	0.05
Minimum	0.09	0.03	0.46	0.90	0.35	0.90	0.46	0.90
5th pct	0.28	0.24	0.58	0.92	0.72	0.92	0.58	0.92
10th pct	0.34	0.33	0.75	0.93	0.88	0.93	0.73	0.93
25th pct	0.77	0.51	1.17	0.96	1.12	0.96	1.26	0.96
50th pct	1.19	0.73	1.50	1.00	1.61	1.00	1.64	1.00
75th pct	1.91	0.88	2.23	1.04	2.41	1.04	2.79	1.04
90th pct	3.82	1.01	3.64	1.07	3.67	1.07	6.10	1.07
95th pct	9.62	1.12	4.76	1.08	4.28	1.08	10.7	1.08
Maximum	31.2	7.96	18.0	1.10	6.67	1.10	13.1	1.10

(4) *in-Vehicle*: These statistics exhibit patterns similar to those observed in the statistics for outdoors-other and outdoors-near road. The median ECP ratio is 1.64; the median APEX ration is 1.00. The APEX ratios values vary over the same narrow range as the outdoor ratios with the same percentile values. The ECP ratios exhibit a much larger variability, with a 10th percentile value of 0.73 and a 90th percentile value of 6.09.

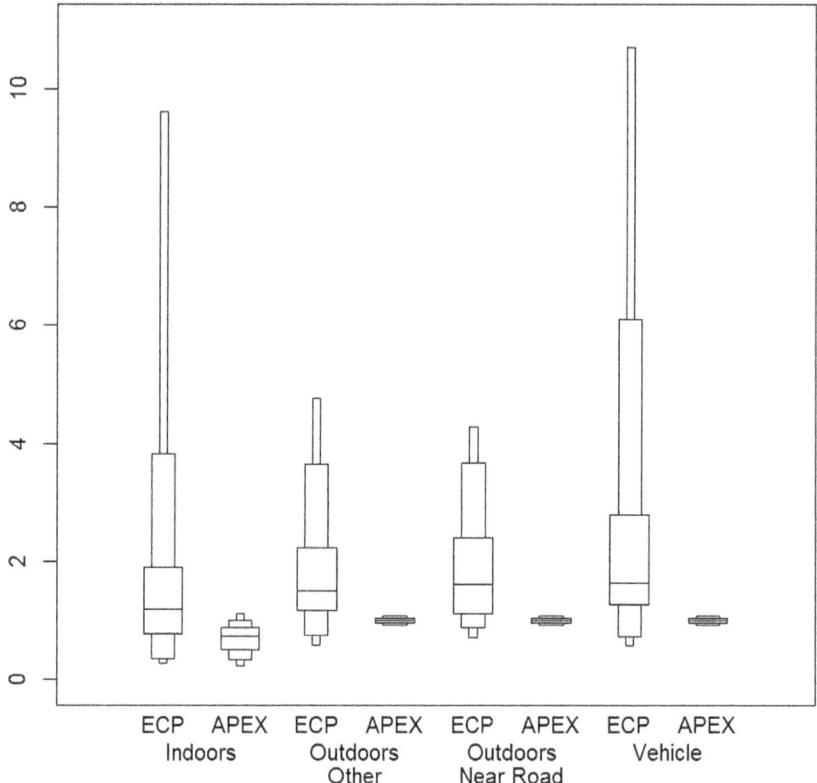

Thickest bars indicate the 25th, 50th, and 75th percentiles. Secondary bars extend
to the 10th and 90th percentiles. Tertiary bars extend to the 5th and 95th percentiles.

Figure 4. Microenvironment-to-monitor PM2.5 ratios.

10. Discussion

Exposure assessment using exposure models is continually evolving and is of growing importance
and complexity within the criteria pollutant NAAQS reviews performed by EPA. There has been growing recognition that people are differentially exposed to outdoor ambient pollution concentrations
depending on where they are and what they are doing. To better estimate health risks there is a
need for an improved methodology for estimating these differential exposures. The evaluation of
APEX presented here will guide further model development and is consistent with the advice to EPA
from stakeholders and scientific advisory committees that calls for more extensive evaluation of
exposure models that inform regulatory decisions.

Personal exposures are a time-weighted average of microenvironment concentrations, weighted
by the time spent in each microenvironment, and modeled personal exposures depend on both the
microenvironment concentrations and the activities of individuals. APEX estimates concentrations
outside a microenvironment by applying a stochastic proximity factor to concentrations measured
at a fixed-site monitor, and for O_3 also adjusting for titration by NO near roadways. The concentrations inside a microenvironment are estimated by APEX using either (1) a compartmental mass-
balance model based on air exchange and deposition/decay rates for indoor microenvironments or
(2) a regression factors model for in-vehicle microenvironments. In applications of APEX, analysts
tend to focus on the simulated distributions of microenvironmental concentrations and associated
exposure estimates rather than the exposures estimated for specific individuals at specific times. For
these reasons, we have compared the distributions of ECP-measured and APEX-estimated concentrations, indoor/outdoor ratios, and indoor/fixed-site ratios for microenvironments typical of those
defined in recent EPA risk assessments for O_3 and CO.

The ratios of microenvironment concentrations to the fixed-site measurements, sometimes re-ferred to as exposure factors, were examined. APEX underestimated the median O_3 microenviron-ment/fixed-site ratios for the near-road microenvironment and underestimated the ratios greater than one for all four of the microenvironment groups (Table 2 defines the groups). For CO, APEX un-derestimated the median ratio for all four groups. With the exception of a few very high ratios for indoor microenvironments predicted by APEX, APEX underestimated the ratios at the percentiles above the medians for all four of the microenvironment groups. For $PM_{2.5}$, APEX underestimated the median ratios and the 75th and higher percentiles by a factor of two or more for all four of the mi-croenvironment groups. The systematic underestimation of the microenvironment/fixed-site ratios is mostly driven by ambient concentrations that were higher at locations where the ECP measure-ments were made than at the locations of the fixed-site monitors. For O_3 at the near-road and in-vehicle microenvironments, the APEX ratios are lowered further by the proximity factors, which are intended to account for titration of O_3 by NO.

The lack of spatial resolution of the ambient (fixed monitor-based) concentrations (Figure 1) con-tributes to the uncertainties of the APEX predictions. Non-representativeness of the APEX distribu-tions of mass-balance and factor model parameters and the proximity factors also lead to differences between the ECP and APEX results. One aspect of these comparisons that stands out is the signifi-cantly higher variability of the ECP outdoor-to-monitor ratios than the comparable APEX ratios (see Figures 2–4). This indicates that the distribution of proximity factors is too narrow for all three pollut-ants. This implies that the variability of estimated exposures is significantly underestimated in this application.

The uncertainty of the mass-balance model and its parameters can be measured by comparing the modeled and measured ratios of indoor concentrations to outdoor concentrations, where the outdoor concentrations are modeled or measured just outside the indoor microenvironment (not at the fixed-site monitor). The concentrations of O_3, CO, and $PM_{2.5}$ in indoor microenvironments were modeled in APEX using a mass-balance model, where the AERs and decay rates are the critical pa-rameters for determining how much outdoor air goes indoors and how long it remains indoors. APEX was set up to estimate concentrations outside the microenvironments by applying a stochastic proximity factor to concentrations measured at a fixed-site monitor, which includes adjustments for titration of O_3 by NO on and near roadways. The ratio of the indoor concentrations to the concentra-tions outside the microenvironment is a measure that is approximately independent of the levels of the concentrations and independent of proximity factors; consequently this ratio is a good measure for evaluating the mass-balance model and the parameters driving it.

11. Results and conclusions

In general, except for the smallest ratios and the highest ratios for CO and $PM_{2.5}$, the indoor/outdoor ratios predicted by APEX are within a factor of two of the measured ratios over most of the ranges of the ratios for all three pollutants. The distributions of the indoor/outdoor ratios for O_3 reflected in the ECP measurements compare fairly well with the APEX predictions, with the APEX median (0.38) slightly higher than the measured median (0.30) and the range of the APEX predictions (0.01–2.71) somewhat wider than the range of the measurement data-set ratios (0.07–1.94) (Table 12, Figure 5). For CO, the medians of the ECP and APEX ratios are quite close, while the range of the APEX ratios is much wider than the ECP ratios. The very high APEX ratios for CO result from conditions when the indoor microenvironment is not well ventilated and the outdoor concentrations rapidly decrease to values close to zero while the indoor concentrations decrease more slowly. This tends to happen in the evening in residences with low AERs. ECP sampled four residences, between 8 am and 8 pm, and did not observe this phenomenon. For $PM_{2.5}$, APEX is overestimating in the central range of the distri-bution and underestimating the highest ratios, which in the ECP samples could be due to pollutant emissions from indoor sources or positive interference pollutant measurement bias, influential at-tributes not controlled for in that study or modeled by APEX. One finding of particular interest is that the variability of estimated exposures is significantly underestimated compared with fixed-site monitors in this application (Figures 2–4). For the grouped microenvironments (Tables 7, 9 and 11),

Table 12. Comparison of distributions of indoor/outdoor ratios: ECP measurements and APEX predictions

Statistic	O_3		CO		$PM_{2.5}$	
	ECP	APEX	ECP	APEX	ECP	APEX
Number of values	25	>1000	25	>1000	26	>1000
Minimum	0.07	0.01	0.55	0.05	0.08	0.03
10th pctile	0.10	0.10	0.67	0.89	0.17	0.34
25th pctile	0.16	0.20	0.87	1.00	0.29	0.54
Median	0.30	0.38	1.10	1.01	0.55	0.76
75th pctile	0.43	0.57	1.60	1.16	0.88	0.90
90th pctile	0.92	0.70	2.00	1.56	12.9	1.00
Maximum	1.94	2.71	19.0	507	24.1	7.41

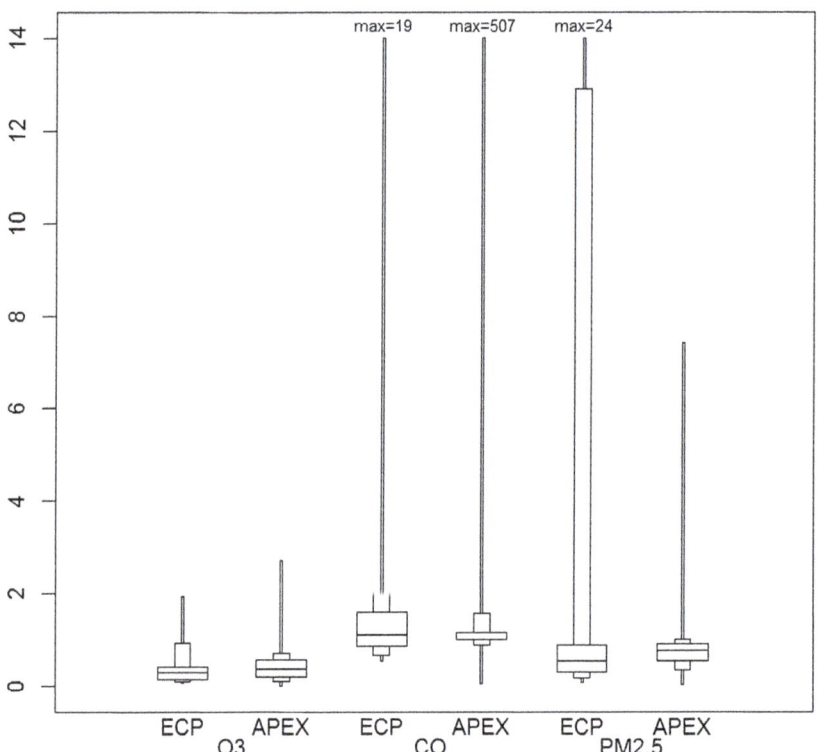

Thickest bars indicate the 25th, 50th, and 75th percentiles. Secondary bars extend to the 10th and 90th percentiles. Tertiary bars extend to the minimum and maximum values.

Figure 5. Comparison of distributions of indoor/outdoor ratios: ECP measurements and APEX predictions.

APEX always underestimates the top 50% of the distributions of ratios to fixed-site monitors, except for O_3 in vehicles, where APEX underestimates the top 10% of the distribution.

The results of this evaluation should be interpreted keeping in mind certain limitations of this study. One geographic area (Los Angeles) was studied, and the relationships between microenvironments and fixed site monitors are unlikely to be representative of a range of geographic areas. The measurements were all made during the day (when people typically experience higher exposures); nighttime microenvironment concentrations are not included in the results. Indoor sources of O_3, CO, and $PM_{2.5}$ were not simulated in APEX, nor were sources of interferences to their measurement methods, whereas some of the microenvironments sampled by ECP may have had concentrations influenced by indoor source emissions. Limited availability of hourly $PM_{2.5}$ fixed-site measurements increased the uncertainties of the APEX microenvironment concentrations. The results of this evaluation will be a useful input to quantitative uncertainty assessments of APEX, and can provide direction in model improvements.

Funding
This work was supported by the US Environmental Protection Agency [contract EP-D-08-100].

Competing interests
The authors declare no competing interest.

Author details
Ted R. Johnson[1]
E-mail: tjohn62124@aol.com
John E. Langstaff[2]
E-mail: Langstaff.John@epa.gov
Stephen Graham[2]
E-mail: Graham.Stephen@epa.gov
Eric M. Fujita[3]
E-mail: Eric.Fujita@dri.edu
David E. Campbell[3]
E-mail: Dave.Campbell@dri.edu
[1] TRJ Environmental, Inc., 713 Shadylawn Rd, Chapel Hill NC 27514, USA.
[2] U.S. Environmental Protection Agency, 109 TW Alexander Drive, Research Triangle Park, NC 27711, USA.
[3] Division of Atmospheric Sciences, Desert Research Institute, 2215 Raggio Parkway, Reno, NV 89512, USA.

References
Allen, R. W., Adar, S. D., Avol, E., Cohen, M., Curl, C. L., Larson, T., ... Kaufman, J. D. (2012). Modeling the residential infiltration of outdoor $PM_{2.5}$ in the multi-ethnic study of atherosclerosis and air pollution (MESA Air). *Environmental Health Perspectives, 120*, 824–830. https://doi.org/10.1289/ehp.1104447

Biller, W., Feagans, T., Johnson, T., Duggan, G., Paul, R., McCurdy, T., & Thomas, H. (1981). A general model for estimating exposure associated with alternative NAAQS. 74th Annual Meeting of the Air Pollution Control, Association, Philadelphia, PA.

Bouilly, J., Karim, L., Claudine, B., & Allard, F. (2005, September). Effect of ventilation strategies on particle decay rates indoors: An experimental and modelling study. *Atmospheric Environment, 39*(27), 4885–4892. https://doi.org/10.1016/j.atmosenv.2005.04.033

Branco, P. T. B. S., Alvim-Ferraz, M. C. M., Martins, F. G., & Sousa, S. I. V. (2014). The microenvironmental modelling approach to assess children's exposure to air pollution—A review. *Environmental Research, 135*, 317–332. https://doi.org/10.1016/j.envres.2014.10.002

Breen, M. S., Long, T. C., Schultz, B. D., Williams, R. W., Richmond-Bryant, J., Breen, M., ... Batterman, S. A. (2015). Air pollution model for individuals (EMI) in health studies: Evaluation for ambient $PM_{2.5}$ in Central North Carolina. *Environmental Science and Technology, 49*, 14184–14194. https://doi.org/10.1021/acs.est.5b02765

Burke, J. M., Zufall, M. J., & Ozkaynak, H. (2001). A population exposure model for particulate matter: Case study results for $PM_{2.5}$ in Philadelphia, PA. *Journal of Exposure Analysis and Environmental Epidemiology, 11*, 470–489. https://doi.org/10.1038/sj.jea.7500188

Challoner, A., & Gill, L. (2014). Indoor/outdoor air pollution relationships in ten commercial buildings: $PM_{2.5}$ and NO_2. *Building and Environment, 80*, 159–173. https://doi.org/10.1016/j.buildenv.2014.05.032

Chaloulakou, A., Mavroidis, I., & Duci, A. (2003). Indoor and outdoor carbon monoxide concentration relationships at different microenvironments in the Athens area. *Chemosphere, 52*, 1007–1019. https://doi.org/10.1016/S0045-6535(03)00263-7

Crist, K. C., Liu, B., Kim, M., Deshpande, S. R., & John, K. (2008). Characterization of fine particulate matter in Ohio: Indoor, outdoor, and personal exposures. *Environmental Research, 106*, 62–71. https://doi.org/10.1016/j.envres.2007.06.008

Dimitroulopoulou, C., Ashmore, M. R., & Terry, A. C. (2017). Use of population exposure frequency distributions to simulate effects of policy interventions on NO_2 exposure. *Atmospheric Environment, 150*, 1–14. https://doi.org/10.1016/j.atmosenv.2016.11.028

Fujita, E., Campbell, D., Arnott, W., Johnson, T., & Ollison, W. (2014). Concentrations of mobile source air pollutants in urban microenvironments. *Journal of the Air and Waste Management Association, 64*(7), 743–758. https://doi.org/10.1080/10962247.2013.872708

Fujita, E., Campbell, D., & Zielinska, B. (2009a). *Exposure classification project, field study results—season 1.* Interim Report. Desert Research Institute, Reno, NV. February 11, 2009.

Fujita, E., Campbell, D., & Zielinska, B. (2009b). *Exposure classification project, field study results—season 2 and 3.* Interim Report. Desert Research Institute, Reno, NV. October 14, 2009.

Georgopoulos, P. G., & Lioy, P. J. (2006). From a theoretical framework of human exposure and dose assessment to computational system implementation: The modeling environment for total risk studies (MENTOR). *Journal of Toxicology and Environmental Health, Part B: Critical Reviews, 9*, 457–483. https://doi.org/10.1080/10937400600755929

Hanninen, O., Kruize, H., Lebret, E., & Jantunen, M. (2003). EXPOLIS simulation model: $PM_{2.5}$ application and comparison with measurements in Helsinki. *Journal of Exposure Science & Environmental Epidemiology, 13*, 74–85. https://doi.org/10.1038/sj.jea.7500260

Isaacs, K., & Smith, L. (2005). *New values for physiological parameters for the exposure model input file 'Physiology.txt.'* Technical memorandum to Tom McCurdy. Office of Air Quality Planning and Standards, U.S. Environmental Protection Agency, Research Triangle Park, NC. December 20, 2005.

Kruize, H., Hanninen, O., Breugelmans, O., Lebret, E., & Jantunen, M. (2003). Description and demonstration of the EXPOLIS simulation model: Two examples of modeling population exposure to particulate matter. *Journal of Exposure Science & Environmental Epidemiology, 13*, 87–99. https://doi.org/10.1038/sj.jea.7500258

Lai, H. K., Kendall, M., Ferrier, H., Lindup, I., Alm, S., Hanninen, O., ... Nieuwenhuijsen, M. J. (2004). Personal exposures and microenvironment concentrations of $PM_{2.5}$, VOC, NO_2 and CO in Oxford, UK. *Atmospheric Environment, 38*, 6399–6410. https://doi.org/10.1016/j.atmosenv.2004.07.013

Long, C. M., Suh, H. H., & Koutrakis, P. (2000). Characterization of indoor particle sources using continuous mass and size monitors. *Journal of the Air & Waste Management Association, 50*, 1236–1250. https://doi.org/10.1080/10473289.2000.10464154

McCurdy, T. (1994). Human exposure to ambient ozone. In D. J. McKee (Ed.), *Tropospheric ozone* (pp. 85–127). Ann Arbor, MI: Lewis Publishers.

McCurdy, T. (1995). Estimating human exposure to selected motor vehicle pollutants using the NEM series of models: Lessons to be learned. *Journal of Exposure Analysis and Environmental Epidemiology, 5*(4), 533–550.

McCurdy, T., Glen, G., Smith, L., & Lakkadi, Y. (2000). The national exposure research laboratory's consolidated human activity database. *Journal of Exposure Science & Environmental Epidemiology, 10*, 566–578. https://doi.org/10.1038/sj.jea.7500114

Meng, Q. Y., Spector, D., Colome, S., & Turpin, B. (2009). Determinants of indoor and personal exposure to $PM_{2.5}$ of indoor and outdoor origin during the RIOPA study. *Atmospheric Environment, 43*, 5750–5758. https://doi.org/10.1016/j.atmosenv.2009.07.066

Milner, J., Vardoulakis, S., Chalabi, Z., & Wilkinson, P. (2011). Modelling inhalation exposure to combustion-related air pollutants in residential buildings: Application to health impact assessment. *Environment International, 37*, 268–279. https://doi.org/10.1016/j.envint.2010.08.015

Mohammed, M. O. A., Song, W. W., Ma, Wan L., Li, W. L., Ambuchi, J. J., Thabit, M., & Li, Y. F. (2015). Trends in indoor-outdoor $PM_{2.5}$ research: A systematic review of studies conducted during the last decade (2003–2013). *Atmospheric Pollution Research, 6*, 893–903. https://doi.org/10.5094/APR.2015.099

Pellizzari, E., Thomas, K., Clayton, C., Whitmore, R. W., & Shores, R. C. (1992). *Particle total exposure assessment methodology (PTEAM): Riverside California pilot study.* Final Report, Volume 1 Prepared for the U.S. EPA by Research Triangle Institute, RTP, NC. RTI/4948/108-02F.

Smith, J. D., Mitsakou, C., Kitwiroon, N., Barratt, B. M., Walton, H. A., Taylor, J. G., … Beevers, S. D. (2016). London hybrid exposure model: Improving human exposure estimates to NO_2 and $PM_{2.5}$ in an urban setting. *Environmental Science and Technology, 50*, 11760–11768. https://doi.org/10.1021/acs.est.6b01817

Soares, J., Kousa, A., Kukkonen, J., Kangas, L., Kauhaniemi, M., … Koskentalo, T. (2014). Refinement of a model for evaluating the population exposure in an urban area. *Geoscientific Model Development, 7*, 1855–1872. https://doi.org/10.5194/gmd-7-1855-2014

U.S. Environmental Protection Agency. (2002). Consolidated human activities database (CHAD) users Guide. Database and documentation. Retrieved from https://www.epa.gov/healthresearch/consolidated-human-activity-database-chad-use-human-exposure-and-health-studies-and

U.S. Environmental Protection Agency. (2008). *Risk and exposure assessment to support the review of the NO_2 primary national ambient air quality standard.* Office of Air Quality Planning and Standards, U.S. Environmental Protection Agency, Research Triangle Park, NC, 27711. EPA-452/R-08-008a Retrieved from http://www.epa.gov/ttn/naaqs/standards/nox/s_nox_cr_rea.html

U.S. Environmental Protection Agency. (2009a). *Risk and exposure assessment to support the review of the SO2 primary national ambient air quality standards.* Office of Air Quality Planning and Standards, U.S. Environmental Protection Agency, Research Triangle Park, NC, 27711. EPA-452/R-09-007. Retrieved from http://www.epa.gov/ttn/naaqs/standards/so2/s_so2_cr_rea.html

U.S. Environmental Protection Agency. (2009b). *Integrated science assessment for particulate matter.* National Center for Environmental Assessment-RTP Division, Office of Research and Development, U.S. Environmental Protection Agency, Research Triangle Park, NC. EPA/600/R-08/139F. Retrieved from http://cfpub.epa.gov/ncea/cfm/recordisplay.cfm?deid=216546

U.S. Environmental Protection Agency. (2010a). *Quantitative risk and exposure assessment for carbon monoxide—amended.* Office of Air Quality Planning and Standards, U.S. Environmental Protection Agency, Research Triangle Park, NC, 27711. EPA-452/R-10-006. Retrieved from https://www.epa.gov/naaqs/carbon-monoxide-co-standards-risk-and-exposure-assessments-current-review

U.S. Environmental Protection Agency. (2010b). *Integrated science assessment for carbon monoxide.* (Final Report). National Center for Environmental Assessment-RTP Division, Office of Research and Development, U.S. Environmental Protection Agency, Research Triangle Park, NC. EPA/600/R-09/019F. Retrieved from http://cfpub.epa.gov/ncea/cfm/recordisplay.cfm?deid=218686

U.S. Environmental Protection Agency. (2010c). *Quantitative health risk assessment for particulate matter.* (Final Report). Office of Air Quality Planning and Standards, U.S. Environmental Protection Agency, Research Triangle Park, NC. EPA-452/R-10-005, June 2010. Retrieved from http://www3.epa.gov/ttn/naaqs/standards/pm/s_pm_2007_risk.html

U.S. Environmental Protection Agency. (2013). Integrated science assessment of ozone and related photochemical oxidants (Final Report), National Center for Environmental Assessment-RTP Division, Office of Research and Development, U.S. Environmental Protection Agency, Research Triangle Park, NC. EPA/600/R-10/076F. Retrieved from http://cfpub.epa.gov/ncea/isa/recordisplay.cfm?deid=247492#Download

U.S. Environmental Protection Agency. (2014). Health risk and exposure assessment for ozone, Office of Air Quality Planning and Standards, U.S. Environmental Protection Agency, Research Triangle Park, NC, 27711. EPA-452/R-14-004a. Retrieved from https://www.epa.gov/naaqs/ozone-o3-standards-risk-and-exposure-assessments-current-review

U.S. Environmental Protection Agency. (2015). The HAPEM user's guide, hazardous air pollutant exposure model, Version 7. Retrieved from https://www.epa.gov/fera/human-exposure-modeling-hazardous-air-pollutant-exposure-model-hapem

U.S. Environmental Protection Agency. (2017). Air pollutants exposure model documentation (APEX, Version 5) volume I: User's guide, volume II: Technical support document. Office of Air Quality Planning and Standards, U.S. Environmental Protection Agency, Research Triangle Park, NC. EPA-452/R-17-001a,b. Retrieved from https://www.epa.gov/fera/human-exposure-modeling-air-pollutants-exposure-model

Weschler, C. (2000). Ozone in indoor environments: concentration and chemistry. *Indoor Air, 10*, 269–288. https://doi.org/10.1034/j.1600-0668.2000.010004269.x

Zhang, Q., & Zhu, Y. (2012). Characterizing ultrafine particles and other air pollutants at five schools in South Texas. *Indoor Air, 22*, 33–42. https://doi.org/10.1111/ina.2012.22.issue-1

8

Diagnosing soil degradation and fertilizer use relationship for sustainable cotton production in Benin

Barthelemy G. Honfoga[1]*

*Corresponding author: Barthelemy G. Honfoga, Faculty of Agronomic Sciences, School of Economics, Socio-Anthropology & Communication for Rural Development, University of Abomey-Calavi (UAC), 06 BP 1892 Cotonou, Akpakpa PK3, Abomey-Calavi, Benin
E-mail: honfogabg@yahoo.fr

Reviewing editor: Isidoro Gómez Parrales, Sevilla University, Spain

Abstract: In Benin and many other cotton-producing countries of West Africa, unsustainable natural resource management is hindering agricultural growth, food security, and poverty reduction. This study addressed the sustainability of fertilizer-based soil fertility management practices in Benin. It diagnosed the relationship between differential soil degradation status over space and fertilizer use in cotton production systems. Referring to sound land use principles, it found that present fertilizer use practices overlook the spatial differences in soil fertility status in export-oriented cotton production systems. Considering more relevant short-run fertilizer needs based on desirable fertilizer doses, the potentials for sustainable fertilizer use were then assessed considering the likelihood of change towards best practices of integrated soil fertility management. More rational fertilizer use practices will be critical in the future to inducing higher cotton yields while preserving the environment. Adjusting current fertilizer recommendations to site-specific soil conditions is urgently required to enhance the sustainability of cotton production systems in Benin. Fertilizer policies will need to rely on updated information on soil and land

ABOUT THE AUTHOR

Barthelemy G. Honfoga is a senior agricultural economist and associate professor from the University of Abomey-Calavi (UAC), Benin, with expertise in agricultural marketing, food security, and policy analysis. He's presently deputy head of the Department of Agricultural Economics, Rural Sociology and Extension. He holds a PhD in economics and business from the University of Groningen (RuG)/Centre for Development Studies (Netherlands) and a degree of Engineer in agricultural economics from the Faculty of Agronomic Sciences/UAC. His research focus is on fertilizer market development and soil fertility management (SFM). He gained about 30 years of experience in agricultural research and development with International Fertilizer Development Center (IFDC), Foundation for Sustainable Food Security in Central West Africa (SADAOC), International Land Coalition (ILC), and UAC. He was consultant for FAO, FANRPAN, FARA, UNDP, Michigan State University, IFDC, UEMOA, etc. This paper adds to a SFM series published in international journals from a research funded by RuG and the International Foundation for Science.

PUBLIC INTEREST STATEMENT

Extensive cotton production and inadequate fertilizer use is leading to land degradation, food insecurity and growing poverty in cotton growing areas of West Africa. Soil fertility maps are quite obsolete because of the lack of high-level modern expertise in the field and related high cost. This study used a socio-environmental approach to diagnose the relationship between soil degradation and fertilizer use in cotton-growing areas of Benin, and provided evidence of an inadequate link. Considering desirable fertilizer doses based on best practices of integrated soil fertility management, short-run fertilizer needs were assessed and poor potentials for sustainable intensification were revealed. Fertilizer policies should urge to monitoring soil dynamics, updating fertilizer recommendations to site-specific soil conditions, and innovation in fertilizer use knowledge management. This is key to enhancing the sustainability of cotton production systems and improving agricultural productivity and net incomes among cotton growers.

use dynamics, and be innovative enough to induce a steady increase in agricultural productivity and improved net incomes cotton growers.

Subjects: Environment & Agriculture; Environmental Studies & Management; Development Studies, Environment, Social Work, Urban Studies; Economics, Finance, Business & Industry

Keywords: environmental degradation; soil fertility; agricultural intensification; cotton production sustainability; potential

1. Introduction

Poor soil fertility management is a subtle form of natural resource misuse which is overlooked whereas it hinders economic development through decreasing agricultural productivity and growing food insecurity (Ndufa, Cadisch, Poulton, Noordin, & Vanlauwe, 2005). Indeed, the diversity and peculiarity of different ecological features associated with the African continent demand that a holistic and all-inclusive approach of soil fertility management be devised instead of a one-size-fits all strategy commonly pursued by national governments and international development partners (Oluwatoyin, Kolawole, Mogobe, & Magole, 2013). The widespread soil nutrient mining in Africa[1] has led to expansion of the agricultural frontier and the opening up of less favorable soils for cultivation. This is a scenario for disaster over the long run, given the difficulty of restoring tropical soils to productive capacity (Morris, Kelly, Kopicki, & Byerlee, 2007). In West Africa, cotton production for export with inadequate soil nutrient management is a typical example (UEMOA, 2013). This includes extensive land use with low fertilizer application or backward intensification where fertilizers are applied with little consideration for the differential soil fertility status of cultivated lands (Smaling, 1993; van Duivenbooden, 1995). Bellwood-Howard (2014) observed that market-oriented green revolution approach with emphasis on inorganic fertilizer and credit among African farmers does not allow room for them to practice site-specific soil fertility management (SFM) and subsistence mechanisms that they have developed as responses to risk. Among other reasons, high levels of indebtedness accounted for most.

Fertilizer use practices in cotton production systems of francophone West Africa are inappropriate and are leading to soil nutrient depletion and rapid land degradation (Saidou, Kossou, Acakpo, Richards, & Kuyper, 2012; UEMOA, 2013) and to the inefficiency and low profitability of fertilizer use. Yet, the terms of trade of cotton-producing countries have been declining since the mid 1990s. Recent trends are quite alarming as average increases in fertilizer prices are about three times the increases in crop price (Ivo, 2008). In Benin, in spite of the injection of government subsidy funds up to 67.1 billion CFA francs in the cotton sub-sector since 2000/2001, farmers' income has been declining since 2003 in relation with growing input prices and low producer prices (MEF, 2010). As a result, food insecurity and poverty have expanded (UNDP-Benin, 2011), leading to environmental degradation, even more rapidly in the cotton-producing zones (Houngbo, 2013). Yet, these zones are usually praised to be better off than others.

This paper is about agricultural sustainability in the Republic of Benin, with reference to soil degradation and fertilizer use in cotton zones.

In Benin, agriculture accounts for 39% in the GDP, employs 70% of the active population and contributes 90% of total export earnings (Benin, 2008). However, its annual growth of 2.7% over the 1990–2004 period (Benin, 2010) and around 3% in 2011 has not yet substantially overtaken population growth which reached 3.25% p.a. in 2013 (DDC, 2013). Actually, low-yield cotton production is the main contributor to agricultural growth while food crops' production is lagging far behind and cannot meet the growing urban food demand. The country turned to cotton as its major export crop after its palm oil's competitiveness in the world market drastically decreased in the late 1980s. A state marketing board (SONAPRA) was then created in 1983 to manage input supply to farmers and cotton ginning and exportation to the world market. In the 1990s, the country implemented structural adjustment programs and economic reforms, including trade liberalization, in response to domestic and international pressures for good governance and macroeconomic stability. As a matter of fact, agricultural sector reforms mainly targeted the cotton sub-sector, which provides about 37% of the country's export revenues and 70% of agricultural exports (AProCA, 2008; Kpadé, 2011). However, low productivity and non-sustainability of agriculture in cotton zones still constitute real concerns for the country's economic development.

The main fertilizers used in Benin include the cotton complex (14-23-14-5S-1B), the cereal complex (15-15-15) and urea (46% N). Following market liberalization in 1992, fertilizer consumption first dropped, then increased rapidly thereafter to reach 114,000 metric tons (MT) in 1999, compared with an average of no more than 20,000 MT between 1966 and 1991 (IFDC, 2005). However, this growth was due to an accelerated extension of cultivated areas in response to attractive cotton world market prices in the early 1990s. Cotton areas increased sharply from 26 800 ha in 1982 to 378 000 ha in 1998 (Kpadé, 2008), and then decreased slightly to 347,000 ha in 2013–2014 (UEMOA, 2014), but this area expansion did not translate into similar increase in fertilizer use intensity (amount/total ha cultivated), and cotton yields remain low. Since 1998 the average yield ranged between 1 and 1.1 ton/hectare (Togbé, 2013) against a potential estimated at around 3 tons/ha.[2] Soil fertility management practices remain unsustainable. Between 2000 and 2005, farmers sharply reduced cotton areas (and fertilizer use) up to 50%, in reaction to declining world cotton prices and a disorganized domestic market. When national and international stakeholders, including the World Bank, urged for a healthier management of the cotton industry farmers rather resumed with area expansion, although the PARFC and PARFCB projects enabled yield increases in a few districts.[3] Cotton production increased to 451,000 MT in 2016/17 and is expected to reach more than 500,000 MT in 2017/18, owing to a new support from the newly elected President/Government to boost cotton production.[4] However, the increased cotton production observed is due more to land expansion than productivity increase. Indeed, although total fertilizer consumption increased (not beyond 100,000 MT/year), land expansion resumed more rapidly leading to only a little increase in fertilizer use intensity.

Overall low cotton yields indicate that current fertilizer recommendations in Benin are inappropriate. They ignore nutrient use efficiency, which is the essence of sustainable fertilizer use (Igué, Gaiser, & Stahr, 2004). They are obsolete, as they overlook market trends of other crops. Although maize has become a commercial crop while remaining a staple food crop (FAO-Bénin, 2015), it often gets only leftovers or residual effects of cotton fertilizers in cotton-cereal rotations (Saidou et al., 2012). As a result, there is an inadequacy between spatial differences in soil fertility status and fertilizer doses applied, and this leads to low efficiency, low profitability and reduction in demand, especially when cotton/fertilizer price ratios sink. The said recommendations are perpetuated by fertilizer supply policies that are anchored on a rigid licensing system held by institutions working under a hidden agenda set by monopolies (Honfoga, 2013). In such conditions, fertilizer use intensity in the complex cotton cropping systems, including food crops at various degrees, will remain low and sustainable agricultural intensification cannot happen. Sinzogan (2006) observed that current pest management recommendations in Benin do not fit well with the range of socioeconomic problems farmers are facing, and that socioeconomic interventions need to complement technical R&D advices. This is also true for existing fertilizer recommendations.

This study aimed to show that current fertilizer use practices in cotton-based cropping systems in Benin are not congruent with the differential land degradation status across the country. The inadequacy of pan-territorial fertilizer recommendations is demonstrated and more environmentally sound practices are suggested. The paper advocates for rational use of fertilizers and related soil fertility-enhancing technologies to promote a sustainable agricultural intensification in Benin.

2. Literature review
This section highlights the theoretical background of the study. A few previous conceptual works on soil degradation vs. soil fertility, fertilizer demand and potential for sustainable fertilizer use are summarized.

2.1. Soil degradation as a proxy of soil fertility depletion

Soil degradation, which is the most important aspect of land degradation in West Africa, refers to a loss of soil productivity. However, the concept of land degradation is valued-laden as it embraces many perceptions that make soil degradation actually very difficult to measure. Indeed, a change in productivity cannot be attributed solely to a change in the quality of the soil. Productivity is at least as much affected by changes in water availability, agricultural, or range management practices, or in the case of cultivated land, factors such as labor input, technology, and crop selection. As a consequence, biomass production (or crop yield) can only serve as a first proxy to soil degradation. It needs to be supplemented by corroborative evidence from actual measurements of the state of the soil (Mazzucato & Niemeijer, 2000, pp. 115–116). These authors have extensively discussed how problematic is the operationalization of the concept of soil degradation. Nonetheless, soils' physical and chemical properties are often correlated with crop yields in the case of cultivated lands, and science-based measurement criteria (lab-measured soil properties) may also agree well with farmers' perceptions. Field extension agents refer to such perceptions and their own observations when asked to assess the soil fertility status of those lands. From farmers' experience, the occurrence of certain weeds on a field (whether cultivated or not) indicate a substantial decrease in soil fertility (Dangbégnon, Nederlof, Tamelokpo, & Mando, 2010). This may be correlated with some basic soil nutrient deficiencies. For example, Saidou, Kuyper, Kossou, Tossou, and Richards (2004) assessed soil fertility status on the basis of dicotyledonous weeds, soil texture and color, and soil fauna (earthworm casting activity). They observed that farmers of the Atacora region of Benin have adapted their cropping systems to the local environment by developing traditional and new strategies and activities that could contribute to maintain or enhance crop productivity. In the present study, the observed soil degradation is assessed based on visual appraisal, as a proxy of soil fertility depletion or proxy opposite of soil fertility. This method has some previous references. ISRIC (1991, 1992) used the "low: 1, average: 2, high: 3" values as weights to factor land areas at those soil degradation levels and assess the relative severity of the phenomenon in Benin and other West African countries.[5] In Togo, Brabant, Darracq, Egué, and Simonneaux (1996) and Abbey and Adou-Rahim-Alimi (2003) also used a similar approach to assess soil degradation (1–2: lowly degraded; 3–4: moderately degraded; 5–6: highly degraded). In Benin, the cotton branch of the national agricultural research institute evaluated soil fertility status by relying on farmers' perceptions on a similar scale (0: poor soils; 1: moderately fertile soils; 2: fertile soils; 3: very fertile soils) (IFDC, 2005, p. 7). In all above cases, the method has the advantage of enabling a rapid appraisal of the soil fertility status of large regions at low cost, compared with the laboratory method. The latter requires expensive soil sampling and analyses while not always leading to a unanimous appraisal of soil fertility (Mazzucato & Niemeijer, 2000). In our study, the visual method was based on ad hoc soil fertility variables (original nature of soils, general state of the landscape, change in vegetation, trends in crop yields) that extension agents and farmers appraised as a whole to estimate average soil degradation levels.

2.2. Determinants of fertilizer demand

The process of fertilizer demand is complex. Many agronomic and socioeconomic factors influence farmers' decisions about fertilizer use (types, methods of application, doses, target crops, etc.) (Tshibaka, Honfoga, Têvi, Houngbo, & Dokoué, 1992). Initial soil fertility status is the most important agronomic factor to start with (Mokwunye, de Jager, & Smaling, 1996). Other major determinants of fertilizer use and sustainable farming systems include weather (in general), water availability/rainfall, cropping systems/crop patterns and their market orientation, livestock endowment, and off-farm income (Ebanyat et al., 2010). Soil fertility/degradation status is a combined effect of these factors and determines the amount of fertilizers to be used now or in the future to compensate soil nutrient extraction by crops and soil nutrient losses through water and wind erosion, leaching, etc. Based on such appraisal, farmers should adopt soil and water conservation techniques that reduce nutrient losses. The fertilizer demand process emphasizes the relationships between fertilizer use intensity and some critical determinants of fertilizer adoption (soil degradation status, farmers' technical skills and cropping systems, land and labor availability, farm household income, credit accessibility, diffusion rates of soil fertility-enhancing technologies, and risks and uncertainties associated with input/output markets). These variables interact with each other at various degrees to determine the agronomic efficiency and profitability of fertilizer use, and at last fertilizer demand (Baum & Heady, 1957; Dudal, 2002; Minot, Kherallah & Berry, 2000). Therefore, focusing on sustainable land use in agriculture, the realistic estimates of country's fertilizer needs should be calculated based on the principles of integrated or sustainable soil fertility management (ISFM) in existing cropping systems, building on soil fertility status before fertilizer application and crop production constraints and objectives.

2.3. Principles of fertilizer use sustainability

The sustainability of agricultural intensification is highly dependent on that of fertilizer use. Poorly designed investments in agricultural intensification are particularly risky in Africa whose environments are fragile (Breman, 2000, p. 17). Therefore, reasonably desirable fertilizer doses should be applied, and fertilizer demand forecasts should be made accordingly. Such doses in cotton-producing zones should incorporate differential soil degradation status, as well as the potential yields that reflect farmers' intensification options[6] and obey to the principles of integrated soil fertility management – ISFM (Breman, 1999). The thrust of the ISFM paradigm is that organic treatments are valuable primarily as a means toward increasing the agronomic efficiency of the "entry point," inorganic fertilizer (Vanlauwe et al., 2010). Recommended fertilizer doses should take into account soil conditions, so that relatively small amounts are applied per ha when soil is not degraded, and vice-versa. Only then would such above-mentioned doses make sense and prompt farmers to adopt them. After several decades of extension services for modern agricultural inputs' adoption in sub-Saharan Africa, diffusion rates and crop yields are still low. The approach proposed here is likely to bring about long-lasting positive changes in fertilizer use practices. For example, ICRISAT's scientists developed an innovation called micro-dosing which consist in placement next to plant roots of the amount of fertilizer which is just needed (ICRISAT 2009, 2012). This technique ensures that even small, yet affordable, doses of fertilizer applied at the right place at the right time vastly benefit the crop (William op. cit.). Therefore, fertilizer use advices should rather rely on micro-doses to enable the majority of farmers to adopt that technology and increase their incomes. This has been recalled in the FAO campaign "Produce much with less" over the last 10 years.

According to ISFM principles, sustainable fertilizer use should take into account both market opportunities (Dudal, 2002, p. 18) and the level of natural resource exploitation. Breman (1999) highlighted that when designing a soil fertility improvement strategy, the situations of under-exploitation, intensification, and over-exploitation should be distinguished, while manure/soil amendments and counter-erosion techniques are recommended in all situations (Table 1). ISFM is aimed at increasing land productivity using cost-effective soil fertility-enhancing technologies, including fertilizers, organic matter, and soil conservation methods (Breman, 1997; Breman & Debrah, 1999; Dudal, 2002; Tshibaka et al., 1992). The pan-territorial recommended dose in Benin assumes that soils are exhausted or degraded everywhere in the country, which is not true. Applying low to moderate doses to maintain the fertility of lowly degraded/fertile soils is justified in the situation of under-exploitation, where ecological or biological/low input agriculture is advised (Breman, 1999, p. 13; Breman, 2000, p. 9).

The ISFM principles are useful for understanding and assessing a region's potential for sustainable fertilizer use and agricultural intensification.

2.4. Potential of sustainable fertilizer use and agricultural intensification

The capacity a region has to achieve maximum crop yields and to keep them over a long period expresses its potential for a sustainable agricultural intensification. It is the capacity to achieve long-run optimum productivity, depending on the time horizon specified, while keeping or improving the quality of natural resources (Dixon, Gulliver, & Gibbon, 2001, p. 12 et p. 35; Schreurs, Maatman, & Dangbégnon, 2002, pp. 67–69). In Benin, Weller (1999) addressed the potential of agricultural intensification as follows: "To what extent agricultural intensification practices will contribute to food supply or what are the agronomic potentials of Benin?" Actually, not only agronomic factors are important. The potential for intensifying the production of a range of crops within complex cropping

Table 1. Principles and technologies for integrated soil fertility management (ISFM) and implications for the dose of fertilizer use

Situations	Probable soil degradation status*	Principles	Technologies	Desirable dose* of mineral fertilizers
General		Prevent soil nutrient losses	Fight against water and wind erosion, and bush fires	
		Prevent the loss in soil structure	Organic matter management	
Under-exploitation	1 (lowly or non-degraded soils)	Compensate nutrient exportations using "green fertilizers"	Fallows, leguminous plants/crops and agro-forestry	Low (1/3 of optimal dose corresponding to potential yields)
Intensification	2 (moderately degraded soils)	Increase water and nutrient availability	Integrated water and nutrient manage-ment	More than moderate (2/3 of optimal dose or more)
Over-exploitation	3 (highly degraded soils)	Recapitalize and improve soil fertility	Integrated use of amendments and mineral fertilizers	High (3/3 of optimal dose)

Source: Adapted from Breman (1999).

*These columns are added by us to Breman's original table. In the field, the situation of intensification becomes quickly that of overexploitation because of high population density. The method for adjusting the recommended dose assumes the following relationship: Desirable dose = (Dose of potential yield) × (relative degree of soil degradation).

systems is associated with the quality of resources (soils/agro-ecological conditions, infrastructure, access to credit and inputs and farm equipment, skills of extension/research services and farmers), and the long-run strategies for using them (Breman & Debrah, 2003; Schreurs et al., 2002, p. 68). These factors reflect the natural potentials and the existing technical and economic opportunities for intensifying agriculture (IFDC, 2006). Then, how to measure the potential for agricultural intensification? From an agronomic perspective alone, Weller (1999) used a complex formula that incorporates soil, climate/weather and crop data to evaluate the agro-ecological potential of maize production in southern Benin. In this paper, the unit of observation is not a crop but rather the soil carrying various cropping systems in different agro-ecological zones, and we focus on a fertilizer-based sustainable agricultural intensification.

3. Materials and methods

3.1. The study approach
In designing fertilizer policies, reliable data on soil fertility status, applicable fertilizer doses, adoption rates and country's fertilizer demand in the short and long run are required. The study wanted to illustrate a socioeconomic approach for revising pan-territorial fertilizer recommendations based on a participatory assessment of differential soil degradation status by district for a wiser policy for fertilizer procurement and use in Benin. That approach is inspired by the Framework for Evaluating Sustainable Land Management (FESLM) which is defined as "a pathway to guide analysis of land use sustainability, and connect all aspects with the multitude of interacting conditions (environmental, economic and social)" (Smyth, Dumanski, Spendjian, Swift, & Thornton, 1993; van Duivenbooden, 1995). Considering the need for cost-effective fertilizer policies, it draws specifically on the principles of sustainable or ISFM and includes the estimation of: (i) adjusted fertilizer doses using average soil degradation scores at district-level based on farmers' perceptions; (ii) the fertilizer needs (short-term demand) at district and at national levels based on adjusted fertilizer doses; and (iii) the potential for sustainable fertilizer use.

3.1.1. Overall data generation and statistical methods used
A thorough description is made in the next sections of the methods for generating data on soil degradation, adjusted recommended fertilizer doses, fertilizer needs (short-run fertilizer demand) and

the potential for sustainable fertilizer use. Descriptive statistics (sample means comparisons) were used for quantitative analysis throughout the paper. Regarding soil degradation in particular, a two-tier statistical method was used. At district-level (extension agents' perceptions) very simple calculus in Excel spreadsheets were done on the data because number of observations is lower than 20 per district. At farm-level appraisal (farmers' perceptions), the SAS software Version 8.1 was used to process and analyze the data because samples were large enough ($N = 26–63$ farmers per district, and $N =$ more than 250 farmers per region). It was, therefore, possible to give the significance levels of means and correlation coefficients.

3.1.2. Method for calculating soil degradation scores

Soil degradation was measured as a proxy opposite of soil fertility, based on extension agents' and farmers' visual appraisal of the nature of the soils (respectively, at district and farm level), general state of the landscape, change in vegetation, and trends in main crops' yields.

At district level, extension field agents were requested to estimate the overall soil degradation level of agricultural lands in each sub-district under their district's agricultural authority, using the scale: 1: lowly or not degraded, 2: moderately degraded, 3: highly degraded. District senior extension officers also participated jointly in the assessment by providing corrections when necessary.[7] Average soil degradation score was then estimated as the weighted average of soil degradation levels (1, 2, 3), the weights being the number of sub-districts represented at each degradation level.[8]

At farm level, farmers in selected districts were requested to say which acreage or percentage of their total cultivable land area is degraded. This means getting soil degradation estimates on a "0–100" scale. In those districts, extension agents' estimation at district-level could thus be compared with farmers' estimation, although the latter refers mostly to cotton fields.

Then the rationale for fertilizer use could be assessed then, as to whether soil degradation status is correlated or not with fertilizer use intensity, and in the "right" direction or not. The "right" direction means relatively low fertilizer use per ha cultivated when there is little or no soil degradation, and vice-versa.

3.1.3. Method for adjusting the recommended fertilizer doses

The pan-territorial dose[9] of 200 kg/ha fertilizers (150 kg NPKSB 14-23-14-5S-1B and 50 kg urea 46% N) recommended to cotton farmers in Benin is not economical, nor is it environmentally sustainable. This study suggests an adjustment of that dose in application of the principle of wiser mineral fertilizer application in decreasing order of soil fertility level, i.e. the more the soil is fertile (or less degraded) before fertilizer application, the less the quantity of mineral fertilizer to be applied. Desirable fertilizer doses are calculated according to that principle. On the "1–3" scale of soil degradation at district level, the logic of fertilizer application proportionately to soil degradation status[10] is proposed as follows:

Desirable dose = $k \times$ (dose for potential yield)

The k coefficients depend on the soil degradation scale that is used. In this study, k is the relative degree of soil degradation at district-level and the dose for potential yield is by default taken as the current recommended dose of 200 kg/ha. Hence:

Desirable or adjusted dose = 200 × (observed absolute soil degradation level)/3.

3.1.4. Method for estimating fertilizer needs (short-run fertilizer demand)

In this paper, we estimated future fertilizer needs (quantities) at the 2008 horizon considering the desirable fertilizer doses for a wise soil fertility improvement and the likely trends in land use (cultivated areas).[11] Against that background, fertilizer needs are estimated based on the following assumptions:

(1) Average cultivated area p.a. during the "2000–2003" period, after the "1995–1999" cotton boom, is likely to be the annual cultivated area in the 5 following years (horizon 2008) if agricultural intensification takes off.[12]

(2) Desirable or adjusted dose = 200 × (observed absolute soil degradation level)/3.

(3) Maximum fertilizer diffusion rate to be expected refers to main fertilized crops (cotton, maize) and to sorghum and cowpea that are showing growing prospects for fertilizer use.[13]

(4) Estimated fertilizer needs or Demand (quantities) = (Desirable dose) × (Fertilized area) = (Desirable dose) × (Maximum expected diffusion rate) × (Average cultivated area, 2000–2003) = (2) × (3) × (1).

The diffusion rate of a technology is the percentage ratio of area on which the technology is applied to total area cultivated. The formulae (4) assume that steady fertilizer demand translates into the application of "reasonably accessible" or desirable doses adopted on a large proportion of cultivated areas according to ISFM principles. Then, the gaps between short-term needs (demand) and actual consumption could be estimated. Considering that cultivated areas increase each year proportionately to population growth, one may need to introduce the latter in the formulae. However, advocating for sustainable land use suggests a non-expansion of cultivated areas and adoption of relatively intensive production techniques.

3.1.5. Method for estimating the potential for sustainable fertilizer use

The *potential for sustainable fertilizer use* (PSFU) was estimated in this study using a systematic method based on farm-level soil degradation rate (% area degraded out of total area available to farm household) and farmers' performance in fertilizer use (% present consumption to potential short-run demand) and other soil fertility management techniques. PSFU's intra-regional variation was then highlighted to guide fertilizer policies. The whole estimation process and methods are presented hereafter.

The potential for sustainable agriculture is expressed as the *potential of sustainable fertilizer use (PSFU)*:

PSFU = f(soil fertility status, current fertilizer use performance, ISFM demand), where:

Soil fertility status = soil quality = f_1(inherent fertility[14], current soil degradation status, climate);

Current fertilizer use performance = f_2(ratio current/optimal dose, adoption rates[15] of recommended fertilizers, dates, and methods of application);

ISFM demand (f_3) = Average diffusion rate of ISFM technologies (fertilizer, organic matter, water and soil conservation techniques, improved seeds).

ISFM is integrated soil fertility management with recommended soil fertility-enhancing technologies. We do not know a priori the functional form of PSFU. Estimating the above complex functions requires a multidisciplinary expertise. For the purpose of simplifying PSFU estimation, let's consider a student's intellectual quotient and his potential of success to exams. When the student usually does not get marks closer to the maximum score in a subject matter or the whole class curriculum, due to his hereditary characteristics, we may say that he has a natural low learning capacity or a low intellectual quotient, which keeps wide the gap between his achieved score and the maximum score. On the opposite, a student with a high intellectual quotient has a small gap and enjoys a great potential of success. In general, any student's potential of success depends on his intellectual quotient (natural ability) and the conditions of his learning environment[16] that determine his willingness to learn or his demand for learning. Therefore, we may intuitively put it simply as follows:

Potential of success = intellectual quotient × (current performance rate) × (index of learning environment)

The intellectual quotient is the hereditary factor (h); and current performance rate = 100 × (current score/maximum score). Hence:

Potential of success (%) = 100 × [h × 100 × (current score/maximum score) × (index of learning environment)]

When we consider that perception for the purpose of estimating an agricultural zone's potential of sustainable fertilizer use, we define:

h = "soil quality" = "soil fertility index" of the zone

Current score = amount of fertilizer actually used (QE)

Maximum score = fertilizer amount required to meet the needs (QE_p)

Index of fertilizer use environment = Index of ISFM technologies' use = (average ISFM diffusion rate)/100.

Therefore: PSFU (%) = 100 × [(soil fertility index) × 100 × (QE/QE_p) × (average ISFM diffusion rate)/100]

Soil fertility index = 1–(percentage rate of soil degradation/100). As mentioned earlier, this is just an indicative measure of soil fertility for the purpose of rapid socioeconomic assessments[17]. Therefore:

PSFU (%) = 100 × [1–(percentage rate of soil degradation/100)] × (QE/QE_p) × [(average ISFM diffusion rate)/100]

$$PSFU\,(\%) = [1 - (\text{percentage rate of soil degradation}/100)] \times \left(QE/QE_p\right)$$
$$\times \left(\text{average ISFM diffusion rate}\right)$$

(5)

ISFM technologies include mineral fertilizer, organic matter, improved seeds, water, and soil conservation methods. Here, improved seeds concern only maize (second fertilized crop after cotton), as cotton is always cultivated with improved seeds. The diffusion rate reflects both the economic capacities of use and the technical skills that determine initial adoption decisions. The fertilizer amount meeting the needs (QE_p) may be: (i) the fertilizer dose to reach potential yield times the achievable crop area, using recommended soil preparation techniques; (ii) the fertilizer dose recommended by extension services times the same achievable crop area; or (iii) the desirable fertilizer dose according to appropriate ISFMtimes the achievable crop area, using current soil preparation techniques.

The achievable crop area is function of existing land tenure regime and farmers' production objectives. For a sustainable agriculture, expansion of cultivated area should be restricted to account for population growth and the consequent land occupation for non-agricultural uses (housing, nature conservation, etc.). Considering that it is difficult to estimate the "achievable crop area using recommended soil preparation techniques", we used the third option by default in the PSFU formulae, thereby approximating the quantity of fertilizer meeting the needs (QE_p) by the estimated needs (QE_b) in formulae (4) in sub-section 3.1.3.

3.2. Study area, sampling, and data collection methods

A PhD research project conducted from 2003 to 2007 on fertilizer distribution in the liberalized cotton sub-sector in Benin. It was carried out in the two main cotton-producing regions of Benin, i.e. Borgou-Alibori (BA) in the north (750–1100 mm/year; 24 inhab/km²); average seed cotton yield around 1300 kg/ha) and Zou-Collines (ZA) in the center (900–1200 mm/year; 23–110 inhab/km²; average seed cotton yield around 900 kg/ha). They are later referred to, respectively, as Northeast region and Central region. In both regions, fertilizer use is cotton-oriented but actually as a hidden front-cover for food crop production. However, the food security basis is weaker in the first region than the second where various food crops (maize, tubers, legumes, etc.) are regularly produced, thanks to a higher and two-season rainfall. But the regional differences in the food security situation may be minimal sometimes, e.g. during bad seasons.

A field survey was done in these regions to collect primary data on fertilizer use and farmers' evaluation of fertilizer marketing services. The present study deals with the first part of the survey. It was conducted in 29 districts of the 2 regions for a first participatory assessment of soil degradation, with a dedicated support of field extension officers. A total of 255 sub-districts covering various soil types and cropping zones over 986 617 ha cultivated were surveyed.

Then an in-depth assessment of farmers' perceptions of soil degradation and fertilizer use practices was done at the farm level with a sample of 577 farmers selected in 191 villages of 14 districts (8 in ZA and 6 in BA). The fertilizer use investigation addressed technological variables (area cultivated, area and crops fertilized, quantities of fertilizers, doses of application, intensity of fertilizer use, rate of diffusion, complementary inputs/technologies, etc.). Districts and villages were chosen purposively taking into account different levels of fertilizer use, crop production diversity and levels of soil degradation. Farmers were "randomly" chosen from lists of members of village-level cotton producers' organizations so as to get small-, medium- and large-scale farmers according to area cultivated. The resulting sample was, therefore, representative of cotton producers in the two zones.

Although the above data may appear, the situation of extensive land use, low fertilizer application and soil degradation has not improved yet.

4. Results and discussion

4.1. Cotton production systems and differential soil degradation status in Benin

In 2003–2004, cotton occupied, respectively, 51 and 32% of total area cultivated in the Borgou-Alibori and Zou-Collines regions. During the cotton boom year (1999), the first region represented about 50% of total cotton area and 45% of fertilizer consumption in the country, and the second one 30 and 33%, respectively. District-level secondary data covering all districts (29) in the 2 regions reveal a high and significant correlation coefficient ($r = 0.867$) between cotton share in total cultivated area and fertilizer use intensity over the 1999–2003 period. Farm-level data in the 14 selected districts also show a high and very significant ($p < 0.0001$) correlation ($r = 0.743$ for both regions together) between the two variables. Indeed, the fertilizer supply policy is mainly cotton-oriented as it is based on cotton area predictions and fertilizer procurement on credit. However, cotton is usually grown in various types of cotton-cereal rotations; production systems are still extensive through the slash-and-burn land clearing technique.

In the central region, about 41% of 332 570 ha cultivated were on highly degraded soils, 35% on moderately degraded, and only 24% on lowly degraded soils, so that average soil degradation score was 2.2. In the northeast region, corresponding figures are 21, 42, and 37% of 654 047 ha cultivated, and 1.9 for the average soil degradation score (Table 2). This slightly lower score may be explained by large areas of fertile lands that are still available in some districts in the north. Farmers' differential accessibility to fertilizers and the large variability in fertilizer use on food crops may also explain the differences within and between the 2 regions. In the central region, little or no fertilizer at all is

Table 2. Differential soil degradation status and fertilizer use intensity in cotton zones of Benin

Districts[a]/Regions	Area cultivated p.a. (ha) 2000–2003	Number and % sub-districts per level of soil degradation[b]						Average score of soil degradation	Fertilizer use intensity (kg/ha)[c]
		1		2		3			
		N	%	N	%	N	%		
Abomey	5,325	0	0	2	29	5	71	2.7	58.1
Agbangninzoun	8,780	0	0	5	50	5	50	2.5	28.5
Bante	30,405	7	78	2	22	0	0	1.2	31.4
Bohicon	6,904	2	25	2	25	4	50	2.3	27.4
Cove	13,017	0	0	1	14	6	86	2.9	9.5
Dassa-Zoume	29,781	0	0	5	56	4	44	2.4	42.9
Djidja	29,664	7	58	2	17	3	25	1.7	99.8
Glazoue	38,770	0	0	6	60	4	40	2.4	26.2
Ouesse	60,700	2	22	2	22	5	56	2.3	10.7
Ouinhi	8,605	0	0	3	75	1	25	2.3	57.3
Save	15,552	4	50	2	25	2	25	1.8	47.4
Savalou	38,693	3	21	7	50	4	29	2.1	38.9
Zagnanado	8,747	0	0	2	33	4	67	2.7	32.8
Za-Kpota	22,782	0	0	3	37	5	63	2.6	57.0
Zogbodomey	14,845	6	55	1	9	4	36	1.8	53.3
Zou-Collines	332,570		24.2		35.1		40.7	2.2	41.4
Tchaourou	40,470	4	57	3	43	0	0	1.4	24.7
Parakou	34,539	0	0	3	100	0	0	2.0	14.9
N'Dali	29,815	2	40	2	40	1	20	1.8	36.6
Perere	27,059	3	50	2	33	1	17	1.7	23.7

(Continued)

Table 2. (Continued)

Districts[a]/Regions	Area cultivated p.a. (ha) 2000–2003	Number and % sub-districts per level of soil degradation[b]						Average score of soil degradation	Fertilizer use intensity (kg/ha)[c]
		1		2		3			
		N	%	N	%	N	%		
Nikki	50,956	3	43	4	57	0	0	1.6	48.9
Kalale[d]	64,723							1.8	64.5
Sinende	36,680	0	0	3	25	9	75	2.8	75.4
Bembereke	52,242	7	47	6	40	2	13	1.7	59.0
Gogounou	32,017	2	14	4	29	8	57	2.4	119.3
Kandi	74,642	11	73	3	20	1	7	1.3	80.6
Banikoara	99,876	7	47	2	13	6	40	1.9	83.6
Segbana	44,862	6	43	6	43	2	14	1.7	84.8
Malanville	33,837	0	0	5	100	0	0	2.0	66.3
Karimama	32,331	0	0	4	80	1	20	2.2	18.9
Borgou-Alibori	*654,047*		*37.3*		*42.1*		*20.5*	*1.9*	*57.2*

Source: Honfoga (2007), PhD thesis, p.147.

[a]Shadowed lines indicate the districts that were selected for farm-level primary data collection.

[b]Soil degradation levels: 1 = lowly or not degraded, 2 = moderately degraded, 3 = highly degraded.

[c]Average fertilizer use intensity 2000–2003: quantity used/area cultivated.

[d]Soil degradation data could not be obtained there and the average degradation level was estimated as the mean value of surrounding districts.

Table 3. Pearson's correlation coefficients (r) between fertilizer use intensity and soil degradation status at district level in Benin, 2000–2003				
	All districts		Selected districts	
	r	N	r	N
Zou-Collines	−0,340	15	−0,599	8
Borgou-Alibori	0,280	14	0,365	6
Both regions	−0,134	29	−0,273	14

Source: Honfoga (2007), PhD thesis, p. 148.

directly used on food crops. Appropriate fertilizer types are not available for food crops and access to cotton fertilizers is restricted to members of cotton producers' cooperatives.

4.2. Rationale for fertilizer use in cotton-producing zones of the study area

In contrary to the "homogenous picture" of cotton zones presented by the cotton board officials, intra-regional differences are so important (see Table 2) that it is worth checking if fertilizer use intensity is congruent with soil degradation status. Correlation coefficients calculated using district-level data are negative for the central region and positive for northeast region (Table 3). The negative relationship is apparently backward, as one would normally expect that greater amounts of fertilizers are applied per ha cultivated on degraded soils to restore fertility, and small amounts on lowly/moderately degraded soils to maintain fertility. In the reality extensive agricultural practices increase soil degradation and lead farmers to apply lower and lower amounts per ha over space. This may be explained by the fact that with little or no application of organic matter or other forms of soil amendments, highly degraded soils do not reward well fertilizer use. It seems to confirm ICRISAT (1987)'s view that such a trend is observed when soils have a low cation exchange capacity and face acidification and aluminum toxicity problems.

The concerned farmers would, therefore, reduce fertilizer use intensity when they start facing such problems. In the central region of Benin, farmers who avail large land areas usually clear new fields where soils are still fertile in order to get greater crop response to fertilizer use. In the northeast region however, the normal relationship seems to dominate (positive correlation coefficients). Yet, fertilizer is not always applied on degraded soils, as complementary soil fertility-enhancing technologies are not practiced, either by lack of resources or by ignorance. The expansion of cotton areas leads to greater fertilizer use in absolute terms but also increases soil degradation through the occupation of new fields via the slash-and-burn land clearing technique.

Farm-level primary data in the selected districts reveal that average land cultivation rate (share of cultivated area in total land area available to the household) is 68% among cotton farmers in the northeast region, and 57% in the central region. This confirms that cotton areas are expanding in the leading cotton-producing districts, especially in the northeast region where animal traction is widely adopted. Farmers' extensive land use strategy was also confirmed by the highly significant positive correlation coefficients between land cultivation rate and soil degradation rate, as declared by farmers themselves (Table 4).

Particularly in the central region, continued extensive slash-and-burn land use practices go with diminishing fertilizer use intensity over space. Five districts out of 8 show negative correlation coefficients between soil degradation rate and fertilizer use intensity at farmer-level, and the coefficients were highly significant 3 times out of 5 (Table 5).

The farm-level data show that fertilizer use intensity decreases with increasing soil degradation over space in the central region. This is alarming, provided that 76% of total cultivated area (ha) belong to moderately or highly degraded soils (see Table 4). On the contrary, the normal relationship

Table 4. Pearson's correlation coefficients between land cultivation rate and soil degradation rate[a] at farmer-level[b]

	Average land cultivation rate(%)	Average rate of soil degradation(%)	Pearson Coefficient r	N	Prob r_N > |r| under H_0: $\rho = 0$[c]
Borgou-Alibori	66.9	34.3	0.102	86	0.3522
Zou-Collines	56.6	28.2	0.341*	105	0.0004
Both regions	61.3	30.9	0.248*	191	0.0005

Note: N = number of observations (villages).

Source: Honfoga (2007), PhD thesis, p. 150.

*1% significance level.

[a]The rate of soil degradation at farm-level is the percentage ratio of degraded fields' area to total cultivable area available to the farm household (according to farmers' own perceptions and evaluations).

[b]Farmers were interviewed in 191 selected villages and the values refer to the 2003/04 and 2004/05 crop seasons. Average village mean values were used here in order to solve the problem of missing values with some farmers.

[c]Probability that observed Pearson's correlation coefficient (r) is zero as hypothesized for population (ρ) and sample (r_N).

Table 5. Correlation between soil degradation rate and fertilizer use intensity at farmer-level in selected districts, 2003/2004–2004/2005 cropping seasons

Regions/ Districts	Average fertilizer use intensity (kg/ ha)	Average soil degradation rate (%)[a]	Correlation between fertilizer use intensity and soil degradation rate		
			Coefficient	N	Prob r_N > |r| under H_0: $\rho = 0$[b]
Borgou-Alibori	154.2	34.3	0.103*	257	0.099
Sinendé	140.8	42.6	0.316*	36	0.060
Bembéréké	139.7	34.4	0.155	45	0.309
Gogounou	169.6	18.8	0.114	42	0.471
Kandi	145.2	11.9	0.027	44	0.860
Banikoara	170.4	79.1	−0.366***	48	0.010
Ségbana	156.9	15.5	−0.074	42	0.638
Zou-Collines	85.6	28.2	−0.045	314	0.422
Dassa-Zoumè	83.3	23.0	0.126	63	0.324
Djidja	106.9	30.3	0.401***	44	0.006
Glazoué	78.0	25.6	−0.227	30	0.227
Ouessè	17.7	27.6	−0.200	32	0.273
Ouinhi	115.6	16.5	−0.391**	26	0.048
Savalou	113.0	30.6	0.309**	42	0.046
Za-Kpota	108.4	27.5	−0.394***	35	0.019
Zogbodomè	58.3	41.3	−0.154	42	0.330
Both regions	116.5	30.9	0.035	571	0.405

Note: N = number of observations (farmers).

Source: Honfoga (2007), PhD thesis, p. 151.

*10% significant level.

**5% significant level.

***1% significant level.

[a]The soil degradation rate is the percentage share of degraded soils' area in total cultivable area available to the surveyed farm households.

[b]Probability that observed Pearson's correlation coefficient (r) is zero as hypothesized for population (ρ) and sample (r_N).

rather dominates (4 districts out of 6) in the northeast region, thereby confirming the results obtained with district-level data, although correlation coefficients are not significant at 5% level.

Differentiation among cotton growers may explain the differences in land- and fertilizer use behavior across regions and within regions in terms of positive versus negative trends in soil fertility management. Indeed, medium class cotton growers have greater access than small-scale growers to agricultural input services when cotton market perspectives improve, especially access to credit and farm equipment. This leads to greater land clearing and fertilizer use capacity. Indeed, credit dispensation by the Cotton Inter-Professional Association (AIC) is based on land area available or cultivated. Moreover, agreements may arise in this regard between cotton ginning companies and a few farmers' organizations under particular contract farming. James and Woodhouse (2017) observed such a trend among black capitalist medium-scale sugar cane growers in Nkomazi, South Africa.

However, the situation in Banikoara—where about half of the country's cotton tonnage is achieved—supports the thesis of slash-and-burn extensive land use, with lower and lower qunatities of fertilizer per ha cultivated as soil degradation increases (correlation coefficient is negative −0,366, significant at 1% level). Some previous studies have found that soil degradation is becoming a real concern because of cotton areas' expansion in the northeast cotton belt of Benin (Alohou & Nelen, 2000; Samba Bio Tobou, 2000, p. 1).

Due to large intra-regional variations in the "soil degradation-fertilizer use" relationship with farm-level data, regional aggregate correlation was quite zero, especially in the central region. But the overall picture is that negative correlation is prevalent in the central region where food production is top priority for farmers but access to appropriate fertilizers is limited. On the contrary, it is positive correlation that dominates in the northeast region where cotton mobilizes most efforts of agricultural production.

However, considering the significance level of correlation coefficients at district-level, it is the negative relationship that reflects more the situation in the field. Cropping techniques rely so much on the slash-and-burn practice of area extension that fertilizer use intensity is less than proportionate to soil degradation rate. This is contrary to the common sense of sustainable soil fertility management. It suggests that farmers' rationale for fertilizer use in cotton-producing zones of Benin is backward. However, the perceived backwardness may be rejected, with the argument that farmers are captured into rapid social and economic transformations that affect input and output demand and supply, especially fertilizers and cotton lint in the international market. Indeed, shifts in the production, sourcing and sales strategies and technologies of transnational manufacturing, and massive new possibilities attendant on information technologies are recent persistent trends in the agrarian world (Bernstein, 2008). Watts (2012) witnessed Bernstein (1977) finding of how capital transformed agriculture regionally through different patterns of agrarian change.

In spite of these transformations and the constraints farmers are facing, the results of the present study among cotton growers in Benin suggest that unless ISFM is adopted, even higher doses of fertilizer on newly colonized cotton fields (leading to greater fertilizer use intensity) would not be enough to curb down the growing soil degradation in the Banikoara district of Benin. If appropriate mechanisms are not designed to supply fertilizers for food crops, soil degradation will continue in spite of apparently sustained fertilizer use on cotton fields. Moreover, in addition to mineral fertilizers, the adoption of other soil-fertility enhancing technologies (organic matter, soil and water conservation, improved seeds, etc.) influence fertilizer use strategies and determine a region's potential for fertilizer-based agricultural intensification. This is illustrated later after future fertilizer needs are estimated according to the method presented in Section 2.

4.3. Short-run fertilizer needs

Short-run fertilizer needs are estimated using formulae (4), and the gaps to present consumption are calculated. Then an extrapolation to the whole country is made considering current consumption patterns over space. The results show very large gaps between short-run demand and present consumption, as about 141% of the latter in the central region and 87% in the northeast region. In the first region, soils are highly degraded because agriculture is food crop-oriented with little access to appropriate fertilizers. Therefore, short-run demand is about 2 to 7 times present consumption. The opposite is observed in the second region where present consumption is closer to the needs (Table 6).

The differences between the 2 regions reflect the regional bias in the cotton-oriented fertilizer supply policy. Yet, some districts in the northeast region are also food crop-oriented and suffer from the same bias: gaps are 1–4 times present consumption. Assuming the northeast region/country ratio of fertilizer consumption in 1999 is kept, the national short-run demand is estimated to about 158,000 MT/year. Adegbidi et al. (2000) estimated the national demand in 2005 to 111,350 MT, based on a scenario deemed economically accessible to farmers (projected area reached, fertilizer dose 200 kg/ha, diffusion rate 30%). Our estimation (158,000 MT/year) rather considers a large range of desirable fertilizer doses (82–181 kg/ha)—depending on the soil degradation status—a higher average diffusion rate (about 70%) for all soil-fertility enhancing technologies, and the likely trends in land use even when the major cash crop (cotton) enjoys a price boom.

Undoubtedly, such short-run future consumption level is environmentally justified and economically accessible. It assumes that the government promotes farmers' access to credit, especially in the zones where soils are highly degraded and where commercial banks may restrict credit supply because of high production risks. If area expansion is allowed in pace with rural population growth, fertilizer demand would be about 163,500 MT/year. This estimated national consumption will become effective if sustainable soil fertility management principles are applied and if appropriate fertilizers become available for food crops (especially cereals).

The above results suggest that in order to increase fertilizer consumption towards the needs, soil fertility management and fertilizer supply policies should better promote desirable or economically accessible doses on large areas cultivated rather than pan-territorial high doses that will be adopted only by a few farmers on a limited area. Future fertilizer demand will be satisfied if food crops become fully involved in reliable commercial streams. According to Rusike, Reardon, Howard, and Kelly (2003), the green revolution in Asia was due mostly to the intensification of cereals' production rather than traditional export crops. In some West African countries, efficient domestic and regional markets, and improved socioeconomic conditions are inducing an intensive food crop production, more than traditional export crops (Breman & Debrah, 2003; Wiggins, 1995).

4.4 Potentials for sustainable fertilizer use in the area

Improved quality of natural resources, farmers' technical knowledge, economic capacity and institutional environment will be determinant to promote a sustainable fertilizer use. Increasing fertilizer consumption alone will not be enough to achieve potential crop yields if sustainable natural resource management is not implemented. In order to illustrate that assertion, the potential for sustainable fertilizer use (PSFU) or fertilizer-based agricultural intensification is calculated according to formula (5) elaborated in sub-section 3.1.4 for that purpose. The results show that both regions together have a very low PSFU (only 13.6%), compared with the maximum potential of 100%. If the central region's low PSFU (8.3%) reflects the limited availability of fertilizers for food crops, the northeast region's PSFU is also low (17.5%) in spite of its higher relative fertilizer use performance (Table 7). The overall picture reflects at least one of the following: very large gaps between fertilizer needs and present consumption, considerable soil degradation rate, and poor diffusion of complementary soil-fertility enhancing technologies. When one looks particularly at the results in individual districts, it appears that a high fertilizer use intensity does not necessarily mean a high potential for

Table 6. Estimated fertilizer needs and gaps to present consumption in cotton regions of Benin

	Average soil degradation status	Desirable fertilizer doses(kg/ha)	Maximum expected diffusion rates(%)	Annual fertilizer needs(MT)	Present annual consumption, 2000–2003(MT)	"Needs-present consumption" gaps(%)
Zou-Collines	2.2	149	67.1	29,918	12,425	141
Abomey	2.7	181	76.6	738	242	205
Agbangnizoun	2.5	167	73.7	1,079	244	342
Bante	1.2	82	58.4	1,447	847	71
Bohicon	2.3	150	80.3	832	184	353
Cove	2.9	191	38.5	956	121	691
Dassa-Zoume	2.4	163	79.0	3,835	1,254	206
Djidja	1.7	111	73.4	2,421	2,955	–18
Glazoue	2.4	160	68.6	4,259	882	383
Ouesse	2.3	156	36.7	3,471	541	541
Ouinhi	2.3	150	74.3	960	486	98
Save	1.8	117	73.4	1,332	749	78
Savalou	2.1	138	63.7	3,407	1,580	116
Zagnanado	2.7	178	68.2	1,062	268	296
Za-Kpota	2.6	175	72.0	2,870	1,298	121
Zogbodomey	1.8	121	69.4	1,249	775	61
Borgou-Alibori	1.9	125	67.8	55,958	40,760	87
Tchaourou	1.4	95	54.9	2,039	980	108
Parakou	2.0	133	54.3	2,491	511	387
N'Dali	1.8	120	65.4	2,410	1,108	117
Perere	1.7	111	70.2	2,209	656	237
Nikki	1.6	105	82.3	4,159	2,516	65
Kalale	1.8	122	77.2	5,854	4,229	38
Sinende	2.8	183	68.1	4,581	2,760	66
Bimbereke	1.7	111	74.5	4,130	3,078	34
Gogounou	2.4	162	84.0	4,266	3,865	10
Kandi	1.3	89	76.9	5,266	6,003	–12
Banikoara	1.9	129	86.7	10,952	8,473	29
Segbana	1.7	114	75.1	3,914	3,817	3
Malanville	2.0	133	55.6	2,516	2,219	13
Karimama	2.2	147	23.3	1,172	544	115

Notes: Fertilizer needs = (Desirable dose) × (Maximum expected diffusion rate) × (Average cultivated area, 2000–2003). Shadowed lines correspond to the districts selected for primary data collection.
Source: Adapted from Honfoga (2007), PhD Thesis, p.172.

Table 7. Potentials for sustainable fertilizer use in cotton-producing zones of Benin

	Farm-level soil degradation rate(%)	Present annual fertilizer consumption(QE, MT)	Annual fertilizer needs(QE$_b$, MT)	Relative fertilizer use performance(100*QE/QE$_b$, %)	Average diffusion rate of ISFM technologies(%)	PSFU (%)
Borgou/Alibori	34.3	40,760	55,998	72.8	36.7	17.5
Sinende	42.6	2760	4581	60.2	36.7	12.7
Bimbereke	34.4	3078	4130	74.5	37.5	18.3
Gogounou	18.8	3865	4266	90.6	36.9	27.1
Kandi	11.9	6003	5266	114.0	34.9	35.1
Banikoara	79.1	8473	10,952	77.4	30.9	5.0
Segbana	15.5	3817	3914	97.5	39.9	32.9
Zou/Collines	28.2	12,425	29,918	41.5	27.7	8.3
Dassa-Zoume	23.0	1254	3855	32.5	31.8	8.0
Djidja	30.3	2955	2421	122.1	25.5	21.7
Glazoue	25.6	882	4259	20.7	19.8	3.0
Ouesse	27.6	541	3471	15.6	9.9	1.1
Ouinhi	16.5	486	960	50.6	38.6	16.3
Savalou	30.6	1580	3407	46.4	38.3	12.3
Za-Kpota	27.5	1298	2870	45.2	41.2	13.5
Zogbodomey	41.3	775	1249	62.0	16.5	6.0
Both regions	30.9	53,185	85,916	61.9	31.8	13.6

Notes: MT = metric tons. PSFU (%) = [1−(percentage rate of soil degradation/100)] × (QE/QE$_p$) × (average ISFM diffusion rate).

Source: Honfoga (2007), PhD Thesis database.

Table 8. Fertilizer use intensity versus potential for sustainable fertilizer use

Cotton-producing districts	Current situation of soil fertility management	PSFU(%)
Kandi	High fertilizer use intensity on lowly degraded soils and high diffusion rate of soil fertility-enhancing technologies	35.1
Banikoara	High fertilizer use intensity on highly degraded soils and low diffusion rate of soil fertility-enhancing technologies	5.0
Glazoué	Low fertilizer use intensity on degraded soils and low diffusion rate of soil fertility-enhancing technologies	3.0
Ouessè	Low fertilizer use intensity on degraded soils and very low diffusion rate of soil fertility-enhancing technologies	1.1

Source: Honfoga (2007), PhD Thesis database.

Table 9. Comparing extension agents and farmers' perceptions of soil degradation in high cotton production districts in Benin

	Extension agents' perceptions(scale: 1–3)*	Farmers' perceptions(scale: 0–100)**
Borgou-Alibori	2.5(n = 86)	34.3(n = 257)
Zou-Collines	2.2(n = 105)	28.2(n = 314)
Both regions	2.3(n = 191)	30.9(n = 571)

Source: Honfoga (2007), PhD Thesis database.

*Soil degradation levels: 1 = lowly or not degraded, 2 = moderately degraded, 3 = highly degraded.

**Soil degradation rate is the percentage share of degraded soils' area in total cultivable area available to the surveyed farm households.

agricultural intensification if soils are highly degraded and average diffusion rate of other complementary technologies is low (See examples in Table 8).

Knowing that fertilizer supply policy in Benin targets mostly the northeast region, very few people can imagine that Banikoara, which is the leading cotton production district (with the highest fertilizer use intensity and more than half of the national cotton tonnage), has a potential of sustainable fertilizer-based intensification as low as that of Glazoué and Ouessè (central region) where fertilizer use is comparatively very low. The situation of soil fertility management is worse in the southern region where food crops are priority but are paradoxically suffering from the lack of specific fertilizers. Overall, the discriminatory cotton-oriented fertilizer policy will not promote a sustainable agricultural intensification in Benin. Therefore, from now on, fertilizer procurement policies should avoid crop or regional biases and rather aim at raising the potentials for sustainable fertilizer use in production systems. This is urgently needed, if food security is to improve significantly in Benin and elsewhere in Africa where similar trends in soil fertility management are observed. Undoubtedly, if one considers the growing international fertilizer prices, local distribution problems, and other economic constraints to the adoption of soil fertility-enhancing technologies, the costs of incorporating soil fertility management into the economic systems will be high. For example, increasing threefold the current average PSFU value in the study regions to about 40% would require far more increases in present investments. But the rewards will be large gains in productivity and sustainability, and food security over at least 1–2 decades. It is worth raising an alert about greater state budget allocation to agriculture, with special focus on farmers' support for sustainable land use, and on private sector's capacity-building for input/output market development.

4.5. Limitations of soil degradation perceptions data used in this study

Using farmers' perceptions to assess soil degradation as a proxy opposite of soil fertility was extensively documented earlier. Although this method may be criticized for its lack of accuracy—with reference to strict soil science methods whereby nutrients and organic matter contents are measured—it is useful for rapid and low-cost socio-environmental assessments of soil fertility and land conservation decision-making, provided that the respondents are on such a move. However, it may be right to doubt about farmers' perceptions. Indeed, the research results reported in this paper indicate that farmers think their lands are not degraded yet (giving them the reason to continue farming with low fertilizer use), whereas field extension agents think land degradation levels are quite high (Table 9). This discrepancy is an evidence of the limitation of the "perceptions" method. Nonetheless, both groups of respondents agree that land degradation is higher in the northern (with about 70% of the country's cotton production) than in the central region. If soil degradation levels, as perceived by farmers, were to be moved higher and closer to extension agents' perceptions, the present capacity or potential of the country for fertilizer-based sustainable soil fertility management would be much lower. This is a strong environmental warning that calls for a swift change in soil fertility policies, anchored on direct support to farmers with less indebtedness as Bellwood-Howard (2014) and Nyantakyi-Frimpong (2015) highlighted.

5. Conclusion

The study addressed the sustainability of fertilizer-based soil fertility management in Benin. It diagnosed the relationship between differential soil degradation status over space and fertilizer use in cotton production systems. It found that present fertilizer use practices overlook the spatial differences in soil fertility status in export-oriented cotton production systems. Evidence was provided of the backwardness of such practices that have led to a very low potential or capacity for sustainable agricultural intensification. Adjusting current fertilizer recommendations to site-specific soil conditions is urgently required to promote the sustainability of cotton production systems. A more rational fertilizer use will be critical to inducing higher land productivity while preserving the environment. Therefore, more informed fertilizer supply and use policies should be implemented, based on updated recommendations that are drawn from prevailing soil fertility status and farmers' adoption capacities. Only then could cotton growers experience a steady increase in agricultural productivity and improved net incomes.

Acknowledgements
Great thanks to Prof Caspar Schweigman, Dr Clemens Lutz, Dr Anselme Adegbidi and Dr Henk Breman for their valuable support, and to anonymous referees.

Funding
This work was supported by the Centre for Development Studies (University of Groningen), International Foundation for Science (IFS, Sweden) and Rijksuniversiteit Groningen respectively through the PhD fellowship programme and the field work grant.

Competing interests
The authors declare no competing interest.

Author details
Barthelemy G. Honfoga[1]
E-mail: honfogabg@yahoo.fr
[1] Faculty of Agronomic Sciences, School of Economics, Socio-Anthropology & Communication for Rural Development, University of Abomey-Calavi (UAC), 06 BP 1892 Cotonou, Akpakpa PK3, Abomey-Calavi, Benin.

Notes

1. In Benin, only 10.4 kg of fertilizers were applied per ha of arable land in 2004, just slightly above the 9 kg/ha average of Sub-Saharan Africa (Honfoga, 2013).

2. Yield potential for rain-grown cotton production systems depends on soil water storage and rainfall but is about 800 kg lint/ha (Constable & Bange, 2015). This corresponds to 2388.05 kg/ha for seed cotton. The potential yield under irrigated conditions is about 4.5 times that of rain-fed cotton, but irrigated cotton is not yet grown in Benin. However, some best practice farmers in Kandi district (Northeast region) were reported to reach 3 tons/ha.

3. In Dassa-Zoumè, yields increased from 0.7 to 1.1 ton/ha (Bessou, 2013).

4. Sources: www.benininfo.com; www.panapress.com.

5. See http://www.un.org/popin/fao/centafriq/frentex3.htm.

6. In a study on the use phosphate rock in cotton and cocoa fields in Togo, three levels of intensification (current practice, fertilizer use introduction, and semi-intensive production) were distinguished with reference to the potential yield. The semi-intensive production was the most promising in terms of sustainable soil fertility management, but in the short run it does not guarantee a value/cost ratio higher than that of fertilizer use introduction (ITRA/IFDC-Africa, 1998, p. 42 et pp. 51–55). Therefore, economic factors are important in the choice of desired intensification level.

7. All the field extension agents (TSPV) who conducted this survey hold a Master's degree in agronomy, and at least 2 years of field experience in the districts investigated. District senior extension officers (RCPA) are of the same education level with greater supervisory experience.

8. Cultivable areas of these sub-districts should have been used as weights, but reliable area data were not available.

9. The doses that correspond to the potential yields of various crops are not known to extension services and farmers in Benin. The pan-territorial dose (200 kg/ha) recommended for cotton by extension services is

considered here as a default reference that reflects the potential aggregate yield in cotton-based cropping systems and the existing capacities to achieve it.

10. Breman (personal communication, October 2005) thinks that higher soil fertility levels should be targeted rather than simply replacing the nutrients lost. This is right but one should be realistic while keeping in mind the principles of soil fertility management (Table 1) and the explanations hereafter provided.

11. A more accurate assessment of needs should consider specific/appropriate fertilizers for each crop or group of crops (cotton, cereals, tubers, legumes, oilseeds, fruits, vegetables, etc.). The method used here is for a rapid appraisal of fertilizer demand.

12. Elsewhere, the likely trends in crop areas/cultivated lands would be referred to, considering recent developments in the agricultural sector.

13. The percentage share of these four crops in total cultivated area in 1999 (peak cotton boom year) is the maximum fertilizer diffusion rate to be expected, assuming that those crops respond equally to price signals to motivate fertilizer demand.

14. The nature of the parent rock of a soil determine of its inherent fertility or original suitability for agriculture.

15. The adoption rate of a technology in a region is the percentage ratio of number of farmers who adopted to the total number of farmers.

16. This includes, e.g. the socioeconomic conditions of his parents/household, cultural inclusion or adaptation of teachers, school density in the neighborhood, access to schoolmate exchange groups, presence of a public library, etc.

17. Inherent soil fertility (parent rock) and climate are ignored here.

References

Abbey, G. A., & Adou-Rahim-Alimi, A. (2003). Selection factors for low-lands crops and productivity improvement techniques in rice cultivation in northern Togo. In SADAOC (Ed.) (Working Paper Series No. 056). Ouagadougou: SADAOC – Executive Secretariat.

Adegbidi, A., Gandonou, E., Padonou, E., Océni, H., Maliki, R., Mègnanglo, M., & Konnon, D. (2000). *Studies of agricultural inputs in Benin (fertilizers, pesticides, seeds, agricultural materials and equipment, organic fertilizers)* Ministry of Rural Development/German Technical cooperation (GTZ), Soil fertility Initiative (IFS, FAO/World Bank).

Alohou, E., & Nelen, J. (2000). *Experiences with a participatory approach to integrated soil fertility management in research and development in Benin. Case of collaboration between PADEC and research-development in six villages in northern Benin.* PADIC-Kandi.

AProCA. (2008, March). *Progress report of the project "Cotton University: Think, share and act at the service of African cotton sub-sectors"* (Working paper).

Baum, E. L., & Heady, E. O. (1957). Overall economic considerations in fertilizer use. In E. L.Baum, E. O. Heady, J. T. Pesek, & C. G. Hildreth (Eds.), *Economic and technical analysis of fertilizer innovations and resource use* (pp. 187–206). Ames, IA: The Iowa State College Press.

Bellwood-Howard, I. R. V. (2014). Smallholder perspectives on soil fertility management and markets in the African Green Revolution. *Agroecology and Sustainable Food Systems, 38*(6), 660–685. doi:10.1080/21683565.2014.896303

Benin. (2008, December 15–17). *National Investment Report.* High-level Conference on Water for Energy and agriculture in Africa: The challenges of climate change. Sirte, Libyan Arab Jamahiriya.

Benin. (2010). *Review of development efforts in the agricultural sector.* Benin, African Union: ECOWAS.

Bernstein, H. (1977). Notes on capital and peasantry. *Review of African Political Economy, 4*(10), 60–73. https://doi.org/10.1080/03056247708703339

Bernstein, H. (2008). Agrarian questions from transition to globalisation. In H. Akram Lodhi & C. Kay (Eds.), *Peasants and globalisation: Political economy, rural transformation and the agrarian question* (pp. 239–261). London: Routledge.

Bessou, M. R. (2013). *Study of the effects of the project of remediation and recovery of the cotton sector in Benin (PARFCB) on cotton production in the municipality of Dassa-Zoumè* (MBA thesis). International Institute of Management (IIM), Cotonou.

Brabant, P., Darracq, S., Egué, K., & Simonneaux, V. (1996). *State of degradation of land resulting from human activities: Explanatory memorandum of the map of Togo to 1/500 000 of the signs of degradation.* Paris: Editions ORSTOM.

Breman, H. (1997). Improvement of soil fertility in West Africa: Constraints and prospects. In G. Renard, K. Becker, & M. Van Oppen (Eds.), *Soil Fertility Management in West African Land Use Systems* (pp. 7–20). Niamey: Proceedings.

Breman, H. (1999). Linking soil fertility improvement to agricultural input and output market development: The key to sustainable agriculture in West Africa. In S. K. Debrah & W. G. Koster (Eds.), *Miscellanous Fertilizer Studies 16* (pp. 52–60). Lome: IFDC-Africa.

Breman, H. (2000, June 6). *Sustainable agricultural Intensification in Africa: The role of financial investment.* Communication to the AfDB conference, Abidjan.

Breman, H., & Debrah, S. K. (1999). Agricultural intensification within sustainable production systems. In FAO (Ed.), *Soil fertility initiative for Sub-Saharan Africa. Proceedings SFI/FAO consultation* (pp. 54–55). World Soil Resources Reports 85, FAO: Rome.

Breman, H., & Debrah, S. K. (2003). Improving African food security. *SAIS Review, XXIII*(1), 153–170.

Constable, G. A., & Bange, M. P. (2015). The yield potential of cotton (*Gossypium hirsutum* L.). *Field Crops Research, 182*, 98–106. doi:10.1016/j.fcr.2015.07.017

Dangbégnon, C., Nederlof, S., Tamelokpo, A., & Mando, A. (2010). Knowledge management in an integrated soil fertility management project in Togo. In F. Pol (van der) & S. Nederlof (Eds.), *Natural Resource Management in West Africa Towards a knowledge management strategy* (pp. 41–55). Karlsruhe: KIT, Bulletin 392-Natural Resource Management.

DDC. (2013). *Swiss cooperation strategy in Benin 2013–2016.* Switzerland: Federal Department of Foreign Affairs FDFA.

Dixon, J., Gulliver, A., & Gibbon, D. (2001). *Agricultural production systems and poverty: Improving the livelihoods of farmers in a changing world—summary.* Rome and Washington: FAO/World Bank.

Dudal, R. (2002). Forty years of soil fertility work in Sub-Saharan Africa. In B. Vanlauwe, J. Diels, N. Sanginga, & R. Merckx (Eds.), *Integrated plant nutrient management in Sub-Saharan Africa* (pp. 7–21). Wallingford: CAB International.

Ebanyat, P., de Ridder, N., de Jager, A., Delve, R. J., Bekunda, M. A., & Giller, K. E. (2010, July). Drivers of land use and determinants of sustainability in smallholder farming systems of Eastern Uganda. *Population and Environment, 31*(6), 474–506. doi:10.1007/s11111-010-0104-2

FAO-Benin. (2015). *Project of intensification and integrated development of the maize value chain in Benin.* Cotonou: FAO Office.

Honfoga, B. G. (2007). *Towards efficient private fertilizer supply and distribution systems for sustainable agricultural intensification in Benin* (PhD Dissertation). University of Groningen, Groningen, CDS Thesis No. 21.

Honfoga, B. G. (2013). Cotton institutions and perverse incentives for fertilizer traders in Benin. *Journal of*

Development and Agricultural Economics, 5(1), 19–34. Retrieved from http://www.academicjournals.org/JDAE https://doi.org/10.5897/JDAE

Houngbo, E. N. (2013). Socio-Economics of the decline of cotton production in Benin: Case of the village Karthik (Centre Benin). *African Agronomy, 25*(2), 1–7.

ICRISAT. (1987). *Sahelian Center Annual Report 1987*. Niamey: Author.

ICRISAT. (2009, January). *Fertilizer Micro-dosing*. Patancheru: Author.

ICRISAT. (2012, July). *Fertilizer Micro-dosing*. Patancheru: Author.

IFDC. (2005). *The state of the market for agricultural inputs in Benin*. Muscle Shoals, AL: IFDC Publications.

IFDC. (2006). *Development and dissemination of sustainable integrated soil fertility management practices for smallholder farmers in Sub-Saharan Africa*. Muscle Shoals, AL: IFDC Publications.

Igué, A. M., Gaiser, T., & Stahr, K. (2004). A soil and terrain digital database (SOTER) for improved land use planning in Central Benin. *European Journal of Agronomy, 21*(1), 41–52. https://doi.org/10.1016/S1161-0301(03)00062-5

ISRIC. (1991). *GLASOD Project*. Wageningen.

ISRIC. (1992). *World map of the status of human-induced soil degradation*. Wageningen: Author.

ITRA/IFDC-Africa. (1998). *Feasibility study of the use of natural phosphate in the coffee-cacao and cotton areas of the plateau region of Togo*. Lomé: ITRA.

Ivo, A. M. (2008). *Food and fertilizer price hypes in Sub-Saharan Africa. It never rains but it pours; Cameroon a caricature of the broader picture*. Cameroon: Africa Centre for Community and Development.

James, P., & Woodhouse, P. (2017). Crisis and Differentiation among Small-Scale Sugar Cane Growers in Nkomazi, South Africa. *Journal of Southern African Studies, 43*(3), 535–549. doi:10.1080/03057070.2016.1197694.

Kpadé, C. P. (2008, December 11–12). *Institutional analysis of the dynamics of the evolution of cotton policies in Benin. 2nd Social Science Research days*. Lille: INRA SFER CIRAD.

Kpadé, C. P. (2011). *Adaptation of coordination and new contradictions among actors of cotton system facing economic liberalization in Benin* (Doctorate Thesis). University of Bourgogne, Dijon.

Mazzucato, V., & Niemeijer, D. (2000). *Rethinking soil and water conservation in a changing society – A case study in eastern Burkina Faso* (Doctoral thesis dissertation). Wageningen University, Wageningen. (Tropical Resource Management Paper, No. 32).

MEF. (2010). *Ex-ante Evaluation of the implementation of the recovery strategies of the cotton-textile pole of Benin*. Cotonou: Ministry of Economics and Finance.

Minot, N., Kherallah, M., & Berry, P. (2000). *Fertilizer market liberalization in Benin and Malawi—A household-level view*. Washington, DC: International Food Policy Research Institute (IFPRI).

Mokwunye, U. A., de Jager, A., & Smaling, E. M. A. (Eds.). (1996). *Restoring the productivity of West African soils: Key to sustainable development*. Lomé: IFDC-Africa.

Morris, M., Kelly, V. A., Kopicki, R. J., & Byerlee, D. (2007). *Fertilizer use in African agriculture. Lessons learned, and good practice guidelines*. Washington, DC: The World Bank. https://doi.org/10.1596/978-0-8213-6880-0

Ndufa, J. K., Cadisch, G., Poulton, C., Noordin, Q., Vanlauwe, B. (2005). *Integrated soil fertility management and poverty traps in Western Kenya*. Original version presented at the "International Symposium on Improving Human Welfare and Environmental Conservation by Empowering Farmers to Combat Soil Fertility Degradation", organized by The

Africa Network for Soil Biology and Fertility (AfNet) of Tropical Soil Biology and Fertility (TSBF) Institute of CIAT, 16–22 May 2004, Yaounde.

Nyantakyi-Frimpong, H. (2015). A political ecology of high-input agriculture in northern Ghana. *African Geographical Review, 34*(1), 1–7.

Oluwatoyin, D., Kolawole, D. O., Mogobe, O., & Magole, L. (2013). Political economy of integrated soil fertility management in the Okavango Delta, Botswana. *World Academy of Science, Engineering and Technology International Journal of Agricultural, Biosystems Science and Engineering, 7*(11), 2013.

Rusike, J., Reardon, T., Howard, J., & Kelly, V. (2003). *Developing cereal-based demand for fertilizer among smallholders in Southern Africa: Lessons learned and implications for other African regions*. Policy Synthesis No. 30, MSU/USAID Food Security II. Retrieved from http://www.aec.msu.edu. agecon/fs2/polsyn/no30.htm

Saidou, A., Kossou, D., Acakpo, C., Richards, P., & Kuyper, W. T. (2012). Effects of farmers' practices of fertilizer application and land use types on subsequent maize yield and nutrient uptake in Central Benin. *International Journal of Biological and Chemical Science, 6*(1), 365–378. Retrieved from http://ajol.info.index.php.ijbcs

Saidou, A., Kuyper, T. W., Kossou, D. K., Tossou, R., & Richards, P. (2004). Sustainable soil fertility management in Benin: Learning from farmers. *Netherlands Journal of Agricultural Sciences (NJAS), 52*(3/4), 349–469.

Samba Bio Tobou, A. (2000). *Identification of the constraints associated with the adoption of organic cotton on farms from the Padic-Kandi pilot tests* (Agricultural Engineer Thesis). FSA/UAC, Abomey-Calavi.

Schreurs, M. E. A., Maatman, A., & Dangbégnon, C. (2002). In for a penny, in for a pound: Strategic site-selection as a key element for on-farm research that aims to trigger sustainable agricultural intensification in West Africa. In B. Vanlauwe, J. Diels, N. Sanginga, & R. Merckx (Eds.), *Integrated plant nutrient management in Sub-Saharan Africa* (pp. 63–74). Wallingford: CAB International.

Sinzogan, A. A. (2006). *Facilitating learning toward sustainable cotton pest management in Benin: The interactive design of research for development* (PhD Thesis). Wageningen University, Wageningen. ISBN 90-8504-432-4.

Smaling, E. M. A. (1993). *An agro-ecological framework for integrated nutrient management, with special reference to Kenya* (Doctoral thesis, pp. (X) + 250). Wageningen Agricultural University, Wageningen.

Smyth, A. J., Dumanski, J., Spendjian, G., Swift, M. J., & Thornton, P. K. (1993). *FESLM: An international framework for evaluating sustainable land management. A discussion paper. World Soil Resources Reports 73* (p. 74). Rome: FAO.

Togbé, E. C. (2013). *Cotton in Benin: Governance and pest management* (PhD thesis, p. 201). Wageningen University, Wageningen. ISBN 978-94-6173-807-3

Tshibaka, B. T., Honfoga, B. G., Têvi, J., Houngbo, A., & Dokoué, J. (1992). *Determinants of the knowledge, adoption, diffusion and effects of modern cultivation techniques and fertilizers in southeast Togo* (Report of the fertilizer Policy Research Project). Lomé: IFPRI/IFDC.

UEMOA. (2013). *Feasibility study of a reliable supply and distribution mechanism for cotton and cereal fertilizers in UEMOA countries and Chad*. Ouagadougou: Author.

UEMOA. (2014). *White gold from West Africa to conquer the world market*. Ouagadougou: Author.

UNDP-Benin. (2011). *National Human Development Report, 2010–2011*. Cotonou, Benin: UNDP.

van Duivenbooden, N. (1995). *Land use systems analysis as a tool in land use planning, with special reference to North* and West African agro-ecosystems (Doctoral thesis). Wageningen Agricultural University, Wageningen.

Vanlauwe, B., Bationo, A., Chianu, J., Giller, K. E., Merckx, R., Mokwunye, U., ... Sanginga, N. (2010). Integrated soil fertility management. Operational definition and consequences for implementation and dissemination. Outlook on Agriculture, 39(1), 17–24. https://doi.org/10.5367/000000010791169998

Watts, M. (2012). Class dynamics of agrarian change. Journal of Peasant Studies, 39(1), 199–204. doi:10.1080/03066150.2012.656235

Weller, U. (1999). Agro-ecological potential of maize in Southern Benin. Retrieved from https://www.uni-hohenheim.de/atlas308/c_benin/projects/c2_2_2/html/english/

Wiggins, S. (1995). Change in African farming systems between the Mid-1970s and the Mid-1980s. Journal of International Development, 7(6), 807–848. https://doi.org/10.1002/(ISSN)1099-1328

9

Assessment of sugarcane grown in wetlands polluted with wastewater

Agnes Oppong[1], David Azanu[2]* and Linda Aurelia Ofori[3]

*Corresponding author: David Azanu, Department of Laboratory Technology, Kumasi Technical University, Kumasi, Ghana
E-mails: azanudavid@gmail.com, david. azanu@kstu.edu.gh

Reviewing editor: Mustafa Ibrahim Khamis, American University of Sharjah, United Arab Emirates

Abstract: Wastewater effluents directly discharged into nearby stream are eventually used to irrigate plants like sugarcane in Ghana. In this study, 24 triplicates sugarcane stems (sugarcane juice was extracted), 24 triplicates soil samples and 8 triplicates water samples were collected from sugarcane farms in four communities in Ashanti region of Ghana. Two of the communities were exposed to wastewater while the other two without wastewater contamination served as control. Metals (Pb, Cd, Cr, Cu, and Fe) concentration were determined in the digested samples using Spectra AA 220 flame atomic absorption spectrometer. The Pb concentration in all sugarcane juice samples ranged between 12.65 and 145.0 µg/L. The mean Cu concentration of the sugarcane juice samples varied between 11.28 and 156.00 µg/L. In general, there were decrease in metals investigated in sugarcane juice as you move away from the stream. However, the reduction was more pronounced in the hotspot sampling areas than control sampling areas. The EDI value was 9.76×10^{-4}, 2.94×10^{-5}, 1.09×10^{-3}, and 9.07×10^{-3} (mg/kg-day) for Pb, Cd, Cu, and Fe, respectively. Mean hazard quotient (HQ) for the metals studied ranged from 0.036 (Fe) to 0.286 (Pb). The results of this study indicate that sugarcane is able to grow in soils where some metals are accumulated. High levels of metals were pronounced in sugarcane originating from wastewater polluted soils as those considered in this study. The consumption of normal quantity of sugarcane juice may not present detrimental health concerns through a lifetime based on the metals contents alone.

ABOUT THE AUTHORS

Agnes Oppong is a senior technician at Department of Chemistry of Kwame Nkrumah University of Science and Technology, Ghana.

David Azanu is a lecturer and Head of Laboratory Services Unit at Kumasi Technical University, Ghana. Linda Aurelia Ofori is a lecturer at Department of Theoretical and Applied Biology of Kwame Nkrumah University of Science and Technology, Ghana.

PUBLIC INTEREST STATEMENT

Plant species cannot avoid taken up metals in soils they are grown. Some of these metals have no use in the body and even poison the body. In this study, sugarcane grown in areas prone to wastewater was analyzed to know the extent of contamination and whether the levels are harmful to humans. The results indicate that sugarcane is able to grow in soils which are contaminated with metal. High levels of metals were pronounced in sugarcane originating from wastewater polluted soils than areas not polluted with wastewater. However, all the concentrations were below levels recommended to be harmful to human. Hence, consumption of normal quantity of sugarcane juice from this study area may not present detrimental health concerns through a lifetime based on the metals contents alone.

Subjects: Environment & Health; Environmental Change & Pollution; Food Chemistry; Chemistry

Keywords: sugarcane; wetlands; cadmium; lead; wastewater; health risk

1. Introduction

Majority of plant species grown in soils that are polluted with metals are incapable of avoiding absorbing them (Baker, 1981). Conversely, human activities that accompany agricultural practices, industrial processes, mineral exploration and waste management greatly contribute to the pollution of natural ecosystems by heavy metals (Alumaa, Kirso, Petersell, & Steinnes, 2002; Bilos, Colombo, Skorupka, & Rodrigues, 2001; Keane, Collier, Shann, & Rogstad, 2001).

Heavy metals and metalloids accumulation in agricultural soils is an alarming subject due to the associated food safety issues and potential health risks, coupled-up with the harm they cause to soil ecosystems (McLaughlin, Parker, & Clarke, 1999).

The harmful effects of metals have extensively been described. For instance, cadmium, mercury, arsenic, nickel, lead, etc., have a wide range of toxicity including hepatotoxic, neurotoxic, teratogenic, mutagenic, and nephrotoxic effects among others (Bucheim, Stoltenburg-Didinger, Lilienthal, & Winnike, 1998; Domingo, 1994; Hudnell, 1999; Kelley, 1999; Lai, Minski, Chan, Leung, & Lim, 1999; McLaughlin et al., 1999). In addition, chromium, cadmium, and arsenic are considered cancer-causing (Costa, 1998).

Contaminating agricultural soils with wastewater that contains heavy metals is a serious matter because it leads to higher heavy metal uptake by crops. This uptake affects the quality and safety of our food, and poses threat to human health (Hough, Young, & Crout, 2003; Mensah et al., 1999). Serious systemic health problems can arise when dietary heavy metals like lead (Pb), cadmium (Cd), and chromium (Cr) accumulate excessively in the human body (Oliver, 1997).

An earlier survey in Kumasi, Ghana, found that rivers or streams in Kumasi are heavily polluted with raw sewage (Cornish, Aidoo, & Ayamba, 2001) as most sewage is either discharged straight into rivers and streams or is collected from septic tanks and then disposed of into waterways (Keraita & Drechsel, 2004)

Routine monitoring of heavy metals concentration in soils and crops are, therefore, essential to know their levels and devise strategies to minimize contamination, in order to reduce risks to human health. The objective of this project is to identify health hazards associated with sugarcane grown in wastewater polluted wetlands by determining Pb, Cd, Cr, Cu, and Fe concentration in water, soil, and sugarcane juice collected from wastewater polluted and unpolluted sugarcane farms (control) where the stream passing through the sugarcane farm serves as the irrigation water for the sugarcane; identifying the correlation between metals concentrations in water, soil and sugarcane juice; estimating the variation in metal concentration due to seasonality, determining metals concentration in sugarcane juice as you move away from the stream and estimating daily intake of metals (EDI) and target hazard quotients (THQ) values for all metals in sugarcane juice.

2. Materials and methods

2.1. Study area

The city of Kumasi is the second largest in Ghana and lies about 150 km northwest of Accra. The population census in 2010 counted about two million inhabitants (Ghana Statistical Service, 2012) Kumasi is located on a drainage divide with 28% of the developed area drain to the west eventually joining the Offin River and 72% of the developed area drain to the Oda River in the south of the city. Most streams originate within the administrative boundaries of Kumasi. The only considerable inflow from outside is noted from Sisa and Wiwi Rivers to the north of Kumasi. Its area comprises 254 km^2

(A)   **(B)**

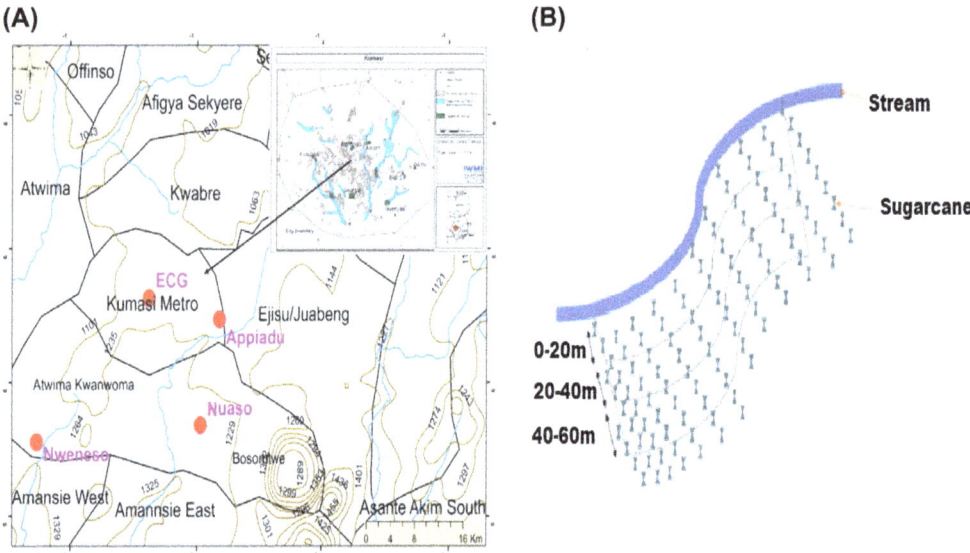

Figure 1. Map of Ashanti region showing areas where sugarcane is grown.

out of which around 80% are developed. In 2005, 75 km^2, or 30%, were either referred to as open space or undeveloped land such as river valleys or other unpopulated areas (Obuobie et al., 2006). Figure 1(a), shows the study area with the rivers, areas where sugarcane is grown, vegetable farms and populated areas. The study areas are wetlands, encircled by farm lands, in which sugarcane has been the planted. These sites are Nweneso, Nuaso, Adum ECG, and Appiadu

2.2. Sampling
Sampling was done in sugarcane farms in four communities in Ashanti region, Ghana. Among the four sugarcane farms under study, two were exposed to wastewater which were chosen as hotspot (Adum ECG and Appiadu) and the other two (Nweneso and Nuaso) had no wastewater inflow into the wetland. The latter served as control areas.

The stem of sugarcane plants and soils were collected along three transects established at the north (N), east (E), west (W), and south (S) directions at each farm, at 0, 20, 40, 60 m distance from the stream shown on a schematic diagram in Figure 1(b). The stream water flowing along the sugarcane farms was sampled. The sampling areas were demarcated using Global Positioning System (Magellan GPS 315). Dry season sampling was done in February 2016, while wet season sampling was done in October 2015. A total of 24 triplicates sugarcane stems, 24 triplicates soil samples and 8 triplicates water samples were collected for investigation. All samples were kept in labeled sample containers and transported to the Chemistry Department of KNUST for analysis.

2.2.1. Sample preparation
Concentrated nitric acid (95%) used was of analytical reagent grade (BDH Chemical Ltd, Poole, England). The nitrate salts of Cd, Pb, Cu, Cr, and Fe (bought from Merck Chemicals, Germany) were used to prepare 1,000 mg/l stock solution in distilled water. Serial dilutions were prepared from this for various elements.

Glassware for digestion and storage of digest were washed under running tap water after been soaked in detergent solution overnight. They were subsequently soaked in 10% nitric acid overnight and then rinsed with distilled water.

Sugarcane stem samples were taken from averagely matured sugarcane farm stands. The sugarcane stems were washed with distilled water to remove soil and dirt. The washed sugarcane stems were peeled and cut to smaller sizes with the help of stainless knife. With the aid of an extractor, the sugarcane juice was squeezed out into plastic vials and refrigerated before analysis (Klein, 1987). The pH and total dissolved sugar (Brix) were determined and are reported as supplementary Table S1.

Sugarcane juice samples were wet digested by technique defined in (AOAC, 2000). One-gram (1 mL) sample was taken into 50 mL digestion tube along with 5 mL 65% nitric acid (Suprapur E. Merck). The contents were heated at 200 °C for 30 min. The digest was cooled, diluted to final volume of 50 ml with distilled water and transferred to 100 mL collection bottle. The soil samples were air-dried in the laboratory for 14 days, grounded with mortar and pestle, sieved with 2 mm nylon sieve. Portion of the soil before grinding was sent to Soil Science Laboratory-KNUST for the determination of its organic carbon content, pH, conductivity, and soil particle size. The soil properties data are presented as supplementary Table S2. The soil samples were digested using aqua regia (1:3) HNO_3:HCl, following a standard procedure (Vowotor et al., 2014). Accurately 0.5 g soil sample was taken into 50 mL digestion tube along with 10 mL aqua regia. The contents were heated at 150 °C for 30 min and cooled and diluted to final volume of 50 ml with deionized water.

A 1000 mL water sample collected at each sampling point, was acidified with 5 mL 65% nitric acid (Suprapur E. Merck) and kept in ice chest containing ice. The samples were kept in the freezer until digestion. Another 1000 mL water sample collected at each sampling point, was not acidified and used to determined physicochemical properties being; pH, conductivity, dissolved oxygen (DO) and turbidity. The result of the physicochemical parameters determined are reported as supplementary Table S3. Digestion of water samples was in accordance to standard methods (American Public Health Association, 2005). Aliquot of 10 mL of 65% nitric acid (Suprapur E. Merck) was added to 1000 mL water sample and boiled at 200 °C on a hot plate in a fume hood. Boiling continued until a solution obtained was about 25 mL. The digest was allowed to cool. It was then filtered into 50 mL volumetric flask and diluted to 50 mL mark with distilled water.

2.2.2. Metal analysis
Metals analysis of the digested sugarcane juice samples were done with (Spectra AA 220 AAS, UK) atomic absorption spectrometer equipped with multiple hollow cathode lamps. The instrument was calibration using different concentrations of standard solutions of Pb, Cd, Cr, Cu, and Fe. The result of metal concentration was expressed in µg/L and presented in Table 2.

2.3. Statistical analysis of data
Statistics was of the data were performed with GraphPad Prism (Version 6.01, San Diego, USA). Metals concentrations for sugarcane juice were subjected to one-way ANOVA and Pearson correlation analysis (Table 3) to test for variation in concentrations found in various sampling points and identify any relationship with physicochemical properties of the sugarcane juice. A probability of 0.05 or less would be considered to be significant. Each experiment was repeated three times and the research findings were presented in the form of mean and standard deviation using Microsoft excel.

2.4. Quality control
Calibration curve was prepared for various elements using triplicate of serial dilution standard solution. Detection limit was determined with blank solution and spike recovery was performed for various element by adding 50 mg of each element to one of the samples and followed the same digestion procedure and determination. The percentage recovery was then calculated. Precision of the instrument was determined by running on sample for eight times and calculating the standard deviation.

2.5. Estimation of daily intake and health risk indices of metals

2.5.1. Estimated daily intake
Estimated daily intake (EDI) of contaminant like metals, depend on exposure frequency and contact time among individuals. This site-specific information was collected from a questionnaire

administered to sugarcane consumers. The estimated daily intake of each heavy metal in this exposure pathway was determined by the equation:

$$EDI\,(mg/kg/day) = (E_F \times E_D \times IR \times C_m)/(W_{AB} \times T_A)$$

where E_F—the exposure frequency; E_D—the exposure duration; IR—sugarcane juice consumption
rate (g/person/day), C_m is the heavy metal concentration in foodstuffs (mg/kg); W_{AB}—average body
weight (average adult body weight was considered to be 70 kg); and T_A—average exposure time for
non-carcinogens (Saha & Zaman, 2012). All the uncertainty parameters were created in Microsoft
Excel 2013 and calculated with Monte Carlo simulation at 10,000 iterations using the @ Risk 7
(Palisade Corporation) software, which is an add-on to Excel. The input parameters used for the estimating hazard quotient are present in supplementary Table S4.

2.5.2. Hazard quotient

Hazard quotients (HQ) were developed by the US Environmental Protection Agency for the estimation of health risks related with long term exposure to chemicals. The non-cancer risks were stated
in terms of a HQ for a single substance as:

$$Hazard\,quotient\,(HQ) = EDI/RfD$$

The HQ has been classified into < 1 being no significant risk or systemic toxicity and HQ > 1 to be
potential risk. A THQ less than 1 means the exposed population is unlikely to experience obvious
adverse effects

2.5.3. Health Index

With exposure involving more than one chemicals, hazard index (HRI) was determined by the sum
of the individual hazard quotients for each chemical. The HRI was used as a measure of the potential
for harm. HRI above 1 means that there is a chance of non- carcinogens effects, with an increasing
probability as the value increases.

3. Results

3.1. Result of quality control analysis

The procedures taken to ensure the validity of the metal analysis data in this study have been described in the method section above. The recoveries, regression co-efficiencies and detection limits
of the elements analyzed are presented in Table 1. The linearity expressed as regression coefficient
values of all the metals ranged from 97 to 99%. The recoveries obtained in this study ranging from
95% to 98% were with the acceptable limits of 95 to 100.4% (Thompson, 2005).

3.2. Metal levels in various environmental samples analyzed

In the water samples, the mean concentration of heavy metals, Pb, Cd, Cr, Cu, and Fe, were 31.1, 9.8,
1.4, 515.3, and 5556.1 ppb, respectively (Table 2). Fe content was the highest and that of Cr was the
lowest in water. The order of accumulation in water was Fe > Cu > Pb > Cd > Cr. In the soil samples
analyzed, maximum concentration of 840.1 µg/L and minimum concentration of 18.4 µg/L were recorded for Pb while Cu concentration determined in the soil samples ranged from 29.7 to 1924.4 µg/L.
Maximum value of 648,500.2 and minimum, 28,497.2 µg/L, was recorded for Fe in soil samples analyzed (Table 2).

In this study, the concentration of heavy metals detected in sugarcane juice were generally lower
to metals in soil. The mean concentration of Pb, Cd, Cu, and Fe, were 55.0, 2.1, 65.9, 567.0, and
931.1 ppb, respectively (Table 2).

Table 1. Recoveries, regression co-efficiencies and detection limits of elements

Element	Detection limits (ppb)	Precision (%CV)	Recovery (%)	Regression coefficient (R^2)
Pb	2	3	97	99
Cd	1	5	95	98
Cr	1	4	95	97
Cu	3	5	98	99
Fe	3	4	96	97

Table 2. Concentration of metals (µg/L) in water, soil, and sugarcane juice and their corresponding recommend limits

Type of sample	Statistical tools/ standard	Pb	Cd	Cr	Cu	Fe
Water	Minimum	7.4	b/d	b/d	38.0	161.2
	Maximum	31.1	9.8	1.4	515.3	5,556.1
	Mean	17.5	4.0	0.3	195.4	1,697.6
	Standard deviation	10.1	3.6	0.6	165.8	1,924.5
	US EPA (2012)	1,000.0	500.0	1,000.0	5,000.0	50,000.0
Soil	Minimum	18.4	b/d	b/d	29.7	28,497.2
	Maximum	840.1	15.1	1.8	1,924.4	648,500.2
	Mean	241.8	6.1	0.6	566.8	211,909.0
	Standard deviation	242.7	3.8	0.6	543.1	171,799.0
	FAO/ISRIC (2004)	150,000.0	5,000.0	250,000.0	100,000.0	5,000,000.0
Sugarcane juice	Minimum	12.7	b/d	b/d	11.3	149.6
	Maximum	145.7	4.0	b/d	156.0	1,205.3
	Mean	55.0	2.1	–	65.9	567.9
	Standard deviation	35.6	0.9	–	47.0	354.9
	EC (2006)	300.0	50.0		500.0	20,000.0

3.3. Correlation of various metals concentration in soil, water and sugarcane juice samples

In general correlation of metal concentrations in soil with water and soil with sugarcane juice were positive (Table 3). Since the Cr concentrations in sugarcane juice were below detection, there was no correlation recorded. There were strong positive correlations of Pb, Cd, Cu, and Fe metals concentration investigated in soil with concentration found in sugarcane juice. The Pearson r^2 values ranged from 0.68 to 0.91. The p-values were all statistically significant being 1.4×10^{-6} for Pb, 0.0002 for Cd, 8.4×10^{-10} for Cu and 2.5×10^{-7} for Fe (Table 3). With the exception of Cu which showed strong positive correlation which is not significant ($p = 0.099$) between metal concentrations in water with metal concentrations in sugarcane juice, the correlation between metal concentrations in water with metal concentrations in sugarcane juice were weak positive correlation for Pb, Cd, and Fe. Additionally, Cu showed a significant ($p = 0.005$) strong positive correlation between metal concentrations in soil with metal concentrations in water, but Cd recorded a strong positive correlation but not significant ($p = 0.054$). Lead, Cr and Fe showed a weak positive correlation between metal concentrations in soil with metal concentrations in water.

3.4. Effect of pollution on metals determined in sugarcane juice

The mean values and standard deviation represent as error bars of Pb, Cd, Cu, and Fe in sugarcane juice samples from the twelve sampling sites are presented in Figure 2.

Table 3: Pearson correlation coefficient of metals between soil, water, and sugarcane juice analyzed

	Pb^s	Cd^s	Cr^s	Cu^s	Fe^s	Pb^w	Cd^w	Cr^w	Cu^w	Fe^w	Pb^J	Cd^J	Cu^J	Fe^J
Pb^s														
Cd^s	0.70*													
Cr^s	0.73*	0.52*												
Cu^s	0.72*	0.90*	0.60*											
Fe^s	0.93*	0.64*	0.52*	0.58*										
Pb^w	0.28	0.84*	0.37	0.78*	0.12									
Cd^w	0.25	0.71*	0.15	0.58	0.11	0.70*								
Cr^w	−0.02	0.69*	−0.05	0.52	−0.15	0.76*	0.91*							
Cu^w	0.14	0.74*	0.12	0.64*	0.00	0.86*	0.94*	0.94*						
Fe^w	0.10	0.68*	0.04	0.53	−0.01	0.76*	0.97*	0.96*	0.98*					
Pb^J	0.81*	0.74*	0.39	0.69*	0.91*	0.38*	0.11	0.00	0.11	0.06				
Cd^J	0.84*	0.68*	0.69*	0.78*	0.79*	0.27	−0.02	−0.11	−0.01	−0.10	0.80*			
Cu^J	0.87*	0.85*	0.57*	0.91*	0.84*	0.65*	0.48	0.30	0.47	0.38	0.90*	0.87*		
Fe^J	0.76*	0.74*	0.37*	0.77	0.84*	0.52	0.20	0.10	0.23	0.14	0.96*	0.84*	0.94*	

Notes: S: soil, W: water, and J: sugarcane juice.

*Correlation is significant at the 95% confidence level.

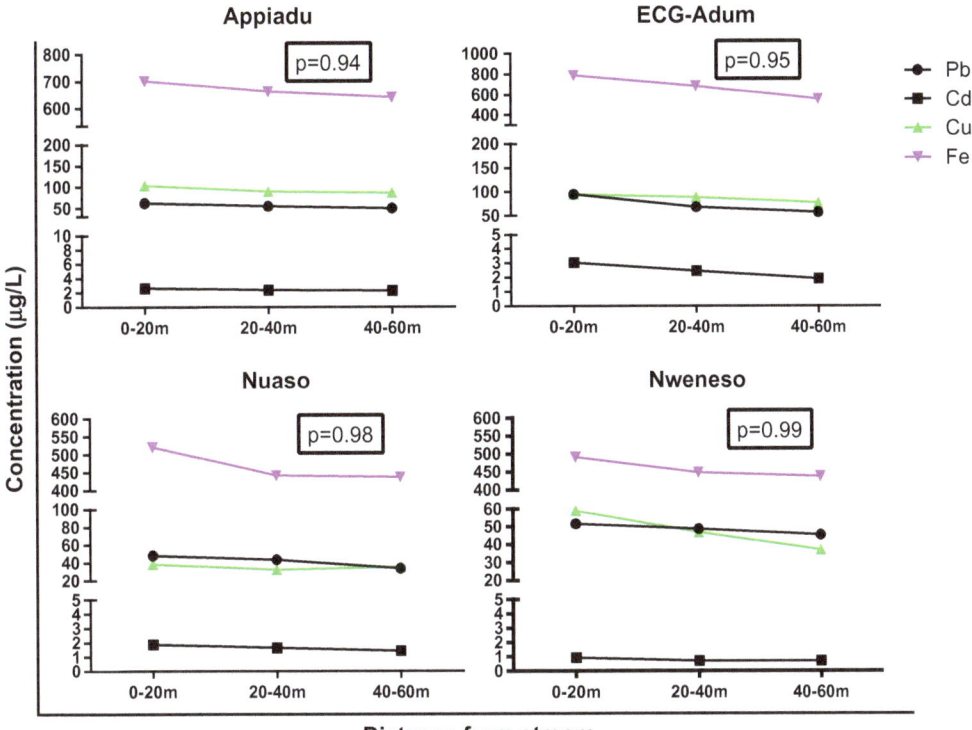

Figure 2. Variation of contamination of metals determined in sugarcane juice away from stream.

The hotspots, being areas polluted with solid and liquid waste (Appiadu and ECG-Adum samples) mean values of Pb, Cd, Cu, and Fe in sugarcane juice samples ranged from 24.00 to 145.00, 1.3–4, 38.38–156.00, and 261.95–1205.00 µg/L, respectively. The control sugarcane samples collected from Nweneso and Nuaso township showed mean values in the sugarcane juice samples ranged from 12.65 to 85.00 µg/L for Pb, b/d – 2.5 µg/L for Cd, 11.28–97.30 µg/L for Cu, and 149.60–855.00 µg/L for Fe. One-way ANOVA, at 95% confidence limit gave p values of 0.19, 0.02, 0.01, and 0.02 for Pb, Cd, Cu, and Fe, respectively.

3.5. Seasonal variation of parameters determined in sugarcane juice

The dry season mean values (bars) and standard deviation (error bars) of Pb (85.33 ± 22.48 µg/L), Cd (2.44 ± 0.98 µg/L), Cu (102.38 ± 38.27 µg/L), and Fe (892.17 ± 159.58 µg/L) are represented in Figure 3. The wet season mean and standard deviation values in the sugarcane juice samples are 24.33 ± 9.01 µg/L for Pb, 1.21 ± 0.56 µg/L for Cd, 29.34 ± 15.69 µg/L for Cu and 243.26 ± 89.18 µg/L for Fe.

The p values for One-way ANOVA, at 95% confidence limit were 0.04, 0.07, 0.001, and 0.002 for Pb, Cd, Cu, and Fe, respectively

3.6. Variation from stream

In general, the concentration of metals investigated in sugarcane juice decreased as you move away from the stream. The mean Pb values of the sugarcane juice samples from the sampling sites ranged between 12.65 and 145.0 µg/L, with the highest value being recorded at ECG-Adum (0–20 m) sample (BDA) while the lowest was recorded at Nuaso (40–60 m) sample (NWC).

ECG-Adum (0–20 m) sugarcane juice sample (BDA) recorded the highest mean value of Cd determined being 4 µg/L and Nweneso and Nuaso samples collected in the wet season were b/d. Chromium values were below detection for all the samples. The mean Cu concentration of the

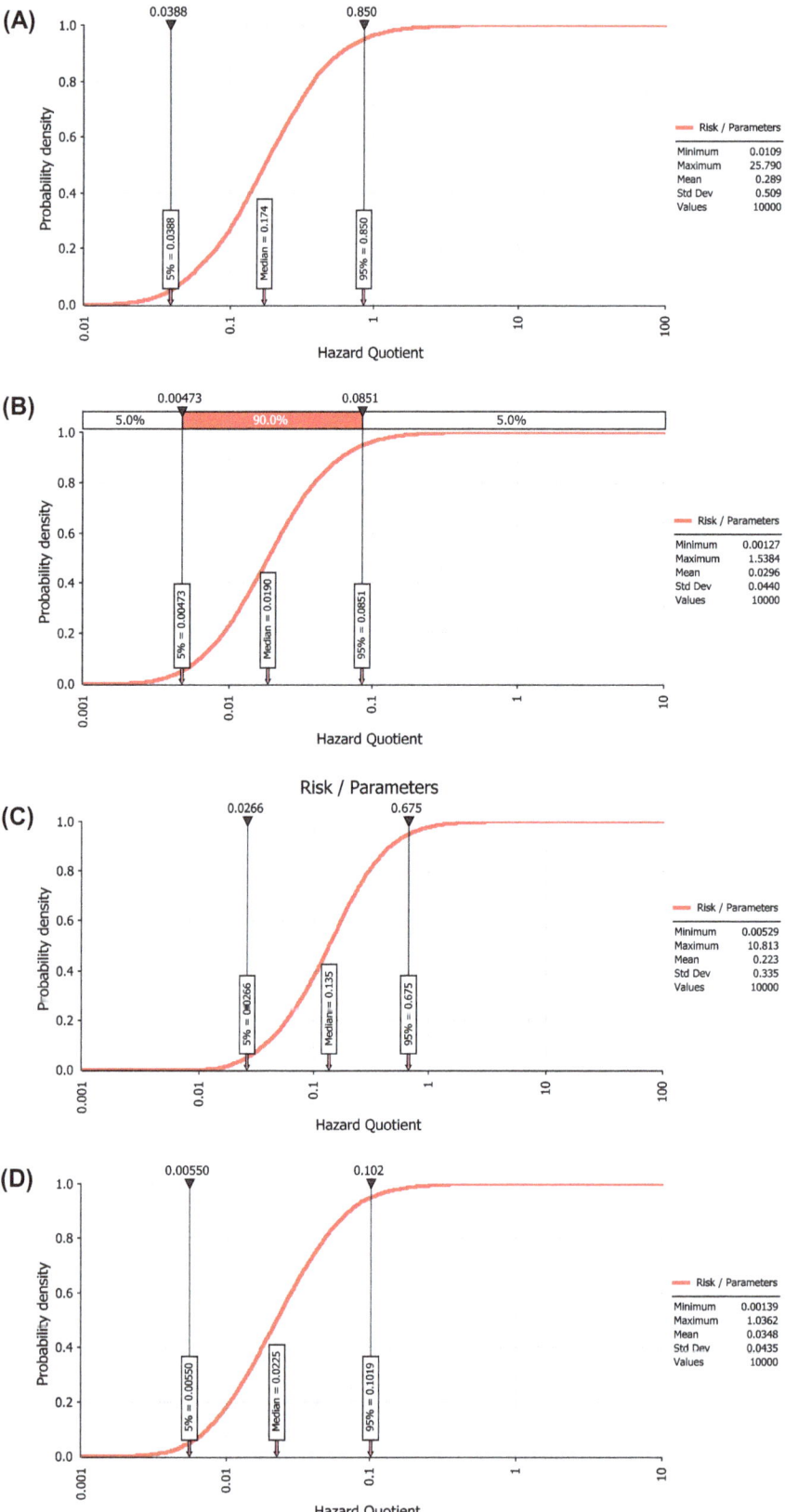

Figure 3. (a) Health quotient graph of lead, (b) Health quotient graph of cadmium, (c) Health quotient graph of copper, and (d) Health quotient graph of iron.

sugarcane juice samples varied between 11.28 and 156.00 µg/L. These values were recorded in Nuaso (20–40 m) (NWB) and Appiadu (0–20 m) (ADA), respectively.

The highest value of Fe in the sugarcane juice samples was 1205.0 µg/L and was recorded at ECG-Adum (0–20 m) (BDA) while the lowest Fe concentration of 149.6 µg/L was recorded at Nweneso (20–40 m) (KWB). The p values for One- way ANOVA, at 95% confidence limit was 0.94 for Appiadu, 0.95 for ECG-Adum, 0.98 for Nuaso and 0.99 for Nweneso, respectively.

3.7. Health risk result

3.7.1. Estimated daily intake
The EDI value was 9.76×10^{-4}, 2.94×10^{-5}, 1.09×10^{-3} and 9.07×10^{-3} (mg/kg-day) for Pb, Cd, Cu, and Fe, respectively. The EDI showed an increasing order of Cd < Pb < Cu < Fe.

3.7.2. Health quotient
The health quotient of lead simulated over the population density shows the median and mean hazard quotient for lead was 0.175 and 0.286, respectively. At 95% probability density, hazard quotient was 0.894 which was still below 1.

Cadmium had a mean hazard quotient of 0.0297 and median hazard quotient of 0.0190. Additionally, hazard quotient at 95% probability density gave a value of 0.0860 which is by far lower than one. The median hazard quotient of copper was 0.137, with a mean of 0.223. Even at 95% probability density, the hazard quotient was 0.665. The hazard quotient at 95% probability density was 0.103 for iron. The median hazard quotient was 0.0233 and mean hazard quotient was 0.036. The health index as a sum of the individual metals hazard quotients was 0.56.

4. Discussion

4.1. Metals discussion
The maximum concentration of Pb, Cd, Cr, Cu, and Fe in the water samples are 31.1, 9.8, 1.4, 515.3, and 5556.1 µg/L, respectively. These were all below their corresponding US EPA (2012) metal thresholds for reuse of wastewater being 10,000.0, 50.0, 1,000.0, 5,000.0, and 50,000 µg/L for Pb, Cd, Cr, Cu, and Fe, respectively.

In Accra, Pb contamination has been recorded in water used for vegetable irrigation to be 80 µg/L. This has been attributed to drains and untreated wastewater (diluted wastewater) joining the streams used for vegetable irrigation (Obuobie et al., 2006). Additionally, Azanu, Jørgensen, Darko, and Styrishave (2016), recorded contamination of Pb (up to 28.7 µg/L) in waste stabilization ponds effluent entering Wiwi River in Kumasi which is used for vegetable irrigation downstream. Based on this result obtained in this study, it is clear that there is metal contamination of the streams in irrigating sugarcane farms. Wastewater used directly or indirectly for irrigation can result in metal accumulation in soils and crops eventually (Hurmanescu, Alda, Bordean, Gogoasa, & Gergen, 2011; Singh & Agrawal, 2010).

Food and Agriculture Organization (FAO) and International Soil Reference and Information Centre (ISRIC) have established recommended values for Pb, Cd, Cr, Cu, and Fe metal concentrations in soils to be 150,000, 5,000, 250,000, 100,000, and 5,000,000 µg/kg (FAO and ISRIC, 2004). The maximum concentration of Pb, Cd, Cr, Cu, and Fe metal concentrations in soils investigated was 840.1, 15.1, 1.8, 1924.4, and 648,500.2 µg/kg respectively and were below the recommended values established by FAO and ISRIC.

Cultivation of food crops in contaminated environments is common in West Africa, as small scale farmers cultivate food crops at dumpsites to maximize yields due to the seemingly high organic contents of waste dumpsite soils. Odai, Mensah, Sipitey, Ryo, and Awuah (2008) reported high levels (mg/kg) of Pb (54.6), Cd (2.87), Cu (1631.67), and Zn (2606.0) in soils used for vegetable cultivation at Kumasi waste dumpsites in Ghana. Soil contamination correlated with high concentrations of Pb (13.50) in onions, Cd (1.78) in cabbages and Cu (90.33) in lettuce grown at the dumpsites. Heavy metal contamination of agricultural soils and crops is particularly worse in developing industrialized countries such as China and India (Keraita, Jimenez, & Drechsel, 2008) due to extensive use of untreated industrial wastewater and Ashanti region is not an exception.

According to JECFA (1983) the provisional tolerable daily intake (PMTDI) of Fe is 0.8 mg/kg and the maximum Fe concentration obtained in this study was 1205.00 ppb. The Fe concentration obtained in this study would only exceed the PMTDI when 1000 mL sugarcane juice is consumed in a day which do not happen. However the maximum Fe concentration obtained in this study was higher than concentration of Fe (352 ppb) in sugarcane juice reported in street vended sugarcane juice in Multan-Pakistan (Akhtar, Ismail, & Riaz, 2015). The variation could be due to pollution of Ghanaian rivers and wetlands (Cornish & Lawrence, 2001; Keraita, 2002), where these sugarcane are grown. In another study, sugarcane juice has been reported with lower Fe contents (0.266 mg/L) (Adekola & Akinpelu, 2002) than observed in present study.

Maximum tolerable limits of copper as contaminants in fresh vegetable juices have been defined by Codex Alimentarius Commission (Codex Alimentarius Commission, 2000) as 5.0 mg/L. and EU as 20 mg/kg (EC, 2006a).

The maximum copper concentration of 156.00 ppb was recorded among all the fresh sugarcane juices sampled. If compared with Codex Alimentarius Commission limits for Cu of 5.0 mg/L (Codex Alimentarius Commission, 2000), present concentrations in this particular study could not have any health risk. Mean copper contents in sugarcane juices of 150 ppb was recorded in street vended sugarcane juice in Multan-Pakistan (Akhtar et al., 2015) and was higher than 65.86 ppb determined in this study. The Cu concentration in street vended sugarcane in Kumasi have been reported to fluctuate between 0.192 mg/kg and dry weight (Azanu, Kyei, & Oppong, 2015). This indicates that sugarcane sold in the street could still contain Cu concentration and could be coming from areas polluted with metals.

Cadmium (Cd) is a toxic element of inert nature with no identified health benefits. European Commission (EC) had defined maximum acceptable level of cadmium in fresh fruit juices as 0.05 mg/L (EC, 2006b). Cadmium levels in all fresh sugarcane juices under study ranged from 0.4 to 4.0 ppb and was approximately 10 times lower in study under discussion to the EC limit. Fruits and beverages derived thereof have been reported with lower level of cadmium as compared to vegetables, cereals and cereals products that readily contribute to cadmium intoxication (EC, 2006b).

Mean concentration of Cd (45 ppb) measured in sugarcane juice from street vended sugarcane juice in Multan-Pakistan (Akhtar et al., 2015) and was higher than 4 ppb determined in this study. However, higher concentration of Cd (41 ppb) has been reported in sugarcane sold on the street of Kumasi (Azanu et al., 2015). Lead carry no health benefit rather extremely toxic if exposed beyond their safe limits that has been identified as 300 ppb for fresh fruits, vegetables and juices derived thereof (EC, 2006b). Codex commission recommends as 0.1 mg/L (Codex Alimentarius Commission, 2000). As compared to the EC limits, maximum Pb contents (145 ppb) of all sugarcane juices was approximately half in study under discussion.

Lead analysis of street vended fresh sugarcane juices of Multan-Pakistan revealed comparative level of 167 ppb (Akhtar et al., 2015). Wetlands in Kumasi under cultivation for sugarcane have often been reported to be contaminated with sewage water and other wastewater supplies. Hence, there could be greater probability of soil contamination with Pb and other toxic trace metals. Hence, toxic metals contamination of sugarcane juice through translocation from wastewater or sewage supplies could not be ignored.

4.2. Variation of contamination of metals determined in sugarcane juice

The hotspots sugarcane sampling areas being Appiadu and ECG-Adum, generally showed higher concentration of metals than the control areas. These differences were significant Cd ($p = 0.02$), Cu ($p = 0.01$), Fe ($p = 0.02$) but not for Pb ($p = 0.19$). In metal polluted soils a lot of vegetable species are not able to avoid the absorption of these elements (Baker, 1981). Concentration of Cd and Cu above EU limit have been reported in sugarcane samples from an area under the influence of a municipal landfill and a medical waste treatment system in Ribeirao, Brazil (Segura-Muñoz et al., 2006).

The maximum Cu concentrations in sugarcane juice were 156.0 and 97.3 µg/L for hotspot and control areas, respectively. It is assessed that in unpolluted soils, Cu concentration in vegetable tissues is between 6 and 25 mg/kg. However, in Cu-polluted soils the concentration of this element in vegetable tissues may reach 80 mg/kg (World Health Organization, 2001). These low values recorded in this study could be due to the use of sugarcane juice instead of the whole stem and the magnitude of pollution.

4.3. Variation from stream

The general decrease in metals concentrations investigated in sugarcane juice as you move away from the stream could be due to attenuation as you move away from point source contamination. This could be the plausible cause because the reduction was more pronounced in the hotspot sampling areas than control sampling areas.

4.4. Seasonal variation of parameters determined in sugarcane juice

Seasonal variation may cause significant difference in the availability of metals to plants. In this study, this fact was confirmed for the metals investigated. Their p values for One-way ANOVA, at 95% confidence limit were generally below 0.05 making them significantly different except Cd with p value of 0.07.

4.5. Estimation of Daily Intake and health risk indices of metals

Evidently, individual EDI's of the various metals investigated in this study ranged from 2.94×10^{-5} to 9.07×10^{-3} mg/kg of body weight were far below RfD values of 0.7 mg/kg recommended by the international regulatory bodies (US EPA, 2009). The average lethal dose of iron is 200–250 mg/kg of body weight, but death has occurred following the ingestion of doses as low as 40 mg/kg of body weight (National Research Council, 1979). Mean hazard quotient (HQ) for each metals studied ranged from 0.036 (Fe) to 0.286 (Pb). The HQ has been defined so that if it is less than 1.0, there should be no significant risk or systemic toxicity and ratios above 1.0 could represent a potential risk. All HQ for the metals studied were below 1, hence there should be no major risk or systemic toxicity when these sugarcane juice studied are consumed.

5. Conclusions

The results of this study has demonstrated that sugarcane is able to grow well in areas where some metals in soils are accumulated. High levels of metals were found in sugarcane originating from soils polluted with wastewater in this field of study. Strong positive correlation was found between Cu concentrations in water and sugarcane juice, but Pb, Cd, and Fe showed weak correlation between metal concentrations in water and sugarcane juice. Additionally, Cu showed a significant ($p = 0.005$) strong positive correlation between metal concentrations in soil with metal concentrations in water, but Cd recorded a strong positive correlation but not significant ($p = 0.054$). The intake of normal amount of sugarcane juice may not pose any detrimental health issue through one lifetime based on the metals contents alone. Further study in this field should be considered with regards to the public health to mainly ascertain the mechanisms of metals integration during sugarcane production and also how use agricultural ways minimize heavy metals translocation in sugarcane. These studies should look on the influence of sugarcane varieties and soil type. Lastly, further studies should be focused on the interactions between metals, fertilizers, and herbicides on sugarcane and not only in aspect of risk, but also in other dimensions.

Funding
The authors received no direct funding for this research.

Competing interests
The authors declare no competing interest.

Author details
Agnes Oppong[1]
E-mail: oppongagnes20@gmail.com
David Azanu[2]
E-mails: azanudavid@gmail.com, david.azanu@kstu.edu.gh
Linda Aurelia Ofori[3]
E-mail: aurelia012001@yahoo.com

[1] Department of Chemistry, Kwame Nkrumah University of Science and Technology, Kumasi, Ghana.

[2] Department of Laboratory Technology, Kumasi Technical University, Kumasi, Ghana.

[3] Department of Theoretical and Applied Biology, Kwame Nkrumah University of Science and Technology, Kumasi, Ghana.

References
Adekola, F. A., & Akinpelu, A. A. (2002). Some trace elements in juice and bagasse of two varieties of sugarcane, location soil and irrigation water from Bacita Sugar Estate, Nigeria. *Bioscience Research Communications, 14*, 175–180.

Akhtar, S., Ismail, T., & Riaz, M. (2015). Safety assessment of street vended juices in Multan-Pakistan: A study on prevalence levels of trace elements. *International Journal of Food and Allied Sciences, 1*(1), 1–10. https://doi.org/10.21620/ijfaas.201511-10

Alumaa, P., Kirso, U., Petersell, V., & Steinnes, E. (2002). Sorption of toxic heavy metals to soil. *International Journal of Hygiene and Environmental Health, 204*, 375–376. https://doi.org/10.1078/1438-4639-00114

American Public Health Association. (2005). *Standard methods for the examination of water and waste water.* (A. D. Eaton, L. S. Clesceri, E. W. Rice, & A. E. Greenberg, Eds.) (22nd ed.). Washington, DC: Author.

AOAC (2000). *Official methods of analysis, official method 999.* Gaithersburg, MD: Author.

Azanu, D., Jørgensen, S. E., Darko, G., and Styrishave, B. (2016). Simple metal model for predicting uptake and chemical processes in sewage-fed aquaculture ecosystem. *Ecological Modelling, 319*, 130–136. Elsevier B.V. doi:10.1016/j.ecolmodel.2015.07.023

Azanu, D., Kyei, S. K., & Oppong, A. (2015). Street sugar cane vendors practices, metals and microbial levels of sugar cane sold in Kumasi, Ghana. *International Journal of Science and Technology, 4*(3), 99–104.

Baker, A. J. M. (1981). Accumulators and excluders: Strategies in the response of plants to heavy metals. *Journal of Plant Nutrition, 3*, 643–654. https://doi.org/10.1080/01904168109362867

Bilos, C., Colombo, J. C., Skorupka, C. N., & Rodrigues, P. M. J. (2001). Sources, distribution and variability of airbone trace metal in La Plata City area, Argentina. *Environmental Pollution, 111*, 149–158.

Bucheim, K., Stoltenburg-Didinger, G., Lilienthal, H., & Winnike, G. (1998). Miopathy: A possible effect of chronic low level lead exposure. *Neurotoxicology, 19*, 539–546.

Codex Alimentarius Commission. (2000). *Revised codex general standards for vegetable juice (CODEX STAN 179-1991).* Rome.

Cornish, G. A., Aidoo, J. B., and Ayamba, I. (2001). Informal irrigation in the Peri-urban zone of Kumasi, Ghana. *An Analysis of Farmer Activities and Productivity. Report OD/ TN 103, February 2001. DFID's Water KAR Project R7132, HR Wallingford, UK,* (February), 33.

Cornish, G. A., & Lawrence, P. (2001). *Informal irrigation in peri-urban areas: A summary of findings and recommendations.* Wallingford: HR Wallingford.

Costa, M. (1998). Carcinogenic metals. *Science Progress, 81*, 329–339.

Domingo, J. L. (1994). Metal-induced developmental toxicity in mammals. *Journal of Toxicology and Environmental Health, 42*, 123–141. https://doi.org/10.1080/15287399409531868

EC (European Commission) (2006a). Commission Regulation (EC) No 1881/2006 Setting maximum levels for certain contaminants in foodstuffs. *Official Journal of the European Union, L, 364*(1881), 5–24.

EC (European Commission) (2006b). Setting maximum levels for certain contaminants in foodstuffs. *Commission Regulation (EC) No 1881/2006 of 19 December 2006, setting, 1*(1881), 5364–5365. doi: 10.1017/ CBO9781107415324.004

FAO and ISRIC. (2004). Guiding principles for the quantitative assessment of soil degradation with a focus on salinization, nutrient decline and soil pollution. Retrieved November 10, 2010, from ftp://ftp.fao.org/agl/agll/docs/ misc36e.pdf

Ghana Statistical Service. (2012). 2010 population and housing census final results. *Ghana Statistical Service* (pp. 1–11). Accra. Retrieved from www.statsghana.gov.gh

Harmanescu, M., Alda, L., Bordean, D., Gogoasa, I., & Gergen, I. (2011). Heavy metals health risk assessment for population via consumption of vegetables grown in old mining area; a case study: Banat County, Romania. *Chemistry Central Journal.* doi:10.1186/1752-153X-5-64

Hough, R. L., Young, S. D., & Crout, N. M. J. (2003). Modelling of Cd, Cu, Ni, Pb and Zn uptake, by winter wheat and forage maize, from a sewage disposal farm. *Soil Use and Management, 19*(1), 19–27. doi:10.1111/j.1475-2743.2003.tb00275.x

Hudnell, H. (1999). Effects from environmental Mn exposure: A review of the evidence from non- occupational exposure studies. *Neurotoxicology, 20*, 379–398.

JECFA. (1983). *Toxicological evaluation of certain food additives and food contaminants* (Vol. 1983). Cambridge.

Keane, B., Collier, M. H., Shann, J. R., & Rogstad, S. H. (2001). Metal content of dandelion (*Taraxacum officinale*) leaves in relation to soil contamination and airborne particulate matter. *Science of The Total Environment, 281*, 63–78. https://doi.org/10.1016/S0048-9697(01)00836-1

Kelley, C. (1999). Cadmium therapeutic agents. *Current Parmaceutical Design, 5*, 229–240.

Keraita, B. (2002). *Wastewater use in urban and peri-urban vegetable farming in Kumasi, Ghana.* The Netherlands.

Keraita, B. N., & Drechsel, P. (2004). Agricultural use of untreated urban wastewater in Ghana. In C. A. Scott et al. (Ed.), *Wastewater use in irrigated agriculture: Confronting the livelihood and environmental realities* (pp. 101–212). Wallingford: CABI Publ. https://doi.org/10.1079/9780851998237.0000

Keraita, B., Jimenez, B., and Drechsel, P. (2008, June 1). Extent and implications of agricultural reuse of untreated, partly treated and diluted wastewater in developing countries. *CAB Reviews: Perspectives in Agriculture, Veterinary Science, Nutrition and Natural Resources, 3*(58), 1–15. doi:10.1079/PAVSNNR20083058

Klein, H. J. (1987). *Reactions to goal setting and feedback*. East Lansing, MI, Michigan State University.

Lai, J. C., Minski, M. J., Chan, A. W., Leung, T. K., & Lim, L. (1999). Manganese mineral interactions in brain. *Neurotoxicology, 20*, 433–444.

McLaughlin, M. J., Parker, D. R., & Clarke, J. M. (1999). Metals and micronutrients—food safety issues. *Field Crops Research, 60*, 43–163.

Mensah, P., Owusu-Darko, K., Yeboah-Manu, D., Ablordey, A., Nkrumah, F. K., & Kamiya, H. (1999). The role of street food vendors in the transmission of enteric pathogens. *Ghana Medical Journal, 33*, 19–29.

National Research Council. (1979). *Iron: A report of the subcommittee on iron, committee on medical and biologic effects of environmental pollutants, division of medical sciences, assembly of life sciences*. Baltimore, MD: University Park Press.

Obuobie, E., Keraita, B., Danso, G., Amoah, P., Cofie, O. O., Raschid-sally, L., & Drechsel, P. (2006). *Irrigated urban vegetable production in Ghana*. Accra: IWMI-RUAF-CPWF.

Odai, S. N., Mensah, E., Sipitey, D., Ryo, S., & Awuah, E. (2008). Heavy metals uptake by vegetables cultivated on urban waste dumpsites: Case study of Kumasi, Ghana. *Research Journal of Environmental Toxicology, 2*(2), 92–99.

Oliver, M. A. (1997). Soil and human health. *European Journal of Soil Science, 48*, 573–592. https://doi.org/10.1046/j.1365-2389.1997.00124.x

Saha, N., & Zaman, M. R. (2012). Evaluation of possible health risks of heavy metals by consumpt ion of foodstuffs available in the central market of Rajshahi City, Bangladesh. *Environmental Monitoring and Assessment, 185*, 3867–3878.

Segura-Muñoz, S. I., da Silva Oliveira, a, Nikaido, M., Trevilato, T. M. B., Bocio, a, Takayanagui, a M. M., and Domingo, J. L. (2006). Metal levels in sugar cane (Saccharum spp.) samples from an area under the influence of a municipal landfill and a medical waste treatment system in Brazil. *Environment International, 32*(1), 52–57. doi: 10.1016/j. envint.2005.04.008

Singh, A., & Agrawal, M. (2010). Effects of municipal waste water irrigation on availability of heavy metals and morpho-physiological characteristics of *Beta vulgaris* L. *Journal of Environmental Biology, 31*(5), 727–736. Retrieved from http://jeb.co.in/journal_issues/201009_sep10_supp/paper_01.pdf

Thompson, M. (2005). *Harmonised guidelines for the in-house validation of methods of analysis*. (Technical Report). Budapest.

US EPA. (2012). *Guidelines for water reuse*. Boston, MA: CDC Smith Inc.

US EPA. (2009). Mercury: Basic information. Retrieved May 4, 2009, from http://www.epa.gov/mercury/about.htm,

Vowotor, M. K., Phil, M., Hood, C. O., Sackey, S. S., Tatchie, E., Osei, D. M., ... Atieomo, S. M. (2014). An assessment of heavy metal pollution in sediments of a tropical Lagoon : A case study of the Benya Lagoon, Komenda Edina Eguafo Abrem municipality (KEEA) – Ghana. *Journal of Health and Pollution, 4*(6), 26–39. https://doi.org/10.5696/2156-9614-4-6.26

World Health Organization. (2001). *Copper. Environmental health criteiavol*. Geneva: International Programme on Chemical Safety., 200.

10

Assessment of pollution levels, potential ecological risk and human health risk of heavy metals/metalloids in dust around fuel filling stations from the Kumasi Metropolis, Ghana

Marian Asantewah Nkansah[1]*, Godfred Darko[1], Matt Dodd[2], Francis Opoku[1], Thomas Bentum Essuman[1] and Joshua Antwi-Boasiako[1]

*Corresponding author: Marian Asantewah Nkansah, Department of Chemistry, Kwame Nkrumah University of Science and Technology, Kumasi, Ghana
E-mail: maan4gr@yahoo.co.uk

Reviewing editor: Peter Fantke, Technical University of Denmark (DTU), Denmark

Abstract: The aim of this study was to evaluate the levels of selected heavy metals/metalloids in filling station dust from the Kumasi Metropolis, Ghana. A total of forty (40) dust samples were analysed for Fe, Ti, Zn, Zr, Mn, Sr, Ba, Cr, Pd, Ni, Cu, As and Mo using X-ray Fluorescence technique. Mean concentrations of Ba, As, Cr, Cu, Fe, Mn, Mo, Ni, Pb, Sr, Ti, Zn and Zr were 92.26, 6.20, 70.41, 50.18, 466.22, 163.68, 4.63, 44.05, 46.93, 106.69, 327.51, 280.32 and 182.05 mg/kg, respectively. The pollution index (PI) and geo-accumulation (I_{geo}) index values were in the order of Ba < Mn < Sr < Zr < Cu < Cr < Ni < Mo < As < Zn < Pb < Fe < Ti. The pollution load index had a mean of 2.20, signifying moderate pollution. Higher PI and I_{geo} value for Pb, Fe and Ti indicated high pollution. The PCA analysis identified anthropogenic inputs and natural origin as the main sources of pollution in filling station dust. The potential ecological risk index decreased as follows: As > Pb > Ni > Cu > Cr > Zn > Mn > Ba. The contribution of hazard quotient via ingestion for most of the heavy metals/metalloids were high with 11.83% for adults and 88.17% for children. For health risk assessment, non-carcinogenic values were below the threshold values, except hazard index via ingestion. The main exposure pathway for both children and adults was ingestion, followed by dermal contact and inhalation.

Subjects: Environment & Agriculture; Earth Sciences; Environmental Studies & Management

Keywords: dust; geo-accumulation index; Kumasi; pollution index; X-ray fluorescence technique

ABOUT THE AUTHORS

The group is involved on the determination of levels and distribution of heavy metals in the environment. The current study on fuel stations forms part of the country-wide study to determine the effects of human activities on heavy metals and the potential health associated with their exposure.

PUBLIC INTEREST STATEMENT

The study looks at the possibility of dust blown from a fuel station causing harm to the health of humans. The likely groups of people to this potential harm are fuel station workers, patrons and residents within the vicinity of these stations. The presence of toxic chemicals in spilled fuel, oils, lubricants and exhaust emissions from vehicles that plough the station have the tendency to pollute the dust in and around these stations. This study gives an overview of the potential health implications of fuel stations on the general public if they come into contact with polluted dust.

1. Introduction

Atmospheric pollution constitutes a major challenge in several countries, especially those under rapid development (He, Yun, Shi, & Jiang, 2013; Schleicher et al., 2011; Teng et al., 2014). During the combustion of wood and fossil fuels, waste incineration, high-temperature industrial processes and traffic, dust containing trace metals are released into the atmosphere (Allen, Nemitz, Shi, Harrison, & Greenwood, 2001). Dust particulates can affect human health (Ruiz-Jimenez et al., 2012), especially the presence of trace metals, which are noxious to humans via inhalation or ingestion (Khairy, Barakat, Mostafa, & Wade, 2011; Lu, Zhang, Li, & Chen, 2014).

Heavy metals/metalloids pollution in dust is a problem due to their non-biodegradability, wide occurrence, toxicity, as well as their ability to accumulate over time (Dong, Yu, Bian, Wang, & Di, 2011). Heavy metals, including Ni, Cd, Pb, Zn, Cu, Hg, Cr and others refer to metals with densities > 5 g/cm^3 (Li, Ma, van der Kuijp, Yuan, & Huang, 2014). As a result of the similarities in chemical properties and fate in the environmental, arsenic is frequently referred to as a heavy metal (Huamain, Chunrong, Cong, & Yongguan, 1999). Heavy metals/metalloids pollution in dust is an irreversible process and very difficult to remove once it occurs (Zhou & Song, 2004). They also have the potential to cause biomagnification and bioaccumulation in the ecosystem (Manahan, 2000). Improper disposal of engine oil, brake fluid, transmission oil and leaded gasoline around the vicinity of fuel contributes to the heavy metals/metalloids load (Dauda & Odoh, 2012; Khorshid & Thiele-Bruhn, 2016; Luo, Yu, & Zhu, 2012; Pant & Harrison, 2013). Heavy metals/metalloids in dust can affect human health when exposed via ingestion, inhalation and dermal contact (Ling, Shen, Gao, Gu, & Yang, 2008; McLaughlin, Hamon, McLaren, Speir, & Rogers, 2000). For example, extreme exposure to Pb can harm the skeletal, nervous, endocrine, circulatory, immune and enzymatic systems (Zhang et al., 2012). In addition, exposure to Cd can cause pulmonary adenocarcinomas, hypertension, kidney dysfunction, lung cancer, prostatic proliferative lesions and bone fractures, while As exposure can cause dermal lesions, skin cancer, peripheral vascular disease and peripheral neuropathy (Chen, Teng, Lu, Wang, & Wang, 2015).

Source identification of heavy metals/metalloids is crucial for effective dust remediation and pollution control (Guo, Huo, Xi, Zhang, & Wu, 2015). To check the potential risk of metals, assessment tools including enrichment factor (EF), geo-accumulation index (I_{geo}), potential ecological risk index (PER), pollution load index (PLI) and human health risk assessment are used to measure the pollution level (Ma, Yang, Li, & Wang, 2016). Multivariate statistical methods have also been used to access the different sources of metals in the dust (Ma et al., 2016).

In relation to several studies, heavy metals/metalloids pollution in the environment mostly originate from anthropogenic sources (Wei & Yang, 2010). In urban road dusts, the anthropogenic sources of heavy metals/metalloids include industrial emission (chemical plant, power plants, metallurgical industry, coal combustion, auto repair shop, etc.), traffic emission (brake lining wear particles, weathered street surface particles, vehicle exhaust particles, tyre wear particles), pavement surface, weathering of building and domestic emission (Ahmed & Ishiga, 2006; Banerjee, 2003; Sezgin, Ozcan, Demir, Nemlioglu, & Bayat, 2004). Lately, dust is assessed as an analytical model with implications on environmental and human health (Shen et al., 2017; Ying, Shaogang, & Xiaoyang, 2016). Therefore, a number of studies on human health risk evaluation of heavy metals/metalloids pollutions in dust are reported (Han, Gao, Wei, Xu, & Gao, 2016; Lin, Fang, Wang, & Xu, 2015; Qing, Yutong, & Shenggao, 2015). Studies on human health risk assessment of heavy metals/metalloids in Ghana have largely focused on food, cosmetic product, soil, wastewater, ground and surface water contamination (Agorku, Kwaansa-Ansah, Voegborlo, Amegbletor, & Opoku, 2016; Akoto, Ephraim, & Darko, 2008, 2009; Asare-Donkor, Kwaansa-Ansah, Opoku, & Adimado, 2015; Atiemo et al., 2011; Boateng, Opoku, Acquaah, & Akoto, 2015; Eze, Udeigwe, & Stietiya, 2010; Guo et al., 2015; Kwaansa-Ansah, Agorku, & Nriagu, 2011; Nkansah, Opoku, & Ackumey, 2016; Obiri, 2007).

In the past decade, several studies on dust have been conducted on metal concentrations, source identification and distribution (Glorennec, Lucas, Mandin, & Le Bot, 2012; Laidlaw & Taylor, 2011; Lu et al., 2010). While there have been some recent studies on road and street dust (Duong & Lee, 2011; Li, Poon, & Liu, 2001), very few studies have been reported in dusts around oil filling stations (Afrifa et al., 2013; Dauda & Odoh, 2012; Ekperusi & Aigbodion, 2015; Emmanuel, Cobbina, Adomako, Duwiejuah, & Asare, 2014). Despite these serious effects of trace metals on human health, particularly for children and adults, information about pollution levels and health risks of heavy metals/ metalloids in the dust around fuel filling stations in Kumasi metropolis are lacking. The metropolis houses several types of highly polluting filling station industries. Thus, the Kumasi metropolis is facing severe threats from contamination induced by the fast growth of filling station activities that adversely can influence human health in the metropolis. This study, therefore, aims: (1) to measure the concentration of Fe, Ti, Zn, Zr, Mn, Sr, Ba, Cr, Pd, Ni, Cu, As and Mo, (2) to ascertain the possible sources and their relationship using multivariate analysis and (3) to evaluate the health risk and pollution levels of heavy metals/ metalloids in dust around the vicinity of fuel filling stations from Kumasi Metropolis.

2. Materials and methods

2.1. Study area

Sampling was done from two filling stations in each of the twenty (20) sub-metros in Kumasi. The Metropolis of Kumasi is centrally situated in the Ashanti Region, Ghana with about 270 km North of Accra. Kumasi lies between longitude 1.30°–1.35° and latitude 6.35°– 6.40° with an elevation above sea level of 250–300 metres. The wet sub-equatorial region is the area where Kumasi Metropolis has a mean maximum and minimum temperature of 30.7 and 21.5°C, respectively. The average humidity is about 84.16 and 60.00%, respectively (Ghana Statistical Services, 2010). The double maxima rainfall regime of Kumasi (214.3 mm in June and 165.2 mm in September) has a direct effect on the environment and population growth (Ghana Statistical Services, 2010). The major rainy season occurs between March/April to July and September to October as the minor rainy season. Furthermore, between November and early February, Kumasi experiences dry season, which affects small streams to dry up. The land area of the Metropolis is about 254 square kilometres. Kumasi is accessible from all zones of the country due to its unique central position. The administrative capital of Ashanti Region is Kumasi, which is the second largest city of Ghana. Kumasi is a rapidly developing metropolis with about two million people and annual growth rate of more than 5.4%. The physical structure of Kumasi is global with a centrally located area for commercial activities, such as the Adum shopping centre, the central market and Kejetia Lorry Park. In addition, there are satellite markets, such as Oforikrom market, Asafo market, Atonsu market and Bantama market in the metropolis. Other economic activities include the Anloga wood market, the Kaase/Asokwa industrial area and Suame Magazine. Most industries, which deal in soap making, logging and food processing, are located at Kaase and Asokwa industrial area. According to the Ghana Statistical Service, about 60, 46 and 48% of the Metropolis are rural, peri-urban and urban, respectively, confirming the rapid rate of urbanisation (Ghana Statistical Services, 2010).

2.2. Sampling

Twenty fuel filling stations were chosen randomly, from each of the 20 sub-metros for study. At the sampling site, the site was zoned out into five parts, namely, the four corners and the centre. The GPS coordinates of all the sampling sites were taken and are presented in Figure 1.

Dust samples were sampled from surfaces by sweeping. The dust samples were gathered from these zones into brown paper envelopes using soft touch brush and plastic dustpan. The paper envelopes were labelled according to the name of the filling station and the sub-metro. Three samples were then sampled from the Kwame Nkrumah University of Science and Technology botanical gardens to serve as pristine samples since the dust location is devoid of any contamination from vehicular activity.

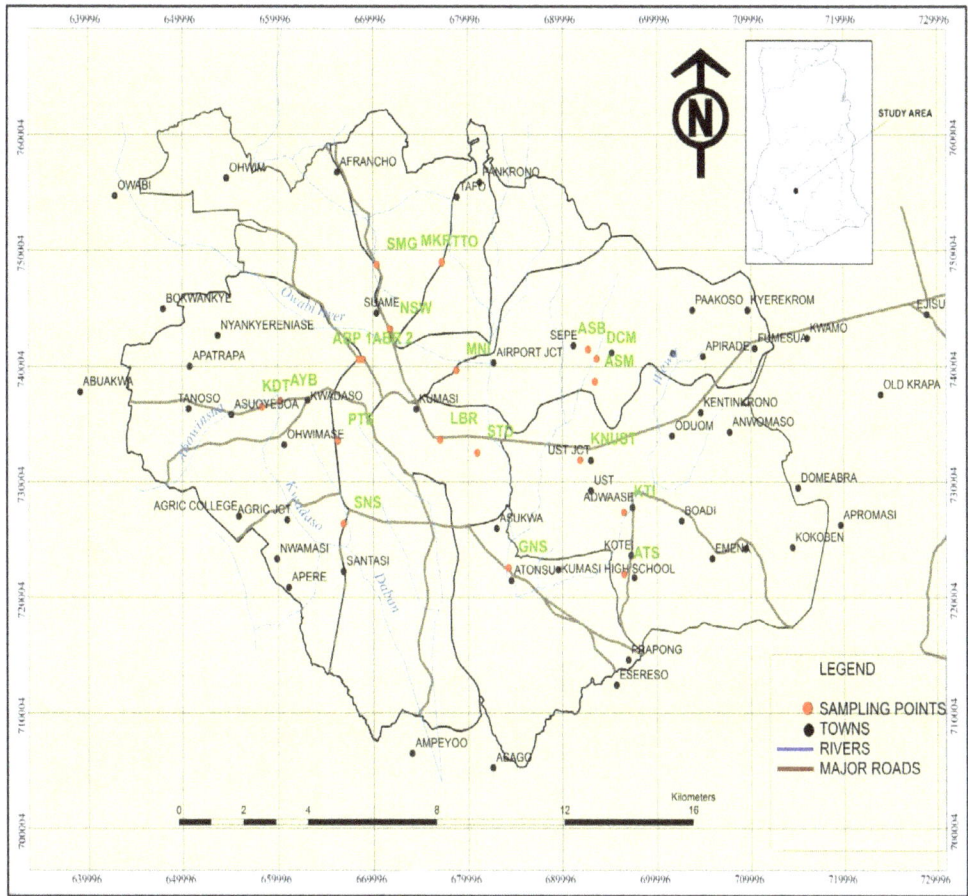

Figure 1. Map of Kumasi displaying the sampling areas.

2.3. Sample preparation

The dust samples were dried at room temperature. The samples were then sieved with a mesh of pore size of 60 and aperture of 20 micrometres. As a measure of avoiding cross-contamination, the sieves were washed intermittently after each sieving. The sieved samples were homogenised and ground with mortar and pestle and then kept in a desiccator prior to analysis.

2.4. Sample analysis

The heavy metals/metalloids content in the dust samples was determined with Thermo Scientific Niton X-ray Fluorescence (XRF) analyser (NDTr-XL3t-86956). The sample holder was filled halfway with the digested sample and covered with a Mylar film. The cupped sample was then placed in the XRF shroud and scanned for 180 s to obtain the desired result. All the samples were treated in the same manner. All the XRF sample analysis was done in triplicate. The equipment was calibrated with reference material of OC USGS SAR-M 180-673. Recovery rates in the range of 90 and 116% were sufficient for analysis. The limits of detection for As, Cr, Ba, Fe, Cu, Mn, Mo, Pb, Sr, Ni, Ti, Zr and Zn were 0.01, 0.05, 0.03, 0.82, 0.28, 0.70, 0.56, 1.60, 0.23, 0.35, 0.64, 1.27 and 0.15 mg/kg, respectively; while limits of quantification were 0.04, 3.24, 0.52, 0.58, 2.13, 0.95, 2.62, 5.12, 3.53, 0.86, 1.56, 0.58 and 4.07 mg/kg, respectively.

2.5. Evaluation of heavy metals/metalloids contamination

Pollution levels of heavy metals/metalloids in filling station dust were investigated using PER, I_{geo} and PI (Pan et al., 2016; Ying et al., 2016). Geo-accumulation index was initially applied to river sediments and has also been used for the evaluation of dust pollution (Salati & Moore, 2010). Geo-accumulation index allows the assessment of environmental contamination between pre-industrial and current concentrations. In this study, I_{geo} was evaluated following the Muller (1969) equation:

$$I_{geo} = I_2\left[\frac{C_n}{1.5B_n}\right] \tag{1}$$

where B_n is geochemical background value in dust and C_n is the estimated concentration of heavy metals/metalloids n. In this study, the background geochemical compositions by Taylor and McLennan (1995) were used as the background values for calculating the I_{geo} values. The constant 1.5 allows us to analyse the natural variations in the concentration of a given substance in the environment and to detect small anthropogenic effects (Wei & Yang, 2010). The I_{geo} classification is given in Table 1 (Chen et al., 2005).

The pollution index was the concentration of heavy metals/metalloids in the filling station dust relative to the background concentration of the equivalent heavy metals/metalloids (Kamani et al., 2015). To evaluate the pollution status of the filling station dust samples, the PLI of each heavy metals/metalloids was calculated according to Equation (2):

$$PLI = \sqrt[n]{PL_1 \times PL_2 \times PL_2 \ldots \times PL_n,} \tag{2}$$

The PLI classification of heavy metals/metalloids contamination is given in Table 1 (Islam, Ahmed, & Habibullah-Al-Mamun, 2015).

2.6. Potential ecological risk

To access the heavy metals/metalloids toxicity, the PER method by Hakanson (1980) was used to estimate their level of pollution in filling station dust. The PER was evaluated according to Equations (3)–(5):

$$PER = \sum_i^n E_r^i \tag{3}$$

$$E_r^i = T_n^i \times C_r^i \tag{4}$$

$$C_r^i = C^i/C_n^i \tag{5}$$

where C_j^i, C^i, C_n^i, E_r^i and T_n^i represent the pollution factor, estimated concentration, background concentration, PER index and toxic response factors for heavy metals/metalloids n, respectively. The toxic response factor for Cu, Zn, Pb, Mn, Ni, Cr, As and Ba were 5, 5, 1, 1, 5, 2, 10 and 1, respectively (Hakanson, 1980).

Table 1. Potential ecological risks, risk category indices and the related classifications of heavy metals/metalloids pollution

I_{geo}	Risk category	PLI	Risk category	E_r^ialue	Risk category	PER value	Risk category
$I_{geo} \leq 0$	Practically unpolluted	PLI ≤ 1	Unpolluted	<40	Low risk	≤ 50	Low risk
$0 \leq I_{geo} \leq 1$	Unpolluted to moderately polluted	$1 \leq PLI \leq 2$	Unpolluted to moderately polluted	40–80	Moderate risk	$50 < PER \leq 100$	Moderate risk
$1 \leq I_{geo} \leq 2$	Moderately polluted	$2 \leq PLI \leq 3$	Moderately polluted	80–160	Considerable risk	$100 < PER \leq 200$	Considerable risk
$2 \leq I_{geo} \leq 3$	Moderately to strongly polluted	$3 \leq PLI \leq 4$	Moderately to highly polluted	160–320	High risk	PER > 200	High risk
$3 \leq I_{geo} \leq 4$	Strongly polluted	$4 \leq PLI \leq 5$	Highly polluted	≥ 320	Significantly risk		
$4 \leq I_{geo} \leq 5$	Strongly to extremely polluted	PLI > 5	Very highly polluted				
$I_{geo} > 5$	Extremely polluted						

As summarised in Table 1, the E_r^i and PER classifications defined by Hakanson (1980) were used in this study.

2.7. Human health risk assessment

Human health risk assessment involves the evaluation of possible human health effect in the contaminated environmental media (Li et al., 2014). The carcinogenic and non-carcinogenic risks through dermal contact, ingestion and inhalation exposure pathways were evaluated using the human health risk assessment (Qing et al., 2015).

2.8. Exposure assessment

The health risk assessment is centred on the exposure factors and guidelines handbook of United States Environmental Protection Agency (USEPA, 2002, 2003). The average daily dose (ADD) via inhalation (ADD$_{inh}$), ingestion (ADD$_{ing}$) and dermal contact (ADD$_{derm}$) for both children and adults were evaluated following Equations (6) and (7):

$$ADD_{ing} = C_{dust} \times \frac{IngR \times EF \times ED}{BW \times AT} \times 10^{-6} \tag{6}$$

$$ADD_{inh} = C_{dust} \times \frac{IngR \times EF \times ED}{PEF \times BW \times AT} \tag{7}$$

$$ADD_{dermal} = C_{dust} \times \frac{SA \times AF \times ABS \times EF \times ED}{BW \times AT} \times 10^{-6} \tag{8}$$

The exposure factors and values (Ying et al., 2016) used in the risk assessment are given in Table 2.

2.9. Non-carcinogenic risk assessment

The carcinogenic and non-carcinogenic adverse effects were evaluated with the hazard quotient (HQ), carcinogenic risk (RI) and hazard index (HI) approaches (Ying et al., 2016). According to (USEPA, 1989), the hazard quotient is the average daily dose of heavy metals/metalloids with reference to its reference dose (RfD):

Table 2. Exposure factors and reference value of parameters used for the human health risk evaluation of heavy metals/metalloids in filling station dust

Factor	Definition	Unit	Value		Reference
			Children	Adults	
C_{dust}	Heavy metal concentration in dust	mg/kg			This study
IngR	Ingestion rate	mg/day	200	100	USEPA (2011)
EF	Exposure frequency	days/year	350	350	Lee-Steere (2009)
ED	Exposure duration	years	6	24	USEPA (2011)
BW	Body weight	kg	15	55.9	Lee-Steere (2009)
AT	Average time	days	365 × ED	365 × ED	USEPA (1989)
InhR	Inhalation rate	m³/day	7.63	12.8	Li et al. (2001)
PEF	Particle emission factor	m³/kg	1.36×10^9	1.36×10^9	USEPA (2011)
SA	Exposure skin surface area	cm²	1600	4350	Lee-Steere (2009)
AF	Skin adherence factor	mg/cm day	0.2	0.7	USEPA (1993)
ABF	Dermal absorption factor	no unit	0.001	0.001	Chabukdhara and Nema (2013)

$$HQ = \frac{ADD}{RfD} \tag{9}$$

If HQ < 1, signifies no adverse effects, whereas HQ > 1, signifies adverse effects (USEPA, 2011).

To assess the overall adverse effects of non-carcinogenic risk, the hazard index approach was applied (USEPA, 1986). The HI is the sum of HQ through the three exposure pathways for heavy metals/metalloids. For a mixture of contaminations, the HI was evaluated according to Equation (10) (USEPA, 1989):

$$HI = \sum HQ_i \tag{10}$$

HI < 1 denotes non-carcinogenic effects, whereas HI > 1 signifies adverse effects.

2.10. Carcinogenic risk assessment

The carcinogenic risk is the possibility of an individual to develop cancer during the lifetime exposure to the carcinogenic threats (Li et al., 2014). According to (USEPA, 1989), the slope factor (SF) directly convert the ADD of contaminant exposed over a lifetime risk of a cancer patient:

$$Risk = ADD \times SF$$

The values for SF in mg/kg day, RfD and other calculated parameters are presented in Table 3.

Risk value < 10^{-6} represents no carcinogenic risk to health from the dust, while a risk value >1×10^{-4} denotes high risk of developing cancer. A risk value ranging from 1×10^{-6}–1×10^{-4} signify an acceptable risk to human health (Hu et al., 2012).

2.11. Statistical analysis

Descriptive statistics and multivariate analysis were calculated using IBM Statistical Package for the Social Sciences (SPSS) *version* 20. The normality tests were analysed using the Shapiro–Wilk test. The Shapiro–Wilk test is based on the correlation between the heavy metals/metalloids concentration and the corresponding normal scores (Peat & Barton, 2008) and offers better power than the Kolmogorov–Smirnov test even after the Lilliefors correction (Steinskog, Tjøstheim, & Kvamstø, 2007). Principal component and Pearson's correlation analysis were used to evaluate and ascertain the possible source of heavy metals/metalloids contamination in the filling station dust. According

Table 3. The toxicity response to heavy metals/metalloids as the oral slope factor and oral reference dose

Heavy metals	Oral RfD (mg/kg/day)	Oral SF[a] (mg/kg/day)$^{-1}$
As	3.0×10^{-4}	1.5
Ba	2.0×10^{-1}	n.d.
Cr	3.0×10^{-3}	0.5
Cu	4.0×10^{-2}	n.d.
Fe	7.0×10^{-1}	n.d.
Mn	1.4×10^{-1}	n.d.
Mo	5.0×10^{-3}	n.d.
Ni	2.0×10^{-2}	n.d.
Pb	3.5×10^{-3}	8.5×10^{-3}
Sr	6.0×10^{-1}	n.d.
Ti	3.0	n.d.
Zn	3.0×10^{-1}	n.d.
Zr	4.0×10^{-4}	n.d.

Notes: n.d.: not determined.
[a]USEPA (2011).

to Lee, Qin and Lee (2006), the principal component is transformed from the original variable via eigen analysis. Herein, varimax rotation with Kaiser normalisation was employed to extract components with eigenvalue >1.

3. Results and discussion

3.1. Heavy metals/metalloids concentration in filling station dust

The descriptive statistics of heavy metals/metalloids concentration in the filling station dust sampled from the Kumasi Metropolis are presented in Table 4.

The filling station dust exhibited distinct variations in the levels of heavy metals/metalloids; with concentration ranges of 2.58–13.28, 33.72–142.63, 21.81–150.04, 89.85–681.87, 14.55–248.58, 2.77–8.93, 56.08–327.06, 6.48–129.87, 9.42–183.94, 347.00–585.53, 12.95–520.68, 21.30–59.91 and 92.02–250.38 mg/kg, for As, Ba, Cr, Fe, Cu, Mo, Mn, Pb, Sr, Ti, Ni, Zn and Zr, respectively. The mean concentrations of As, Ba, Cr, Fe, Cu, Mn, Mo, Pb, Sr, Ti, Zn, Ni and Zr were 6.20, 92.26, 70.41, 466.22, 50.18, 163.68, 4.63, 46.93, 106.69, 327.51, 280.32, 44.05, and 182.05 mg/kg, respectively. The mean concentrations of As, Cr, Cu, Fe, Mn, Mo, Pd, Ti, Zn and Zr were above the corresponding reference value and this indicates pollution of these metals in the filling station dust. Considering the mean concentration, the heavy metals/metalloids in the filling station dust were in the increasing order: Fe > Ti > Zn > Zr > Mn > Sr > Ba > Cr > Cu > Pd > Ni > As > Mo. The normality distributions of heavy metals/metalloids concentrations were checked by the one-sample Shapiro–Wilk normal test. The results showed that the concentrations of Fe, Ti, Zn, Zr, Mn, Sr, Ba, Cr, Cu, Pd, Ni and Mo were normally distributed ($p > 0.05$) in the collected filling station dust (Wang, Markert, Chen, Peng, & Ouyang, 2012), while As concentration showed non-normal distribution due to ($p < 0.05$) (Zhou et al., 2014). According to Karimi Nezhad, Tabatabaii and Gholami (2015), the coefficient of variation, which shows the variability degree of heavy metals/metalloids concentration are classified as low variability with coefficient of variation ≤20%, moderate variability (21% < coefficient of variation ≤ 50%) and high variability (50% < coefficient of variation ≤ 100%) (Qing et al., 2015). Based on this classification, the heavy metals/metalloids in the filling station dusts decreases as: Cu (99.74%) > Pb (68.68%) > Cr (52.24%) > As (43.87%) > Zn (42.74%) > Ba (41.66%) > Mn (40.15%) > Ni (36.98%) > Mo (36.07%) > Sr (34.74%) > Zr (22.18%) > Ti (7.33%) > Fe (6.75%). The coefficient of variation of Pb and Cr specified a moderate degree of variability, which reveals no homogeneous levels. A large coefficient of variation of >70%

Table 4. Descriptive data of heavy metals/metalloids concentration (mg/kg) in filling station dust collected from the Kumasi Metropolis and reference value (Taylor & McLennan, 1995)

	Min	Max	Mean	Std. deviation	Coefficient of variation (%)	Skewness	Kurtosis	Reference value
As	2.58	13.28	6.20	2.72	43.87	1.03	1.22	1.5
Ba	33.72	142.63	92.26	38.44	41.66	−0.09	−1.10	550
Cr	21.81	150.04	70.41	36.78	52.24	0.51	−0.25	35
Cu	14.55	248.58	50.18	50.05	99.74	3.23	12.31	25
Fe	89.85	681.87	466.22	31.46	6.75	0.63	0.68	3.5
Mn	56.08	327.06	163.68	65.72	40.15	0.51	0.14	1.5
Mo	2.77	8.93	4.63	1.67	36.07	1.16	1.28	1.5
Ni	21.30	59.91	44.05	16.29	36.98	−1.18	2.18	20
Pb	6.48	129.87	46.93	32.23	68.68	1.30	1.17	20
Sr	9.42	183.94	106.69	37.06	34.74	−1.07	2.82	350
Ti	347.00	585.53	327.51	24.02	7.33	0.33	−0.24	0.3
Zn	12.95	520.68	280.32	119.81	42.74	−0.43	0.78	71
Zr	92.02	250.38	182.05	40.37	22.18	−0.26	−0.58	190

was found for Cu, an indication that Cu concentration differed greatly in the study region. The skew-ness values of Cu, As, Mo and Pb were >1, indicated that they were positively skewed towards low concentration. Moreover, while the negative kurtosis values of Ba, Cr, Ni, Sr, Zn and Zr, indicated that their distribution in the filling station dust was less steep than normal (Chen, Lu, & Yang, 2012).

3.2. Comparison with other heavy metals/metalloids dust studies
In order to compare the examined heavy metals/metalloids concentration with studies from other countries, their concentrations in filling station dust of Kumasi Metropolis were collated (Table 5).

In Table 5, the mean concentrations of Cr measured agreed with data from Guangzhou and Xiandao, both from China, but lower than those reported in other cities, such as Madrid, Ottawa, Hangzhou, Urumqi. The mean concentrations of Ni measured was comparable with those reported in Oslo, Madrid, Baoji and Urumqi, but was higher in other compared cities except for Shanghai, China. The mean concentration of Cu was lower than cites except for Xiandao. The Zn concentration is similar to Urumqi, while lower than other cities except for Ottawa and Xiandao. The mean concentrations of As and Pb were lower than other compared cites except for Ottawa. However, the mean concentrations of Mn, Fe, Sr, Zr, Ti and Ba were higher in all the compared cities (Table 5).

3.3. Multivariate analysis

3.3.1. Source of heavy metals/metalloids pollution
To check the level of heavy metals/metalloids pollution and identify their possible source in the fill-ing station dust, principal component and correlation coefficient analyses were performed (Facchinelli, Sacchi, & Mallen, 2001). The relationships among the estimated heavy metals/metal-loids were carried out using Pearson's correlation analysis. Table 6 demonstrates that Mn, Cr, Ti and Zr were significantly correlated ($p < 0.01$) with each other, signifying similar sources of these metals.

Pb was positively correlated with Zn, Cu and Fe; Zn was positively correlated with Mn, Cu, Fe and Cr, but negative with Ni. There existed a significant positive correlation between Cu with Fe and Mn. A similar trend was observed for Fe with Mn and Cr; and for Cr with Ti and Zr. In the case of Ti, there was significant positive correlation with Sr, but negative with Zr. Though the significant positive relation-ship cannot always be attributed to a similar source, this correlation analysis can still offer interesting information on the pathway and source of heavy metals/ metalloids (Lu et al., 2010). For instance, Zn, Cu and Fe, and Pd with significant positive correlation could be originating from anthropogenic sourc-es, such as fossil fuel combustion via atmospheric deposition and leaded fuel. Mo, Ba and As might have originated from different sources with reference to the other heavy metals/metalloids.

In addition to the principal component analysis, Bartlett's and KMO test were employed to exam-ine the sampling sphericity and adequacy. The PCA results of heavy metals/metalloids are given in Table 7.

Three components with eigenvalues >1 were obtained, which account for 97.80% of the extracted variance. The first component with positive loading for Mo, As, Mn, Ti, Ba, Zr and Sr, whose coeffi-cients were 0.529, 0.945, 0.534, 0.946, 0.524, 0.734 and 0.928, respectively explained 45.25% of the total variance. The Mo, As, Mn, Ti, Ba, Zr and Sr in PC1 were strongly correlated, signifying their similar source. PC1 consisting of As, Mn and Ba were mainly influenced by natural sources (Lu et al., 2010). The second component showed strong positive loading (>0.75) for Pb, Zn and Cu, whose coefficients were 0.987, 0.999 and 0.862, respectively explained 33.30% of the total variance. PC2 including Pb, Zn and Cu could be associated with anthropogenic inputs, such as industrial and traffic pollution (Chen, Lu, Li, Gao, & Chang, 2014). Moreover, improper disposal of engine oil and a significant amount of emissions influences Pb, Cu and Zn pollution in the filling station dust, as shown in PC2. The third component showed strong loading (>0.75) for Mo (0.830) and Ba (0.851) with 19.25% of the

Table 5. Comparison of heavy metals/metalloids concentrations (mg/kg) observed in this study with those found in other heavy metal dust studies

	Ti	Ba	Cr	Mn	Fe	Ni	Cu	Zn	As	Pb	Sr	Zr	Reference
Hong Kong, China (Street dusts)	2370.00	253.00	124.00	594.00	14,100.00	28.60	110.00	3840.00	66.80	120.00	121.00	378.00	Chen, Wei, Zheng, Wu and Adriano (1991)
Hong kong, China (65 urban street dusts)	–	–	–	–	–	–	173.00	1450.00	–	181.00	–	–	Li et al. (2001)
London (11 street dust)	1465.00	–	112.00	379.00	22,800.00	–	191.00	1176.00	27.00	2008.00	–	–	Fergusson, Forbes, Schroeder and Ryan (1986)
Christchurch, New Zealand (12 street dust)	2117.00	–	103.00	313.00	20,900.00	–	90.80	716.00	14.50	1223.00	–	–	Fergusson et al. (1986)
Oslo, Norway(16 street dust)	7452.00	526.00	–	833.00	51,452.00	41.00	123.00	412.00	–	180.00	344.00	–	de Miguel et al. (1997)
Madrid, Spain (16 street dust)	1100.00	–	61.00	362.00	19,300.00	44.00	188.00	476.00	–	1927.00	–	–	de Miguel et al. (1997)
Ottawa, Canada (50 street dust)	–	576.00	43.30	431.50	18,948.00	15.20	65.84	112.50	1.30	39.05	459.00	–	Rasmussen, Subramanian and Jessiman (2001)
Xi'an, China(157 campus dust)	–	958.90	154.20	546.20	–	32.20	62.10	390.70	11.50	151.60	–	–	Chen et al. (2014)
Xiandao (51 road dust)	–	–	71.60	–	–	–	43.90	171.00	–	66.60	–	–	Li et al. (2016)
Xiandao (51 street dust)	–	–	–	765.19	–	30.74	–	–	24.47	–	–	–	Li et al. (2016)
Xiandao, China (51 road dust)	–	–	80.66	–	–	–	43.90	66.58	–	241.92	–	–	Li et al. (2016)
Kavala, Greece (26 street dust)	–	–	196.00	–	–	57.50	123.90	271.60	16.70	300.90	–	–	Christoforidis and Stamatis (2009)
Hangzhou, China (25 urban dusts)	–	–	51.29	–	–	25.88	116.04	321.40	–	202.16	–	–	Zhang and Wang (2009)
Shanghai, China (273 dusts)	–	–	159.30	–	–	83.98	196.80	733.8	–	294.90	–	–	Shi et al. (2008)
Guangzhou, China (30 urban dusts)	–	–	78.80	–	–	23.00	176.00	586.00	–	240.00	–	–	Duzgoren-Aydin et al. (2006)
Guwahati, India (25 street dusts)	–	–	0.01	0.3	10.8	0.01	0.01	0.03	–	0.01	–	–	Hussain, Rahman, Prakash, and Hoque (2015)
Urumqi, China (169 urban road dusts)	–	–	54.28	–	–	43.28	94.54	294.47	–	53.53	–	–	Wei, Jiang, Li and Mu (2009)
Isfahan, Iran (24 road dusts)	–	–	–	–	–	66.63	182.26	707.19	22.15	393.33	–	–	Soltani et al. (2015)

Table 6. Pearson's correlation matrix for heavy metals/metalloids concentration

	Mo	Pb	As	Zn	Cu	Ni	Fe	Mn	Cr	Ti	Ba	Zr	Sr
Mo	1												
Pb	0.288	1											
As	0.011	0.028	1										
Zn	0.238	0.803[a]	-0.018	1									
Cu	0.357	0.550[a]	-0.027	0.575[a]	1								
Ni	0.201	-0.937	-0.254	-0.973[b]	-0.722	1							
Fe	-0.049	0.488[b]	-0.121	0.655[a]	0.457[b]	0.056	1						
Mn	0.067	0.347	0.134	0.615[a]	0.426[b]	-0.283	0.889[a]	1					
Cr	-0.118	0.374	-0.125	0.455[b]	0.289	0.215	0.940[a]	0.775[a]	1				
Ti	0.025	0.127	0.007	-0.098	0.268	0.037	0.237	0.639[a]	0.714[a]	1			
Ba	0.438	0.075	0.450	0.202	0.460	0.074	0.219	0.612	-0.002	0.204	1		
Zr	-0.296	-0.174	0.193	-0.193	-0.098	0.523	0.320	0.710[a]	0.731[a]	0.723[a]	-0.210	1	
Sr	-0.170	0.032	0.357	0.396	-0.344	0.032	0.338	0.117	-0.030	-0.424[b]	0.265	-0.264	1

[a]Correlation is significant at the 0.01 level.
[b]Correlation is significant at the 0.05 level.

Table 7. Varimax factor loading for heavy metals/metalloids in the filling station dust

	Component		
	PC1	**PC2**	**PC3**
Mo	**0.529**	−0.174	**0.830**
Pb	−0.153	**0.987**	−0.058
As	**0.945**	0.129	0.301
Zn	0.037	**0.999**	0.017
Cu	−0.224	**0.862**	0.455
Ni	−0.187	−0.969	0.159
Fe	−0.906	0.049	−0.421
Mn	**0.534**	0.317	**0.784**
Cr	−0.962	−0.079	−0.262
Ti	**0.946**	−0.177	0.272
Ba	**0.524**	−0.039	**0.851**
Zr	**0.734**	−0.679	0.016
Sr	**0.928**	−0.158	0.337
Eigen value	5.883	4.329	2.788
% of variance	45.252	33.300	19.248
Cumulative %	45.252	78.552	97.800

Note: Bold values represent positive loadings ≥0.5.

extracted variance. The heavy metals/metalloids in the three PC's demonstrate their different sources. The correlation between Mo and Ba in PC1 and PC3, suggests their origin from dust as the parent material. The poor loadings of Ni, Fe and Cr can be attributed to differences in sources of materials and geochemical behaviour of parameters (Edet, Merkel, & Offiong, 2003), as well as quasi-independent behaviour within the group (Lu et al., 2010).

3.4. Heavy metals/metalloids pollution assessment
The estimated I_{geo} result of heavy metals/metalloids in filling station dust is given in Figure 2.

The mean values of I_{geo} increased following the order: Ba < Mn < Sr < Zr < Cu < Cr < Ni < Mo < As < Zn < Pb < Fe < Ti. The mean I_{geo} of Mo, Zr, Sr, Cu, Mn, Cr, Ba and Ni showed that dust from filling stations was practically unpolluted. The mean I_{geo} of As and Zn signify moderately polluted, whereas

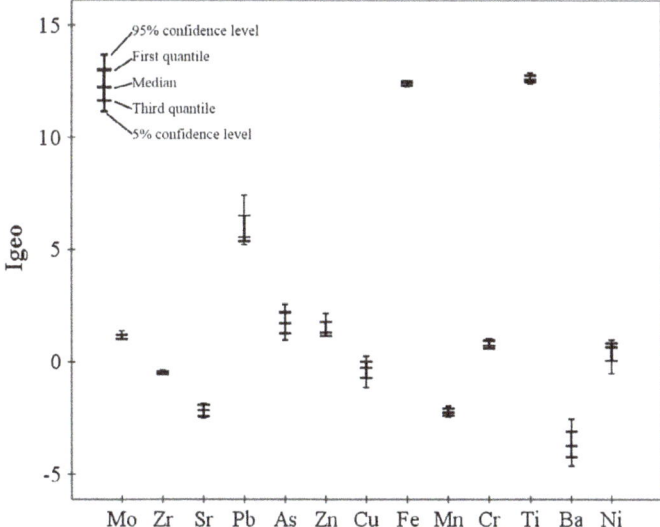

Figure 2. Box-plots of the I_{geo} for heavy metals/metalloids in the filling station dust of Kumasi Metropolis.

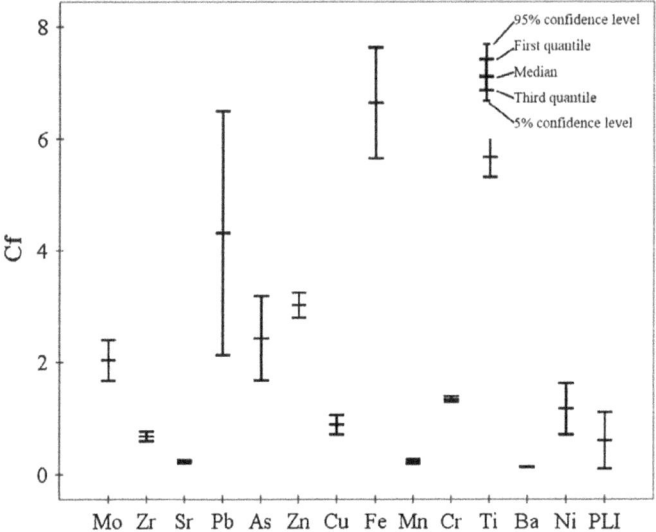

Figure 3. Box-plots of pollution load index and pollution index for heavy metals/metalloids in the filling station dust of Kumasi Metropolis.

the mean I_{geo} for Pb, Fe and Ti designate very highly polluted. In the study area, the I_{geo} values signify "unpolluted" to "highly polluted" of the investigated heavy metals/metalloids. Among the estimated heavy metals/metalloids, Ti and Fe were significantly accumulated in the filling station dust, as showed by their corresponding mean geochemical accumulation values of 12.46 ± 0.41 and 12.06 ± 0.64. The presence of Zn and As in the filling station dust poses an adverse environmental concern. Similarly, Wei and Yang (2010) deduced that for heavy metals in urban soils in China, Cr and Ni appeared to cause the least contamination in the selected cities, while Cu, Pb, Zn and Cd showed the highest I_{geo} values. Cr and Ni appear to be the least contaminated elements in all the cities, while Cu, Pb, Zn and Cd show the highest I_{geo} values for most cities (Wei & Yang, 2010). The PIs results for heavy metals/metalloids in the filling station dust varied significantly across the different heavy metals/metalloids (Figure 3).

The mean PIs of Fe, Pb and Ti were much higher, varying from 16.51 to 97.02, 0.64 to 9.97 and 29.33 to 88.24, with values of 5.75, 46.67 and 58.45, respectively. This showed that Mo, Zn, As, Cu, Cr and Ni of filling station dust were classified as middle PI. Zr, Sr, Mn and Ba exhibited lower mean values and were classified as low PI. The PI value for the calculated heavy metals/metalloids was in the decreasing order: Ba (0.11) < Mn (0.19) < Sr (0.22) < Zr (2.06) < Cr (1.30) < Cu (1.36) < Ni (1.47) < Mo (2.06) < As (2.82) < Zn (2.85) < Pb (5.75) < Fe (6.54) ≤ Ti (6.54). Therefore, Pb, Fe and Ti were the main contaminants of the study area reaching a very strong contamination level. The high level of Pb in filling station dust samples can be credited to leaded fuels, emissions from tyre wear, bearing wear and lubricating oils (Kamani et al., 2015). The PLIs in all filling station dust samples varied between 0.10 and 6.32 with a mean value of 2.20, signifying that the filling station dust was "moderately polluted" in the Kumasi Metropolis.

3.5. Ecological risk index
The PER is a measure of the sensitivity of several biological systems to deadly substances and elucidates the potential risks induced by heavy metals/metalloids (Qing et al., 2015; Ying et al., 2016). The PER results are presented in Table 8.

The mean E_r^i value for all the heavy metals/metalloids except arsenic were below 40, thus indicating low PER. This signifies that arsenic has a moderate risk to the environment and human health. The PER index showed the order: As > Pb > Ni > Cu > Cr > Zn > Mn > Ba. To access the overall PER of the observed heavy metals/ metalloids in the filling station dust in the Kumasi metropolis, PER was evaluated as the summation of the risk factors. The contribution to the overall PER displays that arsenic contributed 32% of the total PER. The PER value for all the sampling sites ranged from 1.34 to

Table 8. Potential ecological risks of heavy metals/metalloids and their ecological index in the filling station dust

Heavy metal	Minimum	Maximum	Mean
Pb	1.62	32.47	11.73
As	17.20	88.53	41.30
Zn	0.18	7.33	3.95
Cu	2.91	18.19	8.22
Ni	5.33	14.98	11.01
Mn	0.09	0.39	0.27
Cr	1.25	8.57	4.02
Ba	0.06	0.26	0.17
PER	1.34	784.73	187.72

784.73 with a mean value of 187.72, indicating low to very high potential ecological risk. Nevertheless, the mean *RI* shows a considerably high risk (PER > 100) in the Kumasi Metropolis, largely due to arsenic contamination. This should be remediated to prevent the possible ecological adverse effect.

3.6. Human health risk assessment

3.6.1. Exposure assessment
The calculated mean daily intake of heavy metals/metalloids for adult and children in the filling station dust via inhalation, dermal contact and ingestion are presented in Table 9.

The trends of ADDs for heavy metals/metalloids via dermal contact, inhalation and ingestion were in the order: Fe > Ti > Zn > Zr > Mn > Sr > Ba > Cr > Cu > Pb > Ni > As > Mo. The exposure route, which on the average brings about the highest risk for adults and children was ingestion followed by dermal contact with inhalation being the lowest. Qing et al. (2015) observed the same order of daily intake of heavy metals.

The health risks assessment results are given in Table 10.

The HQ results for all the heavy metals/metalloids were < 1, signifying no risk for both adult and children. The mean HQ values for adults and children via ingestion were as follows: Zr > Cr > As > Pb

Table 9. Exposure duration (mg/kg-day) for filling station dust in Kumasi Metropolis

	ADD_{ing}		ADD_{inh}		ADD_{dermal}	
	Children	Adult	Children	Adult	Children	Adult
As	7.93E-05	1.06E-05	2.22E-09	1.00E-09	1.27E-07	3.24E-07
Ba	1.18E-03	1.58E-04	3.31E-08	1.49E-08	1.89E-06	4.82E-06
Cr	9.00E-04	1.21E-04	2.53E-08	1.14E-08	1.44E-06	3.68E-06
Cu	6.42E-04	8.61E-05	1.79E-08	8.10E-09	1.03E-06	2.62E-06
Fe	5.96E-03	7.99E-04	1.67E-07	7.53E-08	9.54E-06	2.44E-05
Mn	2.09E-03	2.81E-04	5.87E-08	2.64E-08	3.35E-06	8.55E-06
Mo	5.92E-05	7.95E-06	1.66E-09	7.48E-10	9.47E-08	2.42E-07
Ni	5.63E-04	7.56E-05	1.58E-08	7.11E-09	9.01E-07	2.30E-06
Pb	6.00E-04	8.05E-05	1.68E-08	7.58E-09	9.60E-07	2.45E-06
Sr	1.36E-03	1.83E-04	3.83E-08	1.72E-08	2.18E-06	5.57E-06
Ti	4.19E-03	5.62E-04	1.17E-07	5.29E-08	6.70E-06	1.71E-05
Zn	3.58E-03	4.81E-04	1.01E-07	4.53E-08	5.73E-06	1.46E-05
Zr	2.33E-03	3.12E-04	6.53E-08	2.94E-08	3.72E-06	9.51E-06

Table 10. Non-carcinogenic risk, overall toxic risk and carcinogenic risk ($ADD_{life \times SF}$)							
	HQ_{ing}		HQ_{inh}		HQ_{dermal}		$ADD_{life} \times SF$	
	Children	Adult	Children	Adult	Children	Adult	Children	Adult
As	0.264	0.035	7.41E-06	3.34E-06	3.81E-11	9.72E-11	8.74E-08	1.17E-08
Ba	0.006	0.001	1.65E-07	7.45E-08	3.77E-07	9.64E-07	–	–
Cr	0.300	0.040	8.42E-06	3.79E-06	4.32E-09	1.10E-08	3.31E-07	4.44E-08
Cu	0.016	0.002	4.50E-07	2.03E-07	4.11E-08	1.05E-07	–	–
Fe	0.009	0.001	2.39E-07	1.08E-07	6.68E-06	1.70E-05	–	–
Mn	0.015	0.002	4.19E-07	1.89E-07	4.69E-07	1.19E-06	–	–
Mo	0.012	0.002	3.32E-07	1.49E-07	4.74E-10	1.21E-09	–	–
Ni	0.028	0.004	7.89E-07	3.56E-07	1.80E-08	4.60E-08	–	–
Pb	0.171	0.023	4.81E-06	2.17E-06	3.36E-09	8.57E-09	3.75E-09	5.03E-10
Sr	0.002	0.000	6.38E-08	2.87E-08	1.31E-06	3.34E-06	–	–
Ti	0.001	0.000	3.92E-08	1.76E-08	2.01E-05	5.13E-05	–	–
Zn	0.012	0.002	3.35E-06	1.51E-06	1.72E-06	4.39E-06	–	–
Zr	5.819	0.781	1.63E-08	7.35E-05	1.49E-09	3.80E-09	–	–
HI	6.656	0.893	1.89E-08	8.54E-05	3.07E-05	7.84E-05	–	–

> Ni > Cu > Mn > Mo ≥ Zn > Fe > Ba > Sr > Ti, while that via inhalation were Cr > As > Pb > Zn > Ni > Cu > Mn > Mo > Fe > Ba > Sr > Ti > Zr and via dermal contact were Ti > Fe > Zn > Sr > Mn > Ba > Cu > Ni > Cr > Pb > Zr > Mo > As. In summary, the mean exposure route of heavy metals/metalloids for adults and children decreased according to the following order: inhalation < dermal contact < ingestion. In the present study, cutaneous and oral exposure are more important than inhalation. The findings in this study agree well with other studies (Benhaddya, Boukhelkhal, Halis, & Hadjel, 2016), which also observe ingestion to be the highest exposure pathway to health risk. The contribution of HQ_{ing} for most of the heavy metals/metalloids were the highest with 11.83% for adults and 88.17% for children. This confirms ingestion as the major exposure route that can adversely affect human health. The HI results via inhalation and dermal contact with all the heavy metals/metalloids were < 1, except for ingestion in adult, signifying no non-carcinogenic risk. The HI values in children were observed to develop more non-carcinogenic risk in the filling station dust compared to adults in the Kumasi Metropolis. This can be attributed to the fact that children tend to have significant contact with soil during their outdoor play activities and are more likely to have a direct hand-to-mouth exposure of dust (Luo et al., 2012). Similar findings can be seen from other studies including Qu, Sun, Wang, Huang, and Bi (2012), Zota et al. (2011), and Man et al. (2010). Owing to lack of carcinogenic slope factor for Cu, Ba, Mn, Fe, Ni, Ti, Mo, Zn, Sr and Zr, only the slope factors for As, Cr and Pb were given in the literature. The *RI* values in the filling station dusts of Kumasi Metropolis were 8.74×10^{-8} (As), 3.31×10^{-7} (Cr), and 3.75×10^{-9} (Pb) for children and 1.17×10^{-7} (As), 4.44×10^{-10} (Cr) and 5.03×10^{-9} (Pb) for adults (Table 9). The risk index of Cr, As and Pb for both adults and children were $< 10^{-6}$, showing no plausible carcinogenic risk. However, the HI level via ingestion indicates that the carcinogenic risks of the observed heavy metals/metalloids cannot be overlooked.

4. Conclusion

The concentration, pollution level and health risks assessment of heavy metals/metalloids in dust from filling stations within the Kumasi Metropolis, Ghana were evaluated in this study. This study offers significant information on the concentrations of Fe, Ti, Zr, Mn, Sr, Ba, Zn, Cr, Cu, Pd, As, Ni and Mo in dust impacted by fuel and other related emissions. The PCA analysis identified anthropogenic inputs and natural origin as the main sources of heavy metals/metalloids in the filling station dust of Kumasi Metropolis. The *PI* and I_{geo} revealed the order of Ba < Mn < Sr < Zr < Cu < Cr < Ni < Mo < As < Zn < Pb < Fe < Ti for both I_{geo} and *PI*. The higher *PI* values for Pb, Fe and Ti in the filling station dust indicated a very strong pollution level, which could be attributed to leaded fuels and emissions from bearing wear, lubricating oils and tyre wear. The PLIs in all filling station dust samples varied between 0.10 and 6.32 with an average value of 2.20, signifying moderately pollution in the Kumasi Metropolis. The mean value of E_r^i all the heavy metals/metalloids except As were below 40, indicating low risk. The PER value for all the sampling sites ranged from 1.34 to 784.73 with a mean value of 187.72, The HQ results indicate no risk for both adult and children, as the HQ values were lower than safe limit. The contributions of HQ_{ing} for most of the heavy metals/metalloids were the highest with 11.83% for adults and 88.17% for children. This signifies ingestion as the major exposure route that can adversely affect human health. The HI values were <1, except for HI value for ingestion in adult, signifying no non-carcinogenic risk. Nonetheless, the carcinogenic risks values for adults were lower than the threshold value of 1×10^{-6}, showing no carcinogenic risk. The main exposure pathway for both children and adults is ingestion, followed by dermal contact and inhalation. The health risk assessment revealed ingestion as the major exposure route with children being the most exposed to heavy metals/metalloids pollution compared to adults in the Kumasi Metropolis, Ghana. Therefore, the results in this study will be beneficial for environmental planning authorities in the Kumasi Metropolis to manage and control further heavy metals/metalloids pollution at the various filling station sites in the Metropolis.

Acknowledgement
The authors are grateful to Prof Matt Dodd of the School of Environment and Sustainability of Royal Roads University, Victoria BC, Canada for the provision of the equipment for the analysis of samples in this study.

Authors' contributions
All the authors contributed equally to the preparation of this manuscript. All authors read and approved the final manuscript.

Funding
The authors received no direct funding for this research.

Competing Interests
The authors declare no competing interest.

Author details
Marian Asantewah Nkansah[1]
E-mail: maan4gr@yahoo.co.uk
Godfred Darko[1]
E-mail: godfreddarko@yahoo.com
ORCID ID: http://orcid.org/0000-0001-7157-646X
Matt Dodd[2]
E-mail: matt.dodd@royalroads.ca
Francis Opoku[1]
E-mail: ofrancis2010@gmail.com
ORCID ID: http://orcid.org/0000-0002-8308-9113
Thomas Bentum Essuman[1]
E-mail: thobessman@gmail.com
Joshua Antwi-Boasiako[1]
E-mail: joshuaab81@gmail.com
[1] Department of Chemistry, Kwame Nkrumah University of Science and Technology, Kumasi, Ghana.
[2] School of Environment and Sustainability, Royal Roads University, Victoria, BC, Canada.

References
Afrifa, C. G., Ofosu, F. G., Bamford, S. A., Wordson, D. A., Atiemo, S. M., Aboh, I. J., & Adeti, J. P. (2013). Heavy metal contamination in surface soil dust at selected fuel filling stations in Accra, Ghana. *American Journal of Scientific and Industrial Research, 4,* 404–413.

Agorku, E. S., Kwaansa-Ansah, E. E., Voegborlo, R. B., Amegbletor, P., & Opoku, F. (2016). Mercury and hydroquinone content of skin toning creams and cosmetic soaps, and the potential risks to the health of Ghanaian women. *SpringerPlus, 5,* 1–5. doi:10.1186/s40064-016-1967-1

Ahmed, F., & Ishiga, H. (2006). Trace metal concentrations in street dusts of Dhaka city, Bangladesh. *Atmospheric Environment, 40,* 3835–3844. https://doi.org/10.1016/j.atmosenv.2006.03.004

Akoto, O., Ephraim, J., & Darko, G. (2008). Heavy metals pollution in surface soils in the vicinity of abundant railway servicing workshop in Kumasi, Ghana. *International Journal of Environmental Research, 2,* 359–364.

Akoto, O., Ephraim, J., & Darko, G. (2009). Heavy metals pollution in surface soils in the vicinity of abundant railway servicing workshop in Kumasi, Ghana. *International Journal of Environmental Research, 3,* 359–364.

Allen, A., Nemitz, E., Shi, J., Harrison, R., & Greenwood, J. (2001). Size distributions of trace metals in atmospheric aerosols in the United Kingdom. *Atmospheric Environment, 35,* 4581–4591. https://doi.org/10.1016/S1352-2310(01)00190-X

Asare-Donkor, N. K., Kwaansa-Ansah, E. E., Opoku, F., & Adimado, A. A. (2015). Concentrations, hydrochemistry and risk evaluation of selected heavy metals along the Jimi River and its tributaries at Obuasi a mining enclave in Ghana. *Environmental Systems Research, 4,* 1–14.

Atiemo, M. S., Ofosu, G. F., Kuranchie-Mensah, H., Tutu, A. O., Palm, N., & Blankson, S. A. (2011). Contamination assessment of heavy metals in road dust from selected roads in Accra, Ghana. *Research Journal of Environmental and Earth Sciences, 3,* 473–480.

Banerjee, A. D. (2003). Heavy metal levels and solid phase speciation in street dusts of Delhi, India. *Environmental Pollution, 123,* 95–105. https://doi.org/10.1016/S0269-7491(02)00337-8

Benhaddya, M. L., Boukhelkhal, A., Halis, Y., & Hadjel, M. (2016). Human health risks associated with metals from Urban Soil and road dust in an oilfield area of Southeastern Algeria. *Archives of Environmental Contamination and Toxicology, 70,* 556. https://doi.org/10.1007/s00244-015-0244-6

Boateng, T. K., Opoku, F., Acquaah, S. O., & Akoto, O. (2015). Pollution evaluation, sources and risk assessment of heavy metals in hand-dug wells from Ejisu-Juaben Municipality, Ghana. *Environmental Systems Research, 4*, 1–18. doi:10.1186/s40068-015-0045-y

Chabukdhara, M., & Nema, A. K. (2013). Heavy metals assessment in urban soil around industrial clusters in Ghaziabad, India: Probabilistic health risk approach. *Ecotoxicology and Environmental Safety, 87*, 57–64. doi:10.1016/j.ecoenv.2012.08.032.

Chen, H., Lu, X., Li, L. Y., Gao, T., & Chang, Y. (2014). Metal contamination in campus dust of Xi'an, China: A study based on multivariate statistics and spatial distribution. *Science of the Total Environment, 484*, 27–35. https://doi.org/10.1016/j.scitotenv.2014.03.026

Chen, J., Wei, F., Zheng, C., Wu, Y., & Adriano, D. C. (1991). Background concentrations of elements in soils of China. *Water, Air, and Soil Pollution, 57-58*(1), 699–712. doi:10.1007/BF00282934.

Chen, X., Lu, X., & Yang, G. (2012). Sources identification of heavy metals in urban topsoil from inside the Xi'an Second Ringroad, NW China using multivariate statistical methods. *Catena, 98*, 73–78. https://doi.org/10.1016/j.catena.2012.06.007

Chen, H., Teng, Y., Lu, S., Wang, Y., & Wang, J. (2015). Contamination features and health risk of soil heavy metals in China. *Science of the Total Environment, 512*, 143–153. https://doi.org/10.1016/j.scitotenv.2015.01.025

Chen, T.-B., Zheng, Y.-M., Lei, M., Huang, Z.-C., Wu, H.-T., Chen, H., ... Tian, Q.-Z. (2005). Assessment of heavy metal pollution in surface soils of urban parks in Beijing, China. *Chemosphere, 60*, 542–551. https://doi.org/10.1016/j.chemosphere.2004.12.072

Christoforidis, A., & Stamatis, N. (2009). Heavy metal contamination in street dust and roadside soil along the major national road in Kavala's region, Greece. *Geoderma, 151*, 257–263. https://doi.org/10.1016/j.geoderma.2009.04.016

Dauda, M., & Odoh, R. (2012). Heavy metals assessment of soil in the vicinity of fuel filling station in some selected local government areas of Benue State, Nigeria. *Pelagia Research Library, Der Chemica Sinica, 3*, 1329–1336.

de Miguel, E., Llamas, J. F., Chacón, E., Berg, T., Larssen, S., Røyset, O., & Vadset, M. (1997). Origin and patterns of distribution of trace elements in street dust: Unleaded petrol and urban lead. *Atmospheric Environment, 31*(17), 2733–2740. doi:10.1016/S1352-2310(97)00101-5.

Dong, J., Yu, M., Bian, Z., Wang, Y., & Di, C. (2011). Geostatistical analyses of heavy metal distribution in reclaimed mine land in Xuzhou, China. *Environmental Earth Sciences, 62*, 127–137. https://doi.org/10.1007/s12665-010-0507-5

Duong, T. T., & Lee, B.-K. (2011). Determining contamination level of heavy metals in road dust from busy traffic areas with different characteristics. *Journal of Environmental Management, 92*, 554–562. https://doi.org/10.1016/j.jenvman.2010.09.010

Duzgoren-Aydin, N. S., Wong, C. S. C., Song, Z. G., Aydin, A., Li, X. D., & You, M. (2006). Fate of heavy metal contaminants in road dusts and gully sediments in Guangzhou, SE China: A chemical and mineralogical assessment. *Human and Ecological Risk Assessment: An International Journal, 12*(2), 374–389. doi:10.1080/10807030500538005.

Edet, A., Merkel, B., & Offiong, O. (2003). Trace element hydrochemical assessment of the Calabar Coastal Plain Aquifer, southeastern Nigeria using statistical methods. *Environmental Geology, 44*, 137–149.

Ekperusi, O. A., & Aigbodion, I. F. (2015). Bioremediation of heavy metals and petroleum hydrocarbons in diesel contaminated soil with the earthworm: *Eudrilus eugeniae*. *SpringerPlus, 4*, 493. https://doi.org/10.1186/s40064-015-1328-5

Emmanuel, A., Cobbina, S., Adomako, D., Duwiejuah, A., & Asare, W. (2014). Assessment of heavy metals concentration in soils around oil filling and service stations in Tamale Metropolis, Ghana. *African Journal of Environmental Science and Technology, 8*, 256–266. https://doi.org/10.5897/AJEST

Eze, P. N., Udeigwe, T. K., & Stietiya, M. H. (2010). Distribution and potential source evaluation of heavy metals in prominent soils of Accra Plains, Ghana. *Geoderma, 156*, 357–362. https://doi.org/10.1016/j.geoderma.2010.02.032

Facchinelli, A., Sacchi, E., & Mallen, L. (2001). Multivariate statistical and GIS-based approach to identify heavy metal sources in soils. *Environmental Pollution, 114*, 313–324. https://doi.org/10.1016/S0269-7491(00)00243-8

Ghana Statistical Services. (2010). *2010 population and housing census: Summary report of final results*. Ghana: Accra.

Fergusson, J. E., Forbes, E. A., Schroeder, R. J., & Ryan, D. E. (1986). The elemental composition and sources of house dust and street dust. *Science of The Total Environment, 50*, 217–221. doi:10.1016/0048-9697(86)90363-3.

Glorennec, P., Lucas, J.-P., Mandin, C., & Le Bot, B. (2012). French children's exposure to metals via ingestion of indoor dust, outdoor playground dust and soil: Contamination data. *Environment International, 45*, 129–134. https://doi.org/10.1016/j.envint.2012.04.010

Guo, W., Huo, S., Xi, B., Zhang, J., & Wu, F. (2015). Heavy metal contamination in sediments from typical lakes in the five geographic regions of China: Distribution, bioavailability, and risk. *Ecological Engineering, 81*, 243–255. https://doi.org/10.1016/j.ecoleng.2015.04.047

Hakanson, L. (1980). An ecological risk index for aquatic pollution control. A sedimentological approach. *Water Research, 14*, 975–1001.

Han, L., Gao, B., Wei, X., Xu, D., & Gao, L. (2016). Spatial distribution, health risk assessment, and isotopic composition of lead contamination of street dusts in different functional areas of Beijing, China. *Environmental Science and Pollution Research, 23*, 3247. https://doi.org/10.1007/s11356-015-5535-y

He, B., Yun, Z., Shi, J., & Jiang, G. (2013) Research progress of heavy metal pollution in China: Sources, analytical methods, status, and toxicity. Chinese Science Bulletin, 1-7

Hu, X., Zhang, Y., Ding, Z., Wang, T., Lian, H., Sun, Y., & Wu, J. (2012). Bioaccessibility and health risk of arsenic and heavy metals (Cd Co, Cr, Cu, Ni, Pb, Zn and Mn) in TSP and PM2.5 in Nanjing, China. *Atmospheric Environment, 57*, 146–152. https://doi.org/10.1016/j.atmosenv.2012.04.056

Huamain, C., Chunrong, Z., Cong, T., & Yongguan, Z. (1999) Heavy metal pollution in soils in China: Status and countermeasures. Ambio, 130–134

Hussain, K., Rahman, M., Prakash, A., & Hoque, R. R. (2015). Street dust bound PAHs, carbon and heavy metals in Guwahati city–Seasonality, toxicity and sources. *Sustainable Cities and Society, 19*, 17–25. https://doi.org/10.1016/j.scs.2015.07.010

Islam, S., Ahmed, K., & Habibullah-Al-Mamun, S. (2015). Potential ecological risk of hazardous elements in different land-use urban soils of Bangladesh. *Science of the Total Environment, 512*, 94–102. https://doi.org/10.1016/j.scitotenv.2014.12.100

Kamani, H., Ashrafi, S. D., Isazadeh, S., Jaafari, J., Hoseini, M., Mostafapour, F. K., ... Mahvi, A. H. (2015). Heavy metal contamination in street dusts with various land uses in Zahedan, Iran. *Bulletin of Environmental Contamination and Toxicology, 94*, 382. https://doi.org/10.1007/s00128-014-1453-9

Karimi Nezhad, M. T., Tabatabaii, S. M., & Gholami, A. (2015). Geochemical assessment of steel smelter-impacted urban soils, Ahvaz. *Iran. Journal of Geochemical Exploration, 152*, 91–109. doi:10.1016/j.gexplo.2015.02.005.

Khairy, M. A., Barakat, A. O., Mostafa, A. R., & Wade, T. L. (2011). Multielement determination by flame atomic absorption of road dust samples in Delta Region, Egypt. *Microchemical Journal, 97*, 234–242. https://doi.org/10.1016/j.microc.2010.09.012

Khorshid, M. S. H., & Thiele-Bruhn, S. (2016). Contamination status and assessment of urban and non-urban soils in the region of Sulaimani City, Kurdistan, Iraq. *Environmental Earth Sciences, 16*, 1–15.

Kwaansa-Ansah, E. E., Agorku, S. E., & Nriagu, J. O. (2011). Levels of total mercury in different fish species and sediments from the Upper Volta Basin at Yeji in Ghana. *Bulletin of Environmental Contamination and Toxicology, 86*, 406–409. doi:10.1007/s00128-011-0214-2

Laidlaw, M. A., & Taylor, M. P. (2011). Potential for childhood lead poisoning in the inner cities of Australia due to exposure to lead in soil dust. *Environmental Pollution, 159*, 1–9. https://doi.org/10.1016/j.envpol.2010.08.020

Lee, J.-M., Qin, S. J., & Lee, I.-B. (2006). Fault detection and diagnosis based on modified independent component analysis. *Process Systems Engineering, 52*, 3501–3514. doi:10.1002/aic.10978.

Lee-Steere, C. (2009). *Environmental Risk Assessment Guidance Manual*. Environment Protection and Heritage Council. Australian Environment Agency. Commonwealth of Australia.

Li, F., Zhang, J., Huang, J., Huang, D., Yang, J., Song, Y., & Zeng, G. (2016). Heavy metals in road dust from Xiandao District, Changsha City, China: characteristics, health risk assessment, and integrated source identification. *Environmental Science and Pollution Research, 23*(13), 13100–13113. doi:10.1007/s11356-016-6458-y.

Li, Z., Ma, Z., van der Kuijp, T. J., Yuan, Z., & Huang, L. (2014). A review of soil heavy metal pollution from mines in China: Pollution and health risk assessment. *Science of the Total Environment, 468*, 843–853. https://doi.org/10.1016/j.scitotenv.2013.08.090

Li, X., Poon, C.-S., & Liu, P. S. (2001). Heavy metal contamination of urban soils and street dusts in Hong Kong. *Applied Geochemistry, 16*, 1361–1368. https://doi.org/10.1016/S0883-2927(01)00045-2

Lin, Y., Fang, F., Wang, F., & Xu, M. (2015). Pollution distribution and health risk assessment of heavy metals in indoor dust in Anhui rural, China. *Environmental Monitoring and Assessment, 187*, 1.

Ling, W., Shen, Q., Gao, Y., Gu, X., & Yang, Z. (2008). Use of bentonite to control the release of copper from contaminated soils. *Soil Research, 45*, 618–623.

Lu, X., Wang, L., Li, L. Y., Lei, K., Huang, L., & Kang, D. (2010). Multivariate statistical analysis of heavy metals in street dust of Baoji, NW China. *Journal of Hazardous Materials, 173*, 744–749. https://doi.org/10.1016/j.jhazmat.2009.09.001

Lu, X., Zhang, X., Li, L. Y., & Chen, H. (2014). Assessment of metals pollution and health risk in dust from nursery schools in Xi'an, China. *Environmental Research, 128*, 27–34. https://doi.org/10.1016/j.envres.2013.11.007

Luo, X.-S., Yu, S., & Zhu, Y.-G. (2012). Trace metal contamination in urban soils of China. *Science of the Total Environment, 421–422*, 17–30. https://doi.org/10.1016/j.scitotenv.2011.04.020

Ma, L., Yang, Z., Li, L., & Wang, L. (2016). Source identification and risk assessment of heavy metal contaminations in urban soils of Changsha, a mine-impacted city in Southern China. *Environmental Science and Pollution Research, 23*, 17058–17066. https://doi.org/10.1007/s11356-016-6890-z

Man, Y. B., Sun, X. L., Zhao, Y. G., Lopez, B. N., Chung, S. S., Wu, S. C., ... Wong, M. H. (2010). Health risk assessment of abandoned agricultural soils based on heavy metal contents in Hong Kong, the world's most populated city. *Environment International, 36*, 570–576. https://doi.org/10.1016/j.envint.2010.04.014

Manahan, S. (2000). *Environmenal chemistry* (p. 898). Boca Raton, FL: Lewis.

McLaughlin, M. J., Hamon, R., McLaren, R., Speir, T., & Rogers, S. (2000). Review: A bioavailability-based rationale for controlling metal and metalloid contamination of agricultural land in Australia and New Zealand. *Australian Journal of Soil Research, 38*, 1037–1086. https://doi.org/10.1071/SR99128

Muller, G. (1969) Index of Geoaccumulation in Sediments of the Rhine River. *GeoJournal, 2*, 108–118.

Nkansah, M. A., Opoku, F., & Ackumey, A. A. (2016). Risk assessment of mineral and heavy metal content of selected tea products from the Ghanaian market. *Environmental Monitoring and Assessment, 188*, 1–11. doi:10.1007/s10661-016-5343-y

Obiri, S. (2007). Determination of heavy metals in water from boreholes in Dumasi in the Wassa West District of Western Region of Republic of Ghana. *Environmental Monitoring and Assessment, 130*, 455. https://doi.org/10.1007/s10661-006-9435-y

Pan, L., Ma, J., Hu, Y., Su, B., Fang, G., Wang, Y., ... Xiang, B. (2016). Assessments of levels, potential ecological risk, and human health risk of heavy metals in the soils from a typical county in Shanxi Province, China. *Environmental Science and Pollution Research, 23*, 19330–19340. https://doi.org/10.1007/s11356-016-7044-z

Pant, P., & Harrison, R. M. (2013). Estimation of the contribution of road traffic emissions to particulate matter concentrations from field measurements: A review. *Atmospheric Environment, 77*, 78–97. https://doi.org/10.1016/j.atmosenv.2013.04.028

Peat, J., & Barton, B. (2008) *Medical statistics: A guide to data analysis and critical appraisal*. Hoboken, NJ: John Wiley & Sons.

Qing, X., Yutong, Z., & Shenggao, L. (2015). Assessment of heavy metal pollution and human health risk in urban soils of steel industrial city (Anshan), Liaoning, Northeast China. *Ecotoxicology and Environmental Safety, 120*, 377–385. https://doi.org/10.1016/j.ecoenv.2015.06.019

Qu, C., Sun, K., Wang, S., Huang, L., & Bi, J. (2012). Monte Carlo simulation-based health risk assessment of heavy metal soil pollution: A case study in the Qixia Mining Area, China. *Human and Ecological Risk Assessment: An International Journal, 18*, 733–750. https://doi.org/10.1080/10807039.2012.688697

Rasmussen, P. E., Subramanian, K. S., & Jessiman, B. J. (2001). A multi-element profile of house dust in relation to exterior dust and soils in the city of Ottawa. *Canada. Science of the Total Environment, 267*(1–3), 125–140. doi:10.1016/S0048-9697(00)00775-0.

Ruiz-Jimenez, J., Parshintsev, J., Laitinen, T., Hartonen, K., Petäjä, T., Kulmala, M., & Riekkola, M.-L. (2012). Influence of the sampling site, the season of the year, the particle size and the number of nucleation events on the chemical composition of atmospheric ultrafine and total suspended particles. *Atmospheric Environment, 49*, 60–68. https://doi.org/10.1016/j.atmosenv.2011.12.032

Salati, S., & Moore, F. (2010). Assessment of heavy metal concentration in the Khoshk River water and sediment, Shiraz, Southwest Iran. *Environmental Monitoring and Assessment, 164*, 677–689. https://doi.org/10.1007/s10661-009-0920-y

Schleicher, N. J., Norra, S., Chai, F., Chen, Y., Wang, S., Cen, K., ... Stüben, D. (2011). Temporal variability of trace metal mobility of urban particulate matter from Beijing–A contribution to health impact assessments of aerosols. *Atmospheric Environment, 45*, 7248–7265. https://doi.org/10.1016/j.atmosenv.2011.08.067

Sezgin, N., Ozcan, H. K., Demir, G., Nemlioglu, S., & Bayat, C. (2004). Determination of heavy metal concentrations in street dusts in Istanbul E-5 highway. *Environment*

International, 29, 979–985. https://doi.org/10.1016/S0160-4120(03)00075-8

Shen, F., Liao, R., Ali, A., Mahar, A., Guo, D., Li, R., ... Zhang, Z. (2017). Spatial distribution and risk assessment of heavy metals in soil near a Pb/Zn smelter in Feng County, China. *Ecotoxicology and Environmental Safety, 139*, 254–262. https://doi.org/10.1016/j.ecoenv.2017.01.044

Shi, G., Chen, Z., Xu, S., Zhang, J., Wang, L., Bi, C., & Teng, J. (2008). Potentially toxic metal contamination of urban soils and roadside dust in Shanghai. *China. Environmental Pollution, 156*(2), 251–260. doi:10.1016/j.envpol.2008.02.027.

Soltani, B., Keshavarzi, F., Moore, T., Tavakol, A. R., Lahijanzadeh, N., & Jaafarzadeh, M. K. (2015). Ecological and human health hazards of heavy metals and polycyclic aromatic hydrocarbons (PAHs) in road dust of Isfahan metropolis, Iran. *Science of the Total Environment, 505*, 712–723. https://doi.org/10.1016/j.scitotenv.2014.09.097

Steinskog, D. J., Tjøstheim, D. B., & Kvamstø, N. G. (2007). A cautionary note on the use of the Kolmogorov–Smirnov test for normality. *Monthly Weather Review, 135*, 1151–1157. https://doi.org/10.1175/MWR3326.1

Taylor, S. R., & McLennan, S. M. (1995). The geochemical evolution of the continental crust. *Reviews of Geophysics, 33*, 241–265. https://doi.org/10.1029/95RG00262

Teng, Y., Wu, J., Lu, S., Wang, Y., Jiao, X., & Song, L. (2014). Soil and soil environmental quality monitoring in China: A review. *Environment International, 69*, 177–199. https://doi.org/10.1016/j.envint.2014.04.014

USEPA. (1986). *Guidelines for the health risk assessment of chemical mixtures.* Washington, DC: US Environmental Protection Agency.

USEPA. (1989) *Risk assessment guidance for Superfund. Human health evaluation manual, (part A)* (Vol. 1). Washington, DC: Office of emergency and remedial response.

USEPA. (1993). Assessing dermal exposure from soil: Region 3 technical guidance manual, risk assessment. United States Environmental PROTECTION agency. Hazardous Waste Management Division Office of Superfund Programs. Philadelphia.Retrieved from https://www.epa.gov/risk/assessing-dermal-exposure-soil

USEPA. (2002). *Supplemental guidance for developing soil screening levels for Superfund sites.* Washington, DC: Soild Waste and Emergency Response.

USEPA. (2003). *Example exposure scenarios National Center for Environmental Assessment.* Washington, DC. EPA/600/R-03/036. National Information Service,

Springfield, VA; PB2003-103280. Retrieved from http://www.epa.gov/ncea.

USEPA. (2011). *Exposure factors handbook. National Center for Environmental Assessment.* Washington, DC: United State Environmental Protection Agency.

Wang, M., Markert, B., Chen, W., Peng, C., & Ouyang, Z. (2012). Identification of heavy metal pollutants using multivariate analysis and effects of land uses on their accumulation in urban soils in Beijing, China. *Environmental Monitoring and Assessment, 184*, 5889–5897. https://doi.org/10.1007/s10661-011-2388-9

Wei, B., Jiang, F., Li, X., & Mu, S. (2009). Spatial distribution and contamination assessment of heavy metals in urban road dusts from Urumqi. *NW China. Microchemical Journal, 93*(2), 147–152. doi:10.1016/j.microc.2009.06.001.

Wei, B., & Yang, L. (2010). A review of heavy metal contaminations in urban soils, urban road dusts and agricultural soils from China. *Microchemical Journal, 94*, 99–107. https://doi.org/10.1016/j.microc.2009.09.014

Ying, L., Shaogang, L., & Xiaoyang, C. (2016). Assessment of heavy metal pollution and human health risk in urban soils of a coal mining city in East China. *Human and Ecological Risk Assessment: An International Journal, 22*, 1359–1374. https://doi.org/10.1080/10807039.2016.1174924

Zhang, M. & Wang, H. (2009). Concentrations and chemical forms of potentially toxic metals in road-deposited sediments from different zones of Hangzhou. *China. Journal of Environmental Sciences, 21*(5), 625–631. doi:10.1016/S1001-0742(08)62317-7.

Zhang, X., Yang, L., Li, Y., Li, H., Wang, W., & Ye, B. (2012). Impacts of lead/zinc mining and smelting on the environment and human health in China. *Environmental Monitoring and Assessment, 184*, 2261–2273. doi:10.1007/s10661-011-2115-6

Zhou, Q., & Song, Y. (2004) *Remediation of contaminated soils: Principles and methods* (pp. 1–489). Beijing: Science.

Zhou, L., Yang, B., Xue, N., Li, F., Seip, H. M., Cong, X., ... Li, H. (2014). Ecological risks and potential sources of heavy metals in agricultural soils from Huanghuai Plain, China. *Environmental Science and Pollution Research, 21*, 1360–1369. https://doi.org/10.1007/s11356-013-2023-0

Zota, A. R., Schaider, L. A., Ettinger, A. S., Wright, R. O., Shine, J. P., & Spengler, J. D. (2011). Metal sources and exposures in the homes of young children living near a mining-impacted Superfund site. *Journal of Exposure Science and Environmental Epidemiology, 21*, 495. https://doi.org/10.1038/jes.2011.21

11

Heavy metals concentration and distribution in soils and vegetation at Korle Lagoon area in Accra, Ghana

Benedicta Yayra Fosu-Mensah[1]*, Emmanuel Addae[1], Dzidzo Yirenya-Tawiah[1] and Frank Nyame[2]

*Corresponding author: Benedicta Yayra Fosu-Mensah, Institute for Environment and Sanitation Studies (IESS), University of Ghana, P.O. Box 209, Legon, Accra, Ghana

E-mail: yayramensah@staff.ug.edu.gh

Reviewing editor: Peter Fantke, Technical University of Denmark (DTU), Denmark

Abstract: The call for reclamation of land around Korle Lagoon in Accra, Ghana, where burning of e-waste and cultivation of vegetables takes place, makes risk assessment of heavy metal contaminations important. This study aimed at evaluating the lev-els and risk of heavy metal contamination in soils and vegetation around the Korle Lagoon area in Accra. Geoaccumulation index, enrichment factor and pollution load index were determined to assess the risk of contamination. The levels and distribution of nine heavy metals (Pb, Hg, Cd, As, Zn, Sn, Ni, Cu and Cr) in soil (0–20 cm) and com-mon vegetation (*Panicum maximum, Imperata cylindrica, Lactuca sativa* and *Hibiscus sabdariffa*) from the area using Atomic Absorption Spectrometer (AAS) were assessed. The area was divided into five sites, namely; the e-waste site (S1), gardens area (S2), recreational area (S3), reclaimed area (S4) estuary (S5) and the control (S6) which was about 700 m away. Soil analysis showed that the concentration of Pb (184.44 mg/kg), Cd (103.66 mg/kg), Cu (202.99 mg/kg), Ni (72.00 mg/kg) and Sn (705.32 mg/kg) at S1 exceeded their WHO/FAO thresholds for agricultural soils. Concentrations of heavy metals in soils from the e-waste site were significantly different ($p < 0.01$) from the other sites. High accumulations of heavy metals were also observed in the plants samples collected from the study sites, with the concentrations of Cu, Pb, Ni and Cd

ABOUT THE AUTHORS

Benedicta Yayra Fosu-Mensah is a research fellow of the Institute for Environment and Sanitation Studies (IESS), University of Ghana. She is an Agricultural and Natural Resource scientist (modeller). She has experience in crop modelling, plant nutrition, site-specific nutrient management, climate change impact assessment and environmental pollution assessment.

Emmanuel Addae is an MPhil student at the Institute for Environment and Sanitation Studies (IESS).

Dzidzo Yirenya-Tawiah is a research fellow of the Institute for Environment and Sanitation Studies, University of Ghana. She has expertise in environmental and public health research, community engagement and sanitation issues.

Frank Nyame is an associate professor at the Department of Earth, University of Ghana.

PUBLIC INTEREST STATEMENT

The reclamation of land around the Korle Lagoon in Accra, where burning of e-waste and cultivation of vegetables take place is a national issue in Ghana. The recycling of electronic waste (e-waste) around the Lagoon and the cultivation of vegetables at the adjoining area makes risk assessment of heavy metal contaminations of the area highly important. The open burning of e-waste produce fumes which contain heavy metals and other polyaromatic compounds which end up in soils and vegetation. Heavy metals in food and water in relatively high concentration has serious health implications such as injury to the kidney or kidney dysfunction, among others. Concentrations of heavy metals in soils from the e-waste site were significantly (p < 0.01) higher than the other sites. High accumulations of heavy metals were observed in the plants samples collected, with the concentrations of Cu, Pb, Ni and Cd exceeding their acceptable limits exceeding their acceptable limits. Laws against open burning of e-waste should be enforced and animals should be restricted from grazing on the forage.

Subjects: Environmental Sciences; Agriculture and Food; Soil Sciences; Environmental Impact Assessment; Environment & Health

Keywords: heavy metals; e-waste; pollution; environment; health risks

1. Introduction

The fast technological change occurring with increase in demand for information technology has resulted in large volumes of e-waste globally (Wagner, 2009). e-waste contains many different substances that are classified as hazardous (Cd, Pb, As, Be, Cr, Hg, polycyclic aromatic hydrocarbons (PAHs), chlorofluorocarbons, polybrominated diphenyl ethers and dioxin-like compounds) and non-hazardous (Zn, Cu, Se, precious metals Ag, Au, Pt, among others.) which have significant harmful environmental impacts when not well disposed off (Lim, 2010; Tsydenova & Bengtsson, 2011). However, some non-hazardous chemicals such as Zn, Cu and Se can become hazardous to humans and ecosystems depending on the dose and concentrations, and the target receptors (Itai et al., 2014; Musee, 2011; Tchounwou, Yedjou, Patlolla, & Sutton, 2012).

In West Africa, the informal recycling of e-waste is becoming an emerging problem due to the high use of electric and electronic equipment (EEE) imported from the developed world (Feldt et al., 2014). Currently, West Africa serves as the major route in Africa, with Ghana and Nigeria serving as the main import hubs. It is estimated that about 215, 000 tons of new and used EEE were imported into Ghana and 129, 000 tons were generated in 2009 alone (e-waste Africa Programme, 2011). The methods mostly used in the recycling of these EEE include melting of electronic boards on open fire to extract metals and valuable chips, burning of cable wires to recover copper and open burning of residual waste materials (Feldt et al., 2014). These methods result in the emission of high concentration of toxins into the atmosphere there by polluting the environment, endangering the health of recycling workers and residents in the surrounding communities. Apart from a number of heavy metals and variety of organic toxins, like polychorinated biphenyls (PCBs) and polybrominated dipenyls ethers (PBDEs), research has shown that PAHs constitute an important emission from informal e-waste recycling (Asante et al., 2012; Ma et al., 2009; Wang et al., 2010, 2012).

The Agbogbloshie slum is one of Ghana's largest e-waste dump sites. Obsolete electronic gadgets of no perceived value are disposed off in a large area on the edge of the e-waste market (also located in the Agbogboloshie slum) that is also used for the disposal of a wide range of other types of waste (Brigden, Labunska, Santillo, & Johnston, 2008). The north-eastern part of Agbogbloshie in Accra is where Korle Lagoon is located, with the reclaimed site few metres away. Workers involved in the e-waste recycling, use open fire to extract metals, valuable chips and copper from electronic gadgets as well as cable wires. In addition, residents burn discarded electronic gadgets at a section of the Korle Lagoon reclamation area to extract copper and other valuable metals. The fumes originating from the burning of the e-waste materials might contain heavy metals and other polyaromatic compounds which eventually end up in soils and vegetation of the reclaimed area. Furthermore, herds of cattle from neighbouring communities graze on the vegetation of the reclaimed waste land exposing them to heavy metals. Adjacent to the reclaimed area is a former dump site, which has been converted to an active vegetable garden by the urban poor in the vicinity. These vegetables and vegetation are likely to take up any heavy metals in the soil, hence, engendering the health of the consumers. Research shows that the presence of heavy metals in food and water in relatively high concentration has serious health implications. For example, mercury and cadmium are known to cause injury to the kidney or kidney dysfunction among others. The negative effect of heavy metal exposure in the environment, either through inhalation or ingestion of contaminated food is thus a major concern to researchers, governments and the general public because of its profound effects on humans and wildlife. Little is, however, known about the extent of heavy metal pollution of soil and vegetation in the study area. This paper aims to assess the level of heavy metal pollution in soils and vegetation around the Korle Lagoon reclamation area in Accra.

2. Materials and methods

2.1. Study area

This study was conducted at the Korle Lagoon area in Accra, Ghana (Figure 1). Korle Lagoon is a coastal wetland comprised of beautiful sand dunes, open lagoon, salt pans, marsh and scrubs. The lagoon also provides roosting and nesting grounds for various species of seabirds. It is 104 m above sea level with its coordinates at 5°33′0″ N and 0°13′0″ E. Accra is a coastal savannah zone which experiences bimodal rainfall. The average annual rainfall is about 730 mm, and ranges between 635 mm and 1,140 mm. The mean temperature is about 28°C with a minimum and maximum temperature of 25 and 35°C, respectively.

2.2. Description of the sampling sites

The study area was divided into five (5) sites depending on the activity that was carried out during the time of the research. These comprised of e-waste area (S1) where burning of e-waste, collected along the length and breadth of the capital are brought for burning in order to retrieve essential components for sale abroad; site 2 was the garden area (S2), where vegetables and plants are grown; site 3 was a reclaimed waste dump site (S3), site 4 a recreational site (S4) and site 5 is the estuary area (S5) of the Korle Lagoon. Each site was about 300 m apart. The control (S6) was 700 m away.

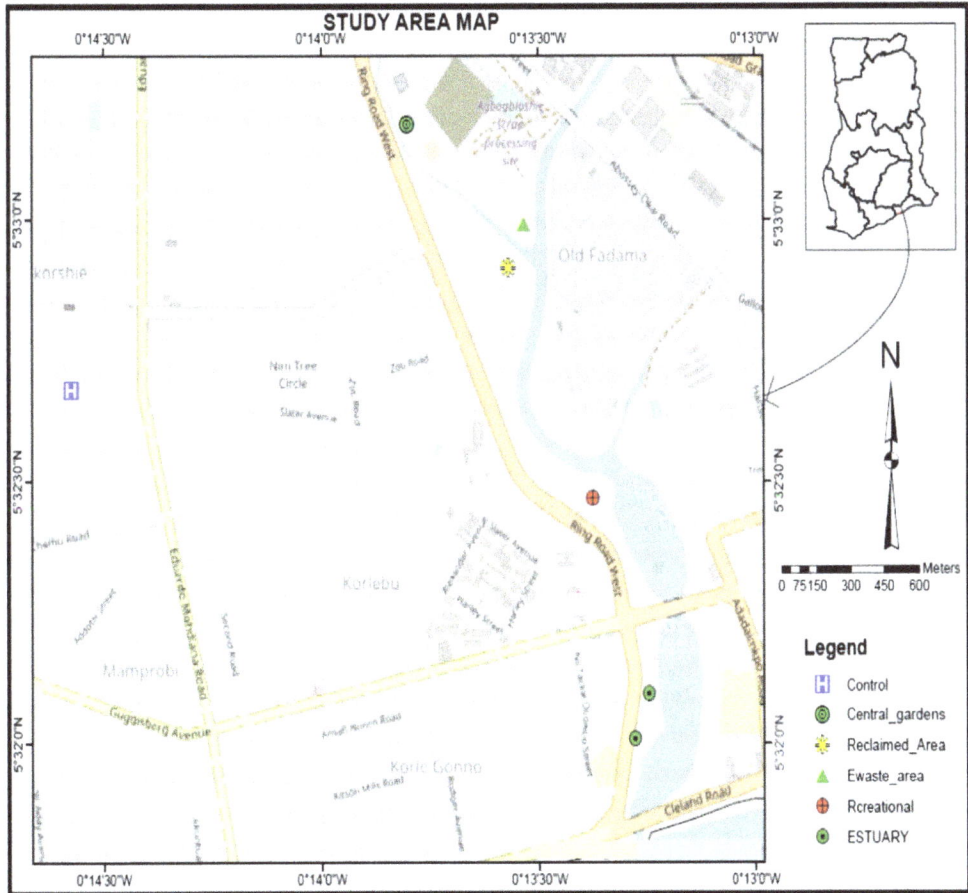

Figure 1. Map of the study area showing sampling sites.

2.3. Sample collection

In each of the selected sites, six quadrats were marked. In each quadrat, four core soil samples were collected randomly at depth 0–20 cm using a soil auger and put together to form a composite sample. The composite samples were well-mixed and sub-samples taken (Qin et al., 2014). Six soil replicates were collected from each site, making a total of 36 soil samples from the study area. The soil samples were kept in zip-locked plastic bags, labelled and transported to the laboratory for further treatment and analysis. The soil samples were air-dried at room temperature for 4 days, and ground to fine powder. They were then sieved with a 2 mm sieve to remove the coarse soil components. Sub-samples of the sieved soils were then taken for soil physicochemical and heavy metals analysis.

In addition, twelve (12) samples each of dominant plant species in each of the sites were collected for analysis. The plants samples collected were lettuce (*Panicum maximum, Imperata cylindrica, Lactuca sativa* and *Hibiscus sabdariffa*). The plant samples were obtained by cutting at a height of 5 cm above the soil surface. The plant samples were freeze-dried using the CHRIST Getriertrocknugsanlagen GmbH Freeze dryer at a temperature of 17°C with a vacuum mbar of 6.110.

2.4. Analysis of soil properties

The soil properties analysed were soil particle size distribution, pH, organic carbon, organic matter and exchangeable cations (K^+, Ca^{2+}, Na^+ and Mg^{2+}). These characteristics were determined using the methods described by Wahabu, Fosu-Mensah, and Nyame (2015) and Fosu-Mensah, Okoffo, Darko, and Gordon (2016). The amount of soil organic matter (SOM) was examined by multiplying the percentage carbon by the factor 1.724 (Walkley & Black, 1934). To determine exchangeable sodium (Na^+), potassium (K^+), Ca^{2+} and Mg^{2+} the methods described by Wahabu et al. (2015) were followed.

2.5. Determination of heavy metals in soil and plant samples

Acid digestion of soil sample was done by weighing 1.5 g of soil sample into a well-labelled 100 mL polytetraflouroethylene (PTFE) Teflon bombs which was previously acid washed. Six millilitre of 65% nitric acid (HNO_3), 3 mL of 35% hydrochloric acid (HCl) and 0.25 mL of 30% hydrogen peroxide (H_2O_2) was added to each sample in a fume chamber. The samples were then loaded onto microwave carousel where they were irradiated for 26 min using milestone microwave lab station ETHOS 900, INSTR: MLS-1200 MEGA (Adama, Esena, Fosu-Mensah, & Yirenya-Tawiah, 2016; Tiimub & Dartey, 2015). Similarly, a 0.5 g plant sample was weighed in an acid-washed Teflon bomb-labelled "100 mL polytetraflouroethylene (PTFE)" and a similar procedure followed. Blanks were prepared to check for background contamination by the reagents used. The digested soil and plant samples were analysed for the heavy metals (Cd, As, Pb, Cu, Ni, Sn, Hg, Cr and Zn) using atomic absorption spectrophotometer (AA 240 FS, Varian) (Kumar, Singh, & Garg, 2012). All the readings were taken in five replicates.

2.6. Methods of assessment of contamination in soils

In this study, geoaccumulation index (I_{geo}), enrichment factor (EF) and pollution load (PL) were calculated to assess the heavy metal contamination levels in soils from the study area.

2.6.1. Geoaccumulation index (I_{geo})

The index of geoaccumulation (I_{geo}) is widely used in the assessment of contamination by comparing the levels of heavy metals obtained to background levels originally used with bottom sediments (Atiemo et al., 2011; Muller, 1969). It is calculated using the equation:

$$I_{geo} = \log_2 \left(\frac{C_n}{1.5B_n} \right)$$

where C_n represents the measured concentration of the elements studied and B_n is the geochemical background value of the element in fossil argillaceous sediment (average shale) (Taylor & McLennan, 1985). The following classification is given for geoaccumulation index: <0 = practically unpolluted,

0-1 = unpolluted to moderately polluted, 1–2 = moderately polluted, 2–3 = moderately to strongly polluted, 3–4 = strongly polluted, 4–5 = strongly to extremely polluted and > 5 = extremely polluted (Lu, Wang, Lei, Huang, & Zhai, 2009).

2.6.2. Enrichment factor

Enrichment factor (EF) of an element in the studied samples was based on the standardization of measured element against a reference element. Al was used as the reference element in this study. The EF is calculated using the equation:

$$EF_x = \frac{[C_x/C_{ref}]\,sample}{[B_x/B_{ref}]\,Background}$$

where C_x is the concentration of the element of interest and C_{ref} is the concentration of reference element for normalization, B_x is the concentration of the element in the crust and B_{ref} is the concentration of the reference element used for normalization in the crust (Ato et al., 2010; Çevik, Göksu, Derici, & Fındık, 2009). Five contamination categories are assigned on the basis of the enrichment factor: EF < 2 = deficiency to minimal enrichment; EF = 2–5 = moderate enrichment; EF = 5–20 = significant enrichment; EF = 20–40 = very high enrichment; and EF > 40 = extremely high enrichment (Yongming, Peixuan, Junji, & Posmentier, 2006).

2.6.3. Pollution load index

The pollution index (PI) is defined as the ratio of element concentration in the study to the background content of the abundance of chemical elements in the continental crust. PI and integrated PI are also commonly used to assess environments quality (dos Anjos et al., 2000). PLI for the soil samples was determined by the equation below, as proposed by Tomilson, Wilson, Harris, and Jeffrey (1980).

$$PLI = (CF_n \times CF_n \times CF_n \times CF_n \times CF_n \times CF_n)^{1/n}.$$

The PI of each element is classified as either low (PI ≤ 1), middle (1 < PI ≤ 3) or high (PI > 3) (Chen et al., 2005).

2.7. Data analysis

The data were analysed using SPSS 16.0. The Tukey test was used for pairwise comparison of means to identify significant differences of heavy metals among the sites.

3. Results and discussion

3.1. Physicochemical properties of soil samples from the study area

Table 1 presents the summary of the physicochemical properties of soil samples from the study sites. The pH of the study sites ranged from 4.67 at S3 to 7.74 at S5 with a mean value of 6.93. The

Table 1. Summary of physicochemical characteristics of soils at different zones in Korle Lagoon area in Accra

Sites	S1	S2	S3	S4	S5	S6	Total means
pH	6.72	7.69	4.67	7.47	7.45	7.27	6.93
%OC	0.85	0.48	0.44	0.42	0.38	0.42	0.50
%OM	2.51	1.42	1.40	1.14	1.13	1.25	1.48
K^+	1.52	0.78	1.52	1.32	0.71	0.58	1.07
Na^+	6.15	2.13	6.07	3.69	3.12	2.01	3.86
Ca^{2+}	97.87	11.70	15.46	18.45	30.31	10.13	30.65
Mg^{2+}	73.40	8.80	11.60	14.00	23.00	7.60	23.07

Note: S1 = e-waste area, S2 = garden area, S3 = a reclaimed waste dump site, S4 = a recreational site, S5 = the estuary area of the Korle Lagoon and S6 = control.

low pH (4.67) at site S3 might probably be due to the dumping of acid-containing waste materials like batteries on the site as the site was once a dumping site. The OC content of the soils ranged from 0.38% at S5 to 0.85% at S1 with a mean value of 0.50%. Similarly, soil organic ranged from 1.13% at S5 to 2.51% at S1 with a mean value of 1.48%. The exchangeable K^+ ranged from 0.58 cmol/kg at S6 to 1.52 cmol/kg at S1 and S3 with a mean value of 1.07 cmol/kg. In addition, the exchangeable Na^+ content of the soils analysed ranged from 2.01 cmol/kg at S6 to 6.15 cmol/kg at S1 with a mean of 3.86 cmol/kg. The exchangeable Ca^{2+} values ranged from 10.13 cmol/kg at S6 to 97.87 cmol/kg at S1 with a mean value of 30.65 cmol/kg. Site S6 recorded the lowest value (7.60 cmol/kg) of exchangeable Mg^{2+} with the highest (73.4 cmol/kg) recorded at S1.

3.2. Heavy metal concentration of soils at the different sites in the study area

Table 2 presents the summary of the mean concentrations of heavy metals (Hg Pb, Cu, Zn, Cr, Cd, Ni, Sn and As) analysed in the soil samples at the various sampling sites at Korle Lagoon area in the Greater Accra region of Ghana. Analysis of variance (ANOVA) revealed a significant ($p < 0.05$) variation in the concentrations of the nine (9) elements among the sites (Table 2) which is an indication of the extent of metal pollution in the soils. The Tukey test revealed that the concentrations of most of the elements were significantly ($p < 0.01$) higher at the e-waste site (S1) compared to the other sites (Table 2). Site 2 generally had the lowest concentrations of most of the metals analysed except for Cr which was higher.

Mercury (Hg) was the least abundant metal recorded in the study area, whereas Sn was the highest among the metals with a mean value of 705.32 mg/kg. Contrary to this finding, Xianjin et al. (2010) reported high concentration of Hg (654.1 mg kg^{-1}) and Sn (660.8 mg kg^{-1}) in simple household e-waste recycling workshop and large-scale e-waste recycling plants, respectively. Mercury in the soil samples analysed ranged from <0.001 mg/kg at S2, S5 and S6 to 0.667 mg/kg at S1. The mean concentration of mercury recorded at the different sites was below the WHO/FAO (2001) limit of 2.00 mg/kg for soils. The low concentration of Hg can be attributed to the fact that Hg easily evaporates into organo-mercury forms (Environmental Health & Safety Manual, 2000).

The concentrations of lead (Pb) in soil samples analysed ranged from 1.28 to 184.44 mg/kg with a mean value of 37.12 mg/kg. The e-waste site (S1) recorded the highest mean concentration of Pb, whereas the garden area (S2) recorded the lowest mean concentration. ANOVA showed significant differences ($p < 0.05$) in the mean concentration of Pb among the sites. The mean concentration of Pb at the e-waste zone (S1) was above the WHO/FAO (2001) permissible limit of 50.00 mg/kg for soils. The higher concentration of Pb recorded at the e-waste site (S1) could be attributed to the

Table 2. Summary of mean of heavy metal concentration of soils within 0–20cm depth in different sites in Korle Lagoon area in Accra

Heavy metals	S1	S2	S3 (mg/kg)	S4	S5	S6
Hg	0.67(0.021)a	ND	ND	0.04(0.006)b	0.003(0.003)b	0.03(0.002)b
Pb	183.66(0.02)a	1.28(0.02)bc	4.39(0.01)b	17.41(0.06)bd	12.65(0.06)b	3.33(0.06)b
Cu	202.99(1.27)a	3.47(0.05)bc	5.20(0.06)be	10.9(1.10)bd	6.39(0.06)bf	3.84(0.06)bg
Zn	37.33(0.58)a	0.83(0.05)bf	2.05(0.03)bc	6.33(0.01)bd	4.51(0.05)be	1.03(0.01)bg
Cr	56.0(1.15)a	3.03(0.10)b	2.28(0.58)b	2.36(0.06)b	20.99(0.06)b	2.63(0.17)b
Cd	103.66(1.73)a	3.55(0.58)b	ND	0.23(0.06)b	0.04(0.01)bc	4.35(0.06)bd
Ni	72(1.16)a	0.91(0.05)b	1.09(0.03)b	4.56(0.12)bc	1.59(0.01)b	1.48(0.003)b
Sn	704.87(2.71)a	27.39(0.86)b	8.77(0.112)bc	16.48(0.23)bd	14.85(0.20)be	24.71(0.12)bf
As	3.67(0.06)a	0.08(0.01)bc	0.47(0.06)bd	1.41(0.00)be	0.88(0.06)bf	0.04(0.012)b

Notes: Values in the parentheses are the standard error, S1 = e-waste area, S2 = garden area, S3 = a reclaimed waste dump site, S4 = a recreational site, S5 = the estuary area of the Korle Lagoon and S6 = control, SE = Standard Error, and nd = Not Detected. Means followed by a different letter(s) in the same row differ significantly ($p = 0.05$) according to Tukey's Multiple Range Test.

burning of e-waste (such as refrigerator, used computers, cables, printers, photocopy machines, automobile tires, batteries, air condition among others). The mean value of Pb recorded in this study was, however, lower than the mean value of 2,645.31 mg/kg reported by Pradhan and Kumar (2014) in an e-waste recycling site soil.

The mean concentration of copper recorded from the various sites ranged from 3.47 mg/kg at S2 to 202.99 mg/kg at S1with a mean value of 38.80 mg/kg. The mean concentrations of copper recorded were below the WHO/FAO (2001) permissible limit of 100 mg/kg for soils except at S1 (202.99 mg/kg) which was above the WHO/FAO (2001) permissible limit. The high concentration of Cu at S1 could be attributed to the burning of electronic gadgets as indicated earlier. This result is similar to the findings of Zhang, Wu, and Simonnot (2012) which recorded extremely high concentrations of copper at e-waste sites, which was beyond the acceptable agricultural soils limits of 50 mg/kg in China.

The mean concentration of zinc (Zn) ranged from 0.83 mg/kg at S2 to 37.33 mg/kg at S5 with a mean value of 8.68 mg/kg. The maximum mean concentration of Zn recorded at the e-waste zone (S1) was below the WHO/FAO (2001) permissible limit of 300.00 mg/kg for soils. The presence of zinc in soil at the various sites could be attributed to the occurrence of dry cells in the municipal waste as reported by Thorpe and Harrison (2008) and the burning of e-waste materials in the area as indicated earlier. In a similar study, Li, Bai, et al. (2011) reported a mean value of 3,500 mg/kg for Zn in e-waste and municipal solid waste dump site in south China. Zinc is an essential microelement which plays a very essential catalytic role in enzyme reactions but its content varies with the type of soil (Knezevic, Stankovic, Krstic, Nikolic, & Vilotic, 2009). High concentration of Zn can, however, pose health threats to humans.

The concentration of chromium Cr ranged from 2.28 mg/kg at S3 to 56.00 mg/kg at S1 with a mean value of 11.55 mg/kg. The high concentration of Cr recorded at S1 could be as a result of the recycling of e-waste such as refrigerator, used computers, cables, printers, photocopy machines, automobile tires and batteries at the site.

Similarly, the concentration of Cd ranged from <0.001 mg/kg at S3 to 103.66 mg/kg at S1 with a mean value of 18.64 mg/kg. The mean concentrations of Cd recorded at all sites were above the WHO/FAO (2001) permissible limit of 3 mg/kg for soils except S3, S4 and S6 which recorded values that were below the WHO/FAO permissible limit. Cadmium is very much connected with non-residual fractions of heavy metals and thus makes them mobile and potentially bio-available for uptake by plants (Zhang et al., 2009).

The mean concentrations of nickel recorded in the soil samples ranged from 0.91 mg/kg at S2 to 72.00 mg/kg at S1 with a mean value of 13.60 mg/kg. The mean concentration of Ni recorded at the various sites were below the WHO/FAO (2001) permissible limit of 50 mg/kg for soils except for S1 which recorded a mean value that was above the permissible limit.

On the other hand, the mean concentration of Sn at the sites ranged from 8.77 mg/kg at S3 to 705.32 mg/kg at S1 with a mean value of 132.86 mg/kg. The mean concentration of Sn obtained at S1 was above the FAO/WHO (1984) permissible value for Sn. The concentration of arsenic (As) recorded in soils ranged from 0.04 mg/kg at (S6) to 3.67 mg/kg at (S1) with mean value of 1.09 mg/kg. The mean concentrations of As at all sites were below the WHO/FAO (2001) permissible limit of 20.00 mg/kg for agricultural soils. This result is in line with the finding of Pradhan and Kumar (2014) who reported the highest mean value of 17.08 mg/kg for as in e-waste dump site in South China.

In general, the e-waste site (S1) recorded the highest concentration of heavy metals in soil samples as compared to the other sites. The burning of the e-waste materials results in the emission of toxins such as heavy metal in high concentration, thus exposing the recycle workers and the

Table 3. Comparison of heavy metal concentration in e-waste dumpsites of some cities of the world with data from e-waste site at the Korle Lagoon area in Accra

Cities	Hg	Pb	Cu	Zn	Cr	Cd	Ni	Sn	As	References
Wenling	0.47	187.30	180.66	343.19	101.29	3.00	48.97	–	–	Tang et al. (2009)
Ibadan	–	427.2	938.2	10,670	72.96	–	34.78	–	–	Timothy and Olajumoke (2014)
Lagos	–	209	–	262.2	27.05	28.032	64.17	–	–	Ofudje et al. (2014)
Guiyu, Guangdong Province	0.21	150	4,800	330	2,600	1.21	480	–	26.03	Li, Duan, et al. (2011)
Guiyu	–	222.8	684.1	572.8	–	1.36	278.4	3,472	–	Quan et al. (2015)
Bangalore, India	<0.05	126	429	192	54	0.478	–	–	–	Ha et al. (2009)
Accra	0.65	183.66	203.0	37.3	56	103.7	72.0	705.3	3.67	This research

communities around to health risk. Li, Duan, and Shi (2011) reported that high concentration of heavy metals is recorded at pollution source.

Table 3 presents comparison of heavy metal concentrations in e-waste dump sites of some cities of the world with data from e-waste site at the Korle Lagoon area in Accra. From the results, it was observed that heavy metal such as Hg, Pb, Zn, Cr, As and Ni were the most prominent heavy metals in the e-waste recycling site compared to Sn and Cu. In the current report, the mean concentration of Pb, Zn, Cr and Ni at the e-waste zone were below the concentration levels of metals reported in other literature (Ha et al., 2009; Li, Duan, et al., 2011; Ofudje, Alayande, Oladipo, Williams, & Akiode, 2014; Tang et al., 2009; Timothy & Olajumoke, 2014; Quan et al., 2015). However, the average concentration of Cd (103.66 mg kg^{-1}) was higher than those (0.478) reported by Ha et al. (2009) in Bangalore, India and 1.21 mg kg^{-1} in Guiyu in Guangdong province by Li, Duan, et al. (2011). The average mean (0.65 mg kg^{-1}) of Hg was higher than the value reported by Ha et al. (2009) in Bangalore, India. While Cd was higher than those reported by Tang et al. (2009), Li, Duan, et al. (2011), Ofudje et al. (2014), Quan et al. (2015).

3.3. Index of geoaccumulation

Soil quality was measured using the I_{geo} index of classification proposed by Muller (1981) (Table 4). The results of I_{geo} analysis indicated that, S1 and S2 were practically uncontaminated by Hg, Zn, Cr, Ni and As, while S1 was strongly polluted by Pb and extremely polluted by Cd and Sn which are the major constituents of e-waste (Pradhan & Kumar, 2014). In S3, S4, S5 and S6, only Sn showed moderate contamination with S3 showing strong contamination. The rest of the elements exhibited no pollution at the various sites. This might be due to the absence of e-waste activity and horizontal spread of metal pollutants in these areas.

Table 4. Geoaccumulation index (I_{geo})

Heavy metals	S1	S2	S3	S4	S5	S6
Hg	−0.60	ND	−5.06	−4.67	ND	ND
Pb	3.30	−3.87	−2.49	−0.11	−0.57	−2.10
Cu	1.30	−4.57	−4.43	−2.93	−3.69	−3.99
Zn	−1.49	−7.00	−6.67	−4.05	−4.54	−5.68
Cr	−1.42	−5.63	−5.84	−5.99	−5.65	−6.14
Cd	8.43	3.56	3.86	−0.51	−2.94	ND
Ni	−0.64	−6.95	−6.25	−4.63	−6.14	−6.70
Sn	7.88	3.19	3.04	2.46	2.31	1.55
As	0.44	−5.08	−6.22	−0.94	−1.62	−2.55

Notes: S1 = e-waste area, S2 = garden area, S3 = a reclaimed waste dump site, S4 = a recreational site, S5 = the estuary area of the Korle Lagoon and S6 = control.

Sites	S1	S2	S3	S4	S5	S6	PLI
Hg	848.72	ND	38.58	51.44	ND	ND	1.534
Pb	9.71	0.07	0.18	0.92	0.67	0.23	2.970
Cu	0.55	0.01	0.01	0.03	0.02	0.01	0.705
Zn	0.06	0.00	0.00	0.01	0.01	0.00	0.124
Cr	0.05	0.00	0.00	0.00	0.00	0.00	0.115
Cd	21,328.05	730.41	895.01	47.32	8.23	ND	93.18
Ni	0.11	0.00	0.00	0.01	0.00	0.00	0.181
Sn	1,450.26	56.35	50.84	33.91	30.55	18.04	66.43
As	9.32	0.20	0.10	3.58	2.24	1.19	0.606

Table 5. Enrichment factor (EF) of heavy metals for soil at different zones in Korle Lagoon area in Accra

Notes: S1 = e-waste area, S2 = garden area, S3 = a reclaimed waste dump site, S4 = a recreational site, S5 = the estuary area of the Korle Lagoon and S6 = control, PLI = pollution load index, nd = Not Detected.

3.4. Enrichment factor (EF) and pollution load index (PLI)

Table 5 presents results of enrichment Factor (EF) and pollution load index of all the heavy metals analysed in the sampled soil at the different sites. From Table 5, S1 exhibited extremely high enrichment of Hg, Cd and Sn, while the same site also showed significant enrichment of Pb and As. The EF for S2, S3, S4, S5 and S6 showed deficiency to minimal enrichment of Pb, Cu, Zn, Cr and Ni, while S2 showed extremely high enrichment for Cd and Sn. Similarly, S3 showed very high enrichment of Hg and extremely high enrichment of Cd and Sn, while S4 showed moderate enrichment of As, very high enrichment of Sn and extremely high enrichment of Hg. Site 5 and 6 showed very high enrichment of Sn. A comparison among different sampling sites of the study area based on EF's of elements showed that S1 is the most contaminated site with the other remaining sites having almost similar EF's. From the results presented, EF's showed that the presence of heavy metals was extremely high in soils of e-waste recycling sites compared to the remaining sites.

The PLI values of Cu, Zn, Cr, As and Ni fell within the PLI ≤ 1 category. This means that the presence of these heavy metals in the earth crust was low (Pradhan & Kumar, 2014) and their pollution in the soil is very minimal. On the other hand, Hg and Pb belong to 1 < PLI ≤ 3 category which means that their presence in the study area fell between low to high pollutants in the soil. However, Cd and Sn were within the high category with PLI > 3. This meant that the presence of Cd and Sn was high in the study area.

3.5. Heavy metal concentration in plants at different sites from the Korle Lagoon area in Accra

Heavy metals with different concentration were detected in common plants in all sites under study. Table 6 presents the summary of the concentrations of nine heavy metals in plants samples collected in the study area. The control site (S6) recorded the least concentration of most of the elements. The trend in heavy metals in the plants samples was similar to what was observed in the soil. There was a significant difference ($p < 0.01$) in the concentration of heavy metals among sites. ANOVA showed significant ($p < 0.01$) higher concentrations of most of the heavy metals (Sn, Cd, Cu, Zn and Ni) in plant samples obtained from the e-waste site (S1) than the other sites. Among the heavy metal analysed, Hg was not observed in any of the sampled plants, while Pb was also not detected at S4. The non-detection of Hg in the plant samples could be attributed to the minimal or no usage of mercury in the activities undertaken at these zones. It might also be the fact that Hg can easily be transformed into other organic forms which are more poisonous (Boening, 2000; Clarkson, 1997). Hg is not essential for plant growth (Lange, Nobel, Osmond, & Ziegler, 2013). The ability of plants to accumulate essential metals equally enables them to acquire other nonessential metals (Djingova & Kuleff, 2000).

Table 6. Summary of heavy metal concentration in vegetation at the Korle Lagoon area in Accra

Heavy metals	S1	S2	S3 (mg/kg)	S4	S5	S6
Hg	<0.001	<0.001	<0.001	<0.001	<0.001	<0.001
Pb	36.72(0.57)a	0.69(0.34)bc	0.48(0.05)bd	<0.001	3.81(0.12)be	0.55(0.80)b
Cu	95.56(0.58)a	10.13(1.04)bc	0.21(0.03)bd	6.92(0.05)be	5.48(0.27)bf	0.21(0.03)b
Zn	34.92(0.59)a	8.04(0.20)bc	14.84(1.13)be	7.88(0.06)bf	7.36(0.40)bg	1.33(0.12)b
Cr	1.56(0.12)a	1.21(0.11)ac	1.08(0.05)a	3.84(0.04)b	3.52(0.29)b	1.23(0.57)a
Cd	1.64(0.12)a	0.83(0.09)bc	0.92(0.04)be	0.44(0.03)b	0.32(0.06)b	0.25(0.06)b
Ni	2.0(0.12)a	5.28(1.72)a	2.25(0.03)ad	2.2(0.12)ae	6.80(0.58)b	0.59(0.11)ac
Sn	22(1.15)a	17.17(1.47)a	3.23(0.06)b	2.52(0.12)b	4.64(0.38)b	22.29(4.89)a
As	0.32(0.05)a	nd	0.12(0.01)b	0.08(0.01)bc	0.16(0.12)be	0.19(0.01)b

Notes: S1 = e-waste area, S2 = garden area, S3 = a reclaimed waste dump site, S4 = a recreational site, S5 = the estuary area of the Korle Lagoon and S6 = control, SE = Standard error, nd = Not Detected. Means followed by a different letter(s) in the same row differ significantly ($p = 0.05$) according to Tukey's Multiple Range Test, values in the parentheses are the standard error.

The concentration of Pb in plant samples ranged from <0.001 mg/kg at S4 to 36.72 mg/kg at S1. Tukey test revealed significant differences ($p < 0.01$) in the mean values of Pb between S1 and the other sites. The mean values recorded at all sites except S4, were above the FAO/WHO (1984) acceptable value of 0.43 mg/kg in edible plants. The high concentration of Pb in plant from the e-waste site could be attributed to the burning of lead-containing products like scrap metals and batteries in the e-waste site. Livestock that graze in the e-waste area are likely to be exposed to health risks with regards to lead toxicity through the consumption of forage grasses growing in this area. Lead is reported to cause liver disorders in livestock especially in cattle. Khan, Khan, Hussain, Marwat, and Ashtray (2008) reported that lead causes both acute and chronic poisoning and thus, poses adverse effects on kidney, liver, vascular and immune system.

Cadmium concentrations recorded ranged from 0.25 to 1.64 mg/kg which was above the WHO/FAO (2007) permissible limit of 0.20 mg/kg for edible plants. Site S6 recorded the least concentration of Cd, while the highest was recorded at S1. The high concentration of Cd at S1 might be as a result of the burning of e-waste containing cadmium–nickel batteries, pigments and paints. Significant concentration of Cd may have gastrointestinal effect and reproductive effect on livestock (Maobe, Gatebe, Gitu, & Rotich, 2012). Jabeen, Shah, Khan, and Hayat (2010) reported that cadmium causes both acute and chronic poisoning, adverse effect on kidney, liver, vascular and the immune system.

Similarly, the concentration of Cu ranged from 0.21 mg/kg at S3 and S6 to 95.56 mg/kg at S1. With the exception of S3 and S6 which recorded mean Cu values that were below the FAO/WHO (1984) permissible limit of 3.0 mg/kg, the other sites recorded values that were above the limit. This could be attributed to continuous dumping of copper-containing electrical gadgets. In a similar study, Pradhan and Kumar (2014) reported a range of Cu values from 11.08 mg/kg to 23.07 mg/kg in a study conducted in China. According to Maobe et al. (2012) high levels of copper can cause metal fumes fever with flu-like symptoms, hair and skin decolouration, dermatitis, irritation of the upper respiratory tract, metallic taste in the mouth and nausea.

Additionally, concentration of Zn ranged from 1.33 mg/kg at S6 to 34.92 mg/kg at S1. However, all the sites recorded mean values of zinc that were below the FAO/WHO (1984) permissible limit of 27.3 mg/kg except for S1. Zinc is required nutrient and becomes toxic to plants only at high concentrations.

Similarly, the concentration of chromium ranged from 1.08 mg/kg at S3 to 3.84 mg/kg at S4, which was within the critical range of 5.00–30.00 mg/kg stated by Radojevic and Bashkin (2006).

The concentration of nickel also ranged from 0.59 mg/kg at S6 to 6.80 mg/kg at S5. With the exception of S6, the remaining sites were above the WHO/FAO permissible limit of 1.63 mg/kg in edible plants. Nickel in plants could be attributed to cadmium–nickel batteries in the electrical gadgets and some paints used to polish the surfaces of the gadgets which might have spread to adjoining sites as it was also detected in the soil samples. High concentration of nickel can lead to health risks. According to Khan et al. (2008), Ni deficiency results in liver disorder.

The concentration of Sn ranged from 2.52 mg/kg at S4 to 22.29 mg/kg at S6 which were all below the recommended FAO/WHO (1984) permissible limit of 200.00 mg/kg in edible plants.

Additionally, As ranged from below detection at S2 to 0.32 mg/kg at S1. The presence of As in the vegetation samples could be attributed to the recycling activities in the area. The concentration of As is in line with the findings of Pradhan and Kumar (2014) on Delhi informal re-cycling site. According to Luo et al. (2011) atmospheric deposition is a major factor for high metal accumulation in plant samples, and this could therefore be the cause of the As in the samples analysed.

The uptake of heavy metals by plants occurs during the vegetative period (Knezevic et al., 2009; Krstić, Stanković, Igić, & Nikolic, 2007). The presence of toxic heavy metals in the soil results in the bio-accumulation and bio-magnification of these toxins into plant tissues (Fagbote & Olanipekun, 2010; Suciu, Cosma, Todică, Bolboacă, & Jäntschi, 2008).

4. Conclusion

The soils and vegetation samples from the Korle Lagoon reclamation area recorded significant levels of heavy metals, especially at the e-waste site (S1). The results showed that all the nine heavy metals (Hg Pb, Cu, Zn, Cr, Cd, Ni, Sn and As) analysed were present in soil samples at the six sites. Similarly, eight heavy metals (except Hg) out of the nine analysed were present in vegetation sampled from the six sites. Generally, the e-waste site (S1) had higher concentration values of all heavy metals in both soil and vegetation samples than the other sites. Pollution assessment revealed high contents of heavy metals in soil samples collected from the e-waste recycling area compared to the other sites. Site S1 exhibited extremely high enrichment of Hg, Cd and Sn, while the same site also showed a significant enrichment of Pb and As. The results from the enrichment factor analysis showed that the presence of heavy metals was extremely high in soils from the e-waste-recycling sites compared to the other sites. The geoaccumulation index showed that S1 was strongly polluted by Pb and extremely polluted by Cd and Sn, with S3 showing strong contamination by Sn. The values for pollution load index of heavy metals recorded low contamination except Hg and Pb which fell between low to high pollutants in the soil, and Cd and Sn which were within the high category which meant that the presence of Cd and Sn was high in the study soils. The contamination of the environment with heavy metals pose hazard to human and animal health, through bioaccumulation and bio-magnification as animals sometimes graze around the area. The informal recycling of e-waste also has the potential to pollute the environment and nearby communities. The need for periodic monitoring of toxic metals in this area is crucial to protect human health and the environment.

5. Recommendation

The Environmental Protection (EPA) Agency of Ghana and the Ghana Customs Excise and Preventive Service (CEPS) should collaborate to educate the public on e-waste separation at the various entry points. The Accra Metropolitan Assembly must enforce laws on open burning and hasten the Korle Lagoon Reclamation Project in order to curtail the activities of these e-waste dealers and the burning of e-waste at the reclamation area. e-waste workers should be educated occasionally to create awareness on the potential health risks they are exposed to. Herds of cattle crossing from nearby communities should be restricted from grazing on the forage grasses in the study area. Children should be restricted from playing within the recreational area since some levels of heavy metals were found in both plant and soil sampled from the area. Farmers cultivating vegetables along the borders of the reclamation zones should be educated on the health risk consumers are expose to by consuming vegetable grown as the plants are likely to pick heavy metals from the polluted soil.

Acknowledgement

The authors thank Elvis Dartey Okoffo and Amoako Ofori for their assistance in some aspect of the data analysis and editing of the manuscript.

Funding

The authors received no direct funding for this research.

Competing Interests

The authors declare no competing interest.

Author details

Benedicta Yayra Fosu-Mensah[1]
E-mail: yayramensah@staff.ug.edu.gh
Emmanuel Addae[1]
E-mail: kwamianane@yahoo.com
Dzidzo Yirenya-Tawiah[1]
E-mail: dzidzoy@staff.ug.edu.gh
Frank Nyame[2]
E-mail: fnyame@ug.edu.gh

[1] Institute for Environment and Sanitation Studies (IESS), University of Ghana, P.O. Box 209, Legon, Accra, Ghana.
[2] Department of Earth Sciences, University of Ghana, Legon, Accra, Ghana.

References

Adama, M., Esena, R., Fosu-Mensah, B., & Yirenya-Tawiah, D. (2016). Heavy metal contamination of soils around a hospital waste incinerator bottom ash dumps site. *Journal of Environmental and Public Health, 2016*, Article ID 8926453. doi:10.1155/2016/8926453

Asante, K. A., Agusa, T., Biney, C. A., Agyekum, W. A., Bello, M., Otsuka, M., ... Tanabe, S. (2012). Multi-trace element levels and arsenic speciation in urine of e-waste recycling workers from Agbogbloshie, Accra in Ghana. *Science of the Total Environment, 424*, 63–73. doi:10.1016/j.scitotenv.2012.02.072

Atiemo, M. S., Ofosu, F. G., Kuranchie-Mansah, H., Tutu, A. O., Palm Linda, N. D. M., & Blankson, A. S. (2011). Contamination assessment of heavy metals in road dust from selected roads in Accra, Ghana. *Research Journal of Environmental and Earth Sciences, 3*(5), 473–480.

Ato, A. F., Samuel, O., Oscar, Y. D., Alex, P., Moi, N., & Akoto, B. (2010). Mining and heavy metal pollution: Assessment of aquatic environments in Tarkwa (Ghana) using multivariate statistical analysis. *Journal of Environmental Statistics, 1*(4), 1–13.

Boening, D. W. (2000). Ecological effects, transport, and fate of mercury: A general review. *Chemosphere, 40*(12), 1335–1351. https://doi.org/10.1016/S0045-6535(99)00283-0

Brigden, K., Labunska, I., Santillo, D., & Johnston, P. (2008). *Chemical contamination at e-waste recycling and disposal sites in Accra and Korforidua, Ghana.* Greenpeace Research Laboratories Technical Note, (10/2008).

Çevik, F., Göksu, M. Z. L., Derici, O. B., & Fındık, Ö. (2009). An assessment of metal pollution in surface sediments of Seyhan dam by using enrichment factor, geoaccumulation index and statistical analyses. *Environmental Monitoring and Assessment, 152*(1–4), 309–317. doi:10.1007/s10661-008-0317-3

Chen, T.-B., Zheng, Y.-M., Lei, M., Huang, Z.-C., Wu, H.-T., Chen, H., ... Tian, Q.-Z. (2005). Assessment of heavy metal pollution in surface soils of urban parks in Beijing, China. *Chemosphere, 60*(4), 542–551. doi:10.1016/j.chemosphere.2004.12.07

Clarkson, T. W. (1997). The toxicology of mercury. *Critical Reviews in Clinical Laboratory Sciences, 34*(4), 369–403. https://doi.org/10.3109/10408369708998098

Djingova, R., & Kuleff, I. (2000). Instrumental techniques for trace analysis. *Trace Metal Environment, 4*, 137–185. https://doi.org/10.1016/S0927-5215(00)80008-9

dos Anjos, M. J., Lopes, R. T., de Jesus, E. F. O., Assis, J. T., Cesareo, R., & Barradas, C. A. A. (2000). Quantitative analysis of metals in soil using X-ray fluorescence. *Spectrochimica Acta Part B: Atomic Spectroscopy, 55*(7), 1189–1194. doi:10.1016/S0584-8547(00)00165-8

Environmental Health and Safety Manual. (2000). *Safe handling of mercury and mercury compounds.* Retrieved October 12, 2017, from https://iaomt.org/TestFoundation/safehandling.htm

e-waste Africa Programme. (2011). *e-waste Africa programme: Where are WEee in Africa? Findings from the basel convention.* Châtelaine: Secretariat of the Basel Convention.

Fagbote, E. O., & Olanipekun, E. O. (2010). Evaluation of the Status of heavy metal pollution of soil and plant (Chromolaena odorata) of Agbabu bitumen deposit area, Nigeria. *American-Eurasian Journal of Scientific Research, 5*(4), 241–248.

FAO/WHO. (1984). *Toxicological evaluation of certain food additives and food contaminants. (Twenty-eight meeting of the Joint FAO/WHO Expert Committee on food additives).* Washington, DC: ILSI Press International Life Sciences Institute.

Feldt, T., Fobil, J. N., Wittsiepe, J., Wilhelm, M., Till, H., Zoufaly, A., ... Göen, T. (2014). High levels of PAH-metabolites in urine of e-waste recycling workers from Agbogbloshie, Ghana. *Science of the Total Environment, 466–467*, 369–376. doi:10.1016/j.scitotenv.2013.06.097

Fosu-Mensah, B. Y., Okoffo, E. D., Darko, G., & Gordon, C. (2016). Assessment of organochlorine pesticide residues in soils and drinking water sources from cocoa farms in Ghana. *Springerplus, 5*(869), doi:10.1186/s40064-016-2352-9

Ha, N. N., Agusa, T., Ramu, K., Tu, N. P. C., Murata, S., Bulbule, K. A., ... Tanabe, S. (2009). Contamination by trace elements at e-waste recycling sites in Bangalore, India. *Chemosphere, 76*(1), 9–15. doi:10.1016/j.chemosphere.2009.02.056

Itai, T., Otsuka, M., Asante, K. A., Muto, M., Opoku-Ankomah, Y., Ansa-Asare, O. D., & Tanabe, S. (2014). Variation and distribution of metals and metalloids in soil/ash mixtures from Agbogbloshie e-waste recycling site in Accra, Ghana. *Science of the Total Environment, 470*, 707–716. https://doi.org/10.1016/j.scitotenv.2013.10.037

Jabeen, S., Shah, M. T., Khan, S., & Hayat, M. Q. (2010). Determination of major and trace elements in ten important folk therapeutic plants of Haripur basin, Pakistan. *Journal of Medicinal Plants Research, 4*(7), 559–566.

Khan, S. A., Khan, L., Hussain, I., Marwat, K. B., & Ashtray, N. (2008). Profile of heavy metals in selected medicinal plants. *Pakistan Journal of Weed Science Research, 14*(1–

2), 101–110. Retrieved from http://www.wssp.org.pk/14,1-2,10.pdf

Knezevic, M., Stankovic, D., Krstic, B., Nikolic, M. S., & Vilotic, D. (2009). Concentrations of heavy metals in soil and leaves of plant species Paulownia elongata and Paulownia fortunei. *African Journal of Biotechnology, 8*(20), 5422–5429.

Krstić, B., Stanković, D., Igić, R., & Nikolic, N. (2007). The potential of different plant species for nickel accumulation. *Biotechnology & Biotechnological Equipment, 21*(4), 431–436. doi:10.1080/13102818.2007.10817489

Kumar, S., Singh, J., & Garg, M. (2012). AAS Estimation of heavy metals and trace elements in Indian herbal cosmetics preparations. *Research Journal of Chemical Sciences, 2*(3), 46–51.

Lange, O. L., Nobel, P. S., Osmond, C. B., & Ziegler, H. (2013). *Physiological plant ecology III: Responses to the chemical and biological environment.* Springer Science & Business Media.

Li, H. Z., Bai, J. M., Li, Y. T., Cheng, H. F., Zeng, E. Y., & You, J. (2011). Short-range transport of contaminants released from e-waste recycling site in South China. *Journal of Environmental Monitoring, 13*(4), 836–843. doi:10.1039/c0em00633e

Li, J., Duan, H., & Shi, P. (2011). Heavy metal contamination of surface soil in electronic waste dismantling area: Site investigation and source-apportionment analysis. *Waste Management and Research, 29*(7), 727–738. doi:10.1177/0734242X10397580

Lim, S. R. S. J. (2010). Toxicity potentials from waste cellular phones, and a waste management policy integrating consumer, corporate, and government responsibilities. *Waste Management, 30*(8–9), 1653–1660. https://doi.org/10.1016/j.wasman.2010.04.005

Lu, X., Wang, L., Lei, K., Huang, J., & Zhai, Y. (2009). Contamination assessment of copper, lead, zinc, manganese and nickel in street dust of Baoji, NW China. *Journal of Hazardous Materials, 161*(2–3), 1058–1062. doi:10.1016/j.jhazmat.2008.04.052

Luo, C., Liu, C., Wang, Y., Liu, X., Li, F., Zhang, G., & Li, X. (2011). Heavy metal contamination in soils and vegetables near an e-waste processing site, south China. *Journal of Hazardous Materials, 186*(1), 481–490. doi:10.1016/j.jhazmat.2010.11.024

Ma, J., Horii, Y., Cheng, J., Wang, W., Wu, Q., Ohura, T., & Kannan, K. (2009). Chlorinated and parent polycyclic aromatic hydrocarbons in environmental samples from an electronic waste recycling facility and a chemical industrial complex in China. *Environmental Science & Technology, 43*(3), 643–649. doi:10.1021/es802878w

Maobe, M. A. G., Gatebe, E., Gitu, L., & Rotich, H. (2012). Profile of heavy metals in selected medicinal plants used for the treatment of diabetes, malaria and pneumonia in Kisii region, Southwest Kenya. *Global Journal of Pharmacology, 6*(3), 245–251.

Muller, G. (1969). Index of geoaccumulation in sediments of the Rhine River. *Geological Journal, 2*, 108–118.

Muller, G. (1981). The heavy metal pollution of the sediments of Neckars and its tributary: A stocktaking. *Chemiker-Zeitung, 105*, 157–164.

Musee, N. (2011). Nanotechnology risk assessment from a waste management perspective: Are the current tools adequate? *Human & Experimental Toxicology, 30*(8), 820–835. https://doi.org/10.1177/0960327110384525

Ofudje, E. A., Alayande, S. O., Oladipo, G. O., Williams, O. D., & Akiode, O. K. (2014). Heavy Metals concentration at electronic-waste dismantling sites and dumpsites in Lagos, Nigeria. *International Research Journal of Pure and Applied Chemistry, 4*(6), 678–690. https://doi.org/10.9734/IRJPAC

Pradhan, J. K., & Kumar, S. (2014). Informal e-waste recycling: Environmental risk assessment of heavy metal contamination in Mandoli industrial area, Delhi, India. *Environmental Science and Pollution Research, 21*(13), 7913–7928. doi:10.1007/s11356-014-2713-2

Qin, F., Ji, H., Li, Q., Guo, X., Tang, L., & Feng, J. (2014). Evaluation of trace elements and identification of pollution sources in particle size fractions of soil from iron ore areas along the Chao River. *Journal of Geochemical Exploration, 138*, 33–49. https://doi.org/10.1016/j.gexplo.2013.12.005

Quan, S. X., Yan, B., Yang, F., Li, N., Xiao, X. M., & Fu, J. M. (2015). Spatial distribution of heavy metal contamination in soils near a primitive e-waste recycling site. *Environmental Science and Pollution Research, 22*(2), 1290–1298. doi:10.1007/s11356-014-3420-8

Radojevic, M., & Bashkin, V. N. (2006). Practical environmental analysis. In *Royal society of chemistry* (2nd ed., pp. 147–170). Cambridge: UK Publishing.

Suciu, I., Cosma, C., Todică, M., Bolboacă, S. D., & Jäntschi, L. (2008). Analysis of soil heavy metal pollution and pattern in central Transylvania. *International Journal of Molecular Sciences, 9*(4), 434–453. doi:10.3390/ijms9040434

Tang, X., Chaofeng, S., Dezhi, S., Sardar, A. C., Muhammad, I. K., Congkai, Z., & Yingxu, C. (2009). Heavy metal and persistent organic compound contamination in soil from Wenling: An emerging e-waste recycling city in Taizhou area, China. *Journal of Hazardous Materials, 173*(1–3), 653–660. doi:10.1016/j.jhazmat.2009.08.134

Taylor, S. R., & McLennan, S. M. (1985). *The continental crust: Its composition and evolution.* Oxford: Blackwell Scientific Publications.

Tchounwou, P. B., Yedjou, C. G., Patlolla, A. K., & Sutton, D. J. (2012). Heavy metal toxicity and the environment. In Luch, A. (Eds.), *Molecular, clinical and environmental toxicology* (vol. 101, pp. 133–164). Springer Basel. doi:10.1007/978-3-7643-8340-44

Thorpe, A., & Harrison, R. M. (2008). Sources and properties of non-exhaust particulate matter from road traffic: A review. *Science of the Total Environment, 400*(1–3), 270–282. doi:10.1016/j.scitotenv.2008.06.007

Tiimub, B. M., & Dartey, E. (2015). Determination of selected heavy metals contamination in water from downstream of the Volta Lake at Manya Krobo district in eastern region of Ghana. *International Research Journal of Public and Environmental Health, 2*(11), 167–173. doi:10.15739/irjpeh.035

Timothy, A., & Olajumoke, M. A. (2014). Heavy metal concentrations around a hospital incinerator and a municipal dumpsite in Ibadan City, South-West Nigeria. *Journal of Applied Science, Environment and Management, 17*(3), 419–422.

Tomilson, D. C., Wilson, J. G., Harris, C. R., & Jeffrey, D. W. (1980). Problems in assessment of heavy metals in estuaries and the formation of pollution index. *Environmental Evaluation, 33*(1), 566–575. doi:10.1007/BF02414780

Tsydenova, O., & Bengtsson, M. (2011). Chemical hazards associated with treatment of waste electrical and electronic equipment. *Waste Management, 31*(1), 45–58. https://doi.org/10.1016/j.wasman.2010.08.014

Wagner, T. P. (2009). Shared responsibility for managing electronic waste: A case study of Maine, USA. *Waste Management, 29*(12), 3014–3021. https://doi.org/10.1016/j.wasman.2009.06.015

Wahabu, S., Fosu-Mensah, B. Y., & Nyame, F. K. (2015). Impact of charcoal production on physical and chemical properties of soil in the Central Gonja District of the Northern Region. Ghana. *Environment and Natural Resources Research, 5*(3), 11–18. doi:10.5539/enrr.v5n3p11

Walkley, A., & Black, I. A. (1934). An examination of the Degtjaref method for determining soil organic matter, and a proposed modification of the chronic acid titration method. *Soil Science, 37*(1), 29–38. doi:10.1097/00010694-193401000-00003

Wang, J., Chen, S., Tian, M., Zheng, X., Gonzales, L., Ohura, T., ... Massey Simonich, S. L. (2012). Inhalation cancer risk associated with exposure to complex polycyclic aromatic hydrocarbon mixtures in an electronic waste and urban area in south China. *Environmental Science and Technology, 46*(17), 9745–9752. doi:10.1021/es302272a

Wang, X. Y., Li, Q. B., Luo, Y. M., Ding, Q., Xi, L. M., Ma, J. M., ... Cheng, C. L. (2010). Characteristics and sources of atmospheric polycyclic aromatic hydrocarbons (PAHs) in Shanghai, China. *Environmental Monitoring and Assessment, 165*(1–4), 295–305. https://doi.org/10.1007/s10661-009-0946-1

WHO/FAO. (2001). *Codex alimentarius commission. Food additives and contaminants. Joint FAO/WHO Food Standards Programme, ALINORM 10/12A*. Retrieved from www.transpaktrading.com/static/pdf/research/ achemistry/introTofertilizers.pdf

WHO/FAO. (2007). *Joint FAO/WHO Food Standard Programme Codex Alimentarius Commission 13th Session. Report of the Thirty Eight Session of the Codex Committee on Food Hygiene*. Houston, TX, ALINORM 07/30/13.

Xianjin, T., Shen, C., Shi, D., Cheema, S. A., Khan, M. I., Zhang, C., & Chen, Y. (2010). Heavy metal and persistent organic compound contamination in soil from Wenling: An emerging e-waste recycling city in Taizhou area, China. *Journal of Hazardous Materials, 173*(1–3), 653–660. doi:10.1016/j.jhazmat.2009.08.134

Yongming, H., Peixuan, D., Junji, C., & Posmentier, E. (2006). Multivariate analysis of heavy metal contamination in urban dusts of Xi'an, Central China. *Science of the Total Environment, 355*(1–3), 176–186. doi:10.1016/j.scitotenv.2005.02.026

Zhang, W. H., Wu, Y. X., & Simonnot, M. O. (2012). Soil contamination due to e-waste disposal and recycling activities: A review with special focus on China. *Pedosphere, 21*(4), 434–455. doi:10.1016/S1002-0160(12)60030-7

Zhang, X., Yang, F., Luo, C., Wen, S., Zhang, X., & Xu, Y. (2009). Bioaccumulative characteristics of hexabromocyclod-odecanes in freshwater species from an electronic waste recycling area in China. *Chemosphere, 76*(11), 1572–1578. doi:10.1016/j.chemosphere.2009.05.031

12

First Nations wastewater treatment systems in Canada: Challenges and opportunities

Mofizul Islam[1] and Qiuyan Yuan[1]*

*Corresponding author: Qiuyan Yuan, Department of Civil Engineering, University of Manitoba, Winnipeg, MB, R3T 5V6 Canada
E-mail: qiuyan.yuan@umanitoba.ca
Reviewing editor: Murat Eyvaz, ebze Technical University, Turkey

Abstract: The Government of Canada has prioritized the availability of water and wastewater services for the Canadian First Nations Communities (CFNC) and introduced the First Nations Water and Wastewater Action Plan. Several studies explore that many wastewater treatment systems (WWTS) in the CFNC do not meet the effluent discharge limits. The objectives of this study were to examine the existing WWTS in CFNC, investigate the progress and improvement opportunities, evaluate the risk levels, encapsulate the financial condition, and provide recommendations for the overall improvement of the WWTS in CFNC. The authors found significant improvement in 2011 when 98% of the Canadian First Nations houses received wastewater services in comparison to only 50% in 1978. However, 1,777 First Nations houses did not receive any wastewater services. In 2011, 21% of the wastewater systems were operated exceeding the facilities' design capacities. The overall high-risk and medium-risk wastewater systems have reduced from 14 and 51% in 2011

ABOUT THE AUTHORS

Mofizul Islam is a master's student in the CREATE H2O program, which is designed to address the research and training gaps of the Canadian First Nations' water and sanitation security, under the supervision of Qiuyan Yuan, an assistant professor at the University of Manitoba. Yuan's research goals include development of sustainable technologies for water and waste treatment processes. Her research team unravels a variety of environmental engineering topics including nutrient removal and recovery, landfill leachate treatment, biomass fermentation, and anaerobic digestion. Mofizul Islam's research is focused on improving biological treatment efficiencies of landfill leachate using selected fungal strains, which produce and secrete enzymes, to degrade complex structure of recalcitrant compounds of leachate. Since CREATE H2O program is the first science engineering research training program working towards the improvement of Canadian First Nations communities, researchers and especially the First Nations community personnel would like to access the knowledge and research gaps from this article to mitigate the issues related to the wastewater treatment facilities.

PUBLIC INTEREST STATEMENT

Canadian First Nations Communities (CFNC) are facing greater challenges to reduce the high and medium risk level wastewater treatment systems. The severe lack of research in this area makes it harder to understand the real scenario. This leads the authors to compile the available information to get a clear overall scenario of the WWTS in CFNC. It was explored that 1,777 houses did not receive any wastewater services in 2011. According to the latest survey in 2014–2015, 6% wastewater systems were identified as high-risk level. The most concerning figure was related to the medium-risk level systems (41%). It is concluded that more funding is required to eliminate or reduce the high and medium risk level systems. The information from this article would be useful to the government and non-government agencies, policy-makers, CFNCs, and researchers for the future programs and policies related to WWTS.

to 6 and 41% in 2014–2015, respectively. The Government of Canada committed to provide $4.2 billion for the 10-year period (2011–2021) against the estimated cost of $6.3 billion. Increasing and proper utilization of the allocated budget is recommended to fill up the financial gaps.

Subjects: Environmental Management; Environment & Economics; Environmental Change & Pollution

Keywords: sanitary system; First Nations; wastewater treatment systems; wastewater lagoon; risk level

1. Introduction

The Government of Canada is committed to ensure that the wastewater treatment systems (WWTS) in Canadian First Nations Communities (CFNC) are safe, meet the community necessities with appropriate monitoring systems, and also meet the federal and provincial effluent discharge limits (Government of Canada, 2016). Efficient and effective wastewater management process has become not only a concern but also an emerging challenge for the CFNC due to the lack of resources, remoteness, and extreme northern climate condition. The assessment report, published in 2003, of Indigenous and Northern Affairs Canada (INAC) (previously referred to as Aboriginal Affairs and Northern Development Canada) by the Department of Indian Affairs and Northern Development (DIAND) identified six major deficiency areas of WWTS in CFNC: treatment technology design, operation and maintenance, operator skills and knowledge levels, monitoring system, dysfunctional equipment, and protecting the receiving watershed (INAC, 2003). The additional major challenges are the availability of equipment and supplies, high operator turnover rate, and economic constraint (Mallett, 2010).

CFNC receive wastewater treatment services through centralized and decentralized wastewater systems. Centralized system is a communal WWTS in which wastewater is transported to a central treatment facility, for example, wastewater stabilization lagoon and activated sludge plant, by collecting through a piped collector system. On the other hand, decentralized system is an on-site WWTS, for instance septic system, managed by a group or groups of Band Councils. Providing wastewater treatment services to the CFNC in a centralized system is a shared responsibility among the First Nations' Band Councils, Environment and Climate Change Canada (previously known as Environment Canada), Health Canada, and INAC (INAC, 2016; Simeone, 2010). The Band Council is responsible for 20% of financial costs as well as ensuring the proper design, construction, operation, and maintenance of WWTS following the more stringent federal and provincial government standards (INAC, 2010a; White, Murphy, & Spence, 2012). They are also accountable for organizing training sessions to ensure that the operators are properly trained and certified (Heinke, Smith, & Finch, 1991; INAC, 2010a). Environment and Climate Change Canada regulates the wastewater effluent discharge into the receiving watershed and is also accountable for providing advice and technical expertise (AANDC, 2012a; INAC, 2016; Simeone, 2010). Environmental and public health assessment, plant inspections, responding to complaints, and planning reviews are under the job scope of Health Canada. In addition, mass people awareness programs such as community-based education pertaining to the wastewater issues are developed and communicated to the concerned communities by the Health Canada (AANDC, 2012b; Health Canada, 2016; White et al., 2012). INAC affords financial support for wastewater treatment facilities (i.e. 80% of design, construction, operation, and maintenance) along with staff (operator) training, setting standards, developing regulations to implement standards, and reviewing designs and facility performance in collaboration with Environment and Climate Change Canada (AANDC, 2012a; Simeone, 2010).

The roles and responsibilities of First Nations, INAC, Environment Canada, and Health Canada in decentralized on-site WWTS are different from the centralized systems. Within the First Nations, the responsibilities of managing decentralized systems are divided between Band Councils and system operators. Band Councils are responsible to ensure that the wastewater systems are designed, constructed, and upgraded by following the Protocol for Decentralized Water and Wastewater Systems in First Nations Communities (INAC, 2010b). Systems operators are responsible for operating, maintaining, sampling, testing, monitoring, and keeping records. AANDC provides full or partial financial assistance and facilitates training. Environment Canada administrates and enforces the acts and regulations as well as provides information regarding the regulatory and environment protection requirements. On the request of individual resident, Health Canada may sample and test the wastewater in the close proximity but they do not have any obligatory sampling and testing requirements (INAC, 2010b).

The Government of Canada has set a goal for the WWTS in CFNC to meet the more stringent federal wastewater systems effluent regulations (White, Murphy, & Spence, 2012) and to address the concerns through the Canadian Environmental Sustainability Indicators (CESI) program under the Federal Sustainable Development Strategy (2013–2016).

Ontario Clean Water Agency (2001) stated in the assessment report prepared for the Shibogama First Nations Council that effluent discharge exceeded the *Escherichia coli (E. coli)* concentration to 12,500/100 ml (*E. coli* discharge limit for Ontario: 400/100 ml). The report also mentioned numerous reasons for the increase of *E. coli* in effluent such as overloading of the plant's capacity, equipment malfunctioning, overflowing, and floor flooding. This information presents challenges to the Canadian Government to achieve the commitment of mitigating First Nations needs of wastewater services and eventually necessitates a complete review of the WWTS in CFNC.

The objectives of this study were to: (1) summarize the literature and technical reports pertaining to WWTS in CFNC for the better understanding of the existing systems, (2) identify and track down the progress and improvement opportunities in accordance with the previous records, (3) evaluate the critical risk levels of different WWTS in CFNC, (4) do a comparison study between the actual investments by the Government of Canada and the estimated cost by the national assessment to meet the need of First Nations wastewater services as well as to fulfill the commitment undertaken by the Government of Canada, and finally (5) provide recommendations to address the current challenges and improvement opportunities.

2. First Nations wastewater treatment systems

Indian and Northern Affairs Canada (INAC) reported that there were 617 First Nations communities in Canada in 2014 (INAC, 2014). This number was about 570 in 2011, out of which only 418 First Nations communities received wastewater services from the 532 treatment systems (Neegan Burnside, 2011i) in the form of wastewater stabilization lagoons (aerated and facultative lagoons), mechanical treatment plants, Municipal Type Agreements (MTA), and individual septic systems (INAC, 2003; Neegan Burnside, 2011i). The wastewater system has been defined as "an organized process and associated structures for collecting, treating, and disposing of wastewater", which was specified as a system serving five or more houses for the purposes of the national assessment report (Neegan Burnside, 2011i). The number of wastewater systems and the corresponding percentages in terms of the relative significance of each WWTS used in CFNC have been captured in Figure 1.

According to the national assessment of Neegan Burnside (2011i), there were a total of 484,321 on-reserve First Nations' population residing in 112,836 houses with an average of 4.3 persons per household. Sixty-three percent of the dwellings were served by the communal wastewater systems. Many CFNC receive wastewater treatment services from the neighboring municipalities, First Nations,

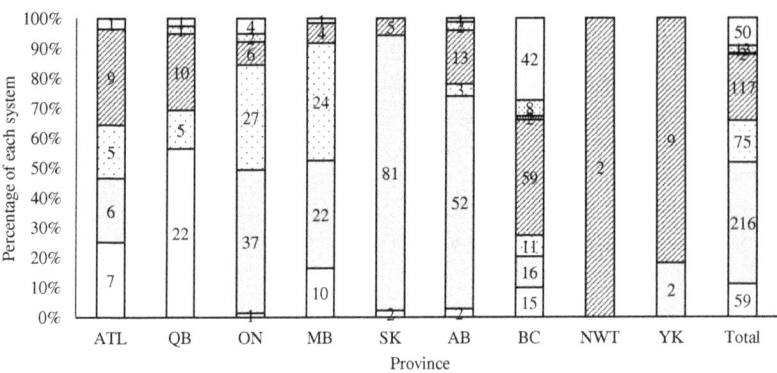

□ Aerated Lagoon □ Facultative Lagoon □ Mechanical Treatment ▨ MTA ▥ None □ Other □ Septic System

Figure 1. The overall categorization of the WWTS in CFNC.

Notes: The value inside the bar indicates the number of WWTS and the height of each bar represents the corresponding relative percentage of that system in comparison with the total systems. Data adapted from Neegan Burnside (2011i).

or other private sectors through the MTA (FCM, 2011; INAC, 2003). MTAs were used as a servicing arrangement for 22% of WWTS in 2011.

Different types of WWTS that are implemented in CFNC are described in the following sections and the relevant concerns related to a particular WWTS are also highlighted.

3.1. Wastewater stabilization Lagoon

Wastewater treatment in the lagoon is performed in an engineered constructed wetland or a pond with a preventive measure of avoiding liquid drainage into the ground water or to the surrounding fields (Spellman & Drinan, 2014). Both aerobic and anaerobic decomposition of waste matter occurs naturally in the lagoon without any mechanical equipment (aeration) and it is commonly known as a facultative lagoon. Some lagoons have been designed with the provision of mechanical aeration systems known as an aerated lagoon, which facilitates higher depth of the lagoon cells for the same level of treatment performance as a facultative lagoon, and results in a smaller footprint (Spellman & Drinan, 2014). Lagoon provides both treatment and storage facilities of wastewater, sludge, and other contaminants. Wastewater treatment proceeds through a series of physical (UV radiation, settling, and mixing), chemical (precipitation and nutrients reaction), and biological processes (algal and bacterial activity) (Spellman & Drinan, 2014), which removes contaminants from the wastewater.

Wastewater stabilization lagoon is the most commonly used centralized WWTS in CFNC to treat sewage wastes reported in the national assessments (INAC, 2003; Neegan Burnside, 2011i). More than 50% of the WWTS in CFNC are lagoon based, whereas 11% are aerated lagoons and 41% are facultative lagoon systems (Figure 1). The First Nations communities of Alberta (72%) and Saskatchewan (92%) mostly use facultative lagoons following the province of Ontario (48%). On the other hand, aerated lagoons are highly used in Quebec (56%).

Although the lagoon-based WWTS are suitable for the small communities of First Nations, some serious health and environmental issues have emerged due to the lack of resources, proper operations, and monitoring or site selection. For instance, nutrient pulses (Nitrogen and Phosphorus) exceeded its recommended threshold limit in the Dead Horse Creek, Manitoba, Canada during the discharge period (Carlson et al., 2013). The worst-case scenario was found in the Kashechewan reserve's First Nations community, where the Minister of Aboriginal Affairs, Ontario, had to evacuate 1,000 residents due to the high level of E. coli in the community's drinking water (Senate of Canada, 2007). According to the final reports of the Standing Senate Committee on Aboriginal Peoples (2007) and CBC News (2007), the main contamination source was identified as the poor-quality source water because of the installation of drinking water treatment plant downstream of the community's sewage stabilization lagoon. The national assessment report revealed that 18% of the wastewater stabilization ponds (both aerated and facultative lagoons) in CFNC have failed to meet the federal effluent discharge limits due to the lack of proper operation (Neegan Burnside, 2011i) and 24% aerated lagoons and 47% facultative lagoons do not have effluent discharge data (Figure 3).

To keep the nutrient discharge limits of unionized ammonia, total nitrogen, and phosphorous within the threshold limit in the extreme cold conditions, wastewater management strategies like constructed wetlands and/or natural wetlands (secondary treatment options) have been considered as a proven technology. On the other hand, some of the researchers have successfully restricted the nutrient discharge using limestone rock filters (Strang & Wareham, 2006), active slag filters (Pratt, Shilton, Pratt, Haverkamp, & Bolan, 2007), or some form of chemical treatment (for instance, the utilization of aluminum salts or iron salts such as alum, sodium aluminate, poly-aluminum chloride, ferric chloride, ferrous chloride, and ferrous sulfate) (US EPA, 2010) in the wastewater stabilization ponds.

3.2. Mechanical treatment plant

Mechanical wastewater treatment plant, a centralized WWTS, for the CFNC is designed in small capacity range, which consists of the main treatment units of Rotating Biological Contactors (RBC), Sequencing Batch Reactors (SBR), Extended Aeration (EA), or Moving Bed Biofilm Reactor (MBBR) (Pictou Landing First Nation, 2012). Most of the mechanical treatment plants are designed with the provision of aeration and agitation arrangement, in addition to a preliminary treatment (screening, grit removal) unit and a final stabilization-disinfection process such as a sand filter or UV disinfection (Neegan Burnside, 2011i). Mechanical treatment plants are not very common in the CFNC due to the small community size, complexity in operation and maintenance, and community remoteness. In Canada, among the 532 WWTS in CFNC, 14% are mechanically designed treatment units, while Manitoba and Ontario have the highest percentages of 40% and 35% mechanical WWTS, respectively (Figure 1).

Table 1 exhibits that CFNC have 75 mechanical WWTS, 20 systems of which have not been categorized and are thus mentioned as Mechanical (unidentified) in the National Assessment report published in 2011.

3.2.1. Rotating biological contactor (RBC)

The RBCs are considered as the mechanical treatment plants and are designed to accelerate the natural biological wastewater treatment process by facilitating the growth and accumulation of micro-organisms on the surface of the rotating contactor disk. The report from the Assembly of First Nations (2008) expressed that 15% of the total WWTS in CFNC in the province of Ontario had the technological basis of RBC or SBR in 2005, which increased to 26% in 2011 (Neegan Burnside, 2011e). As shown in Table 1, the First Nations in the province of Ontario have the highest number of RBC systems (18 out of 27 RBC systems, 67% of Canadian First Nations' RBC systems and entire

Table 1. Number of mechanical WWTS in CFNC (Neegan Burnside, 2011a–2011f)

Treatment type/Region	BC	AB	AT	MB	ON	QB	Total	Percentage (%)
RBC	5	1			18	3	27	36
SBR	2	1		18	2		23	31
Activated sludge plant	2				1		3	4
Trickling filter plant		1		1			2	3
Mechanical (unidentified)	2		5	5	6	2	20	27
Total	11	3	5	24	27	5	75	

Notes: BC : British Columbia, AB: Alberta, AT: Atlantic Provinces (Nova Scotia, New Brunswick, Prince Edward Island, Newfoundland and Labrador), MB: Manitoba, ON: Ontario, QB: Quebec, RBC: Rotating Biological Contactor, SBR: Sequencing Batch Reactor.

province's mechanical systems) as compared to other regions. Nationally, among the total mechanical WWTS in CFNC, RBC-based systems are used the most at 36% (27 out of the total of 75 systems).

3.2.2. Sequencing batch reactor (SBR)
SBR is another form of mechanical treatment plant, which is designed based on the modification of the conventional activated sludge process. According to the Manitoba regional roll-up report of INAC prepared by Neegan Burnside (2011d), there are 18 SBR WWTS (30% of the total systems) in the Manitoban First Nations communities, which is the highest number across all other Canadian provinces of First Nations' SBR systems. SBR is the second highest technologically used system, accruing 31% of all mechanical systems, following RBC in CFNC (Table 1).

3.3.3. Activated sludge plant
Activated sludge plant is the conventional WWTS having the secondary treatment options and more economical for a large community. There are a total of three activated sludge wastewater treatment plants in the CFNC, one of them is in the province of Ontario (Independent First Nations' community) and other two are in British Columbia (Beecher Bay and Nanoose First Nations' community), where the community population size is very large (Table 1).

3.3.4. Trickling filter plant
Trickling filter plant is a type of mechanical treatment system that consists of a fixed bed (rocks, coke, gravel, slag, polyurethane foam, ceramic, or plastic media) where the wastewater flows downward over the bed and causes growth of a layer of biofilm (microbial slime) to cover the bed of media. Trickling filters are used for the treatment of very small rural sewage systems. In the CFNC, there are only two WWTS based on this technology, one is in Manitoba (Brokenhead First Nations' community), and other one is in Alberta (Stoney Nakoda Tribal Nation).

3.3.5. Moving bed biofilm reactor (MBBR)
An MBBR is a relatively new mechanical wastewater treatment technology in the CFNC. Biofilm, biofilter, and activated sludge processes have been integrated together to design an MBBR system that enables the entire reactor volume for biomass growth (Rusten, Eikebrokk, Ulgenes, & Lygren, 2006; Tomaszek & Koszelnik, 2015). Recently, God's Lake First Nations community in Manitoba had planned to set up an MBBR system and had placed an invitation to tender for the community's sewage treatment plant (God's Lake First Nations's, 2013).

3.4. Septic systems
Septic system, a decentralized WWTS, is used to treat small-scale sewer wastes for an individual or a small group of houses. In the national assessment report, Neegan Burnside (2011i) mentioned that 153 First Nations communities, which contain about 40,800 houses (36% of total First Nations homes), have individual septic systems in Canada. Nationally, 9% wastewater systems (50 of 532 systems) are based on septic systems in CFNC. The First Nations communities in British Columbia have the highest number of septic systems at 27% (42 of 153 systems) (Figure 1). A serious health and environmental concern has surfaced that accounted for about 20% of the systems discharging septic wastes directly into the ground surface and the higher rates of 40 and 42% were recorded in Alberta and Saskatchewan, respectively. The assessment also showed that nationally 47% septic systems had operational concerns of limited maintenance, aging of the systems, not pumping out septage waste on a regular basis, and inappropriate leaching bed. The maximum operational concerns, associated with over 60% of the septic systems, were recorded in Manitoba, Saskatchewan, and Alberta (Neegan Burnside, 2011i).

4. Evolution of wastewater treatment systems in Canadian First Nations communities
An improvement is noticeable for the essential sewer and wastewater services received by First Nations communities in the past 30 years. For instance, in 1978, almost 50% on-reserve First Nations

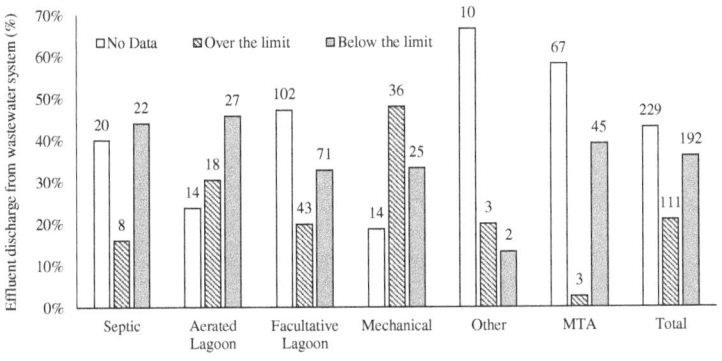

Figure 2. The effluent quality data of wastewater treatment systems in First Nations' of Canada.

Notes: The value above the bar represents the corresponding number of wastewater treatment systems. Data adapted from Neegan Burnside (2011a–2011h).

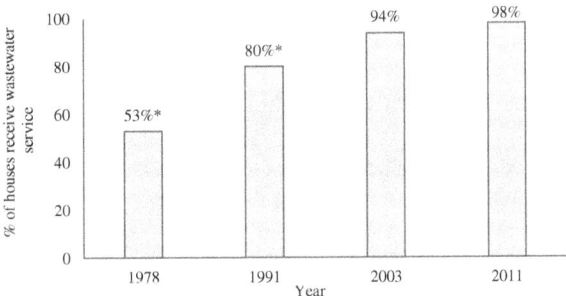

Figure 3. Progressive trend of wastewater services in the CFNC in last 30 years.

Notes: Adapted from Canadian Encyclopedia (2015), INAC (2003) and Neegan Burnside (2011i).*Based on the number of First Nations' houses on reserve that received minimum water and wastewater services.

houses did not have any water and wastewater facilities; whereas, in 2011, 98% CFNC received wastewater services by sewage pipe, truck-haul, or an individual septic system (Figure 2). Although the progress is perceptible, still 2% of First Nations homes that comprises of 1,777 houses in Canada do not have any wastewater services at all. The report also mentioned that 21% of the First Nations wastewater systems (113 systems) provided service at or exceeding the facilities' design capacities (Neegan Burnside, 2011i). The assessment also stated that 5% of the First Nations homes in the province of Manitoba do not have minimum wastewater services, which is the highest among the all Canadian provinces (Neegan Burnside, 2011d).

The regional reports of Neegan Burnside (2011a–2011g) showed that 21% of all WWTS in CFNC exceed the federal effluent discharge limits of either biological oxygen demand (BOD), total suspended solids (TSS), or total fecal coliform (Figure 3) in 2011, which was almost the same as reported by the first national assessment in 2003 of 22% wastewater systems exceeding the effluent discharge guideline (INAC, 2003). The most recent data show that this figure is about 23% in 2014–2015 (AANDC, 2015). Therefore, the discharge limits of several effluents have not been regulated since 2003 and require concrete actions. The matter of more concern is that nationally about 45% wastewater systems do not have any effluent discharge data (Figure 3).

Quebec and Saskatchewan possessed the most vulnerable WWTS in terms of exceeding the effluent discharge limits (41 and 35%, respectively). On the other hand, Alberta, Atlantic, Yukon, and Ontario had the highest number of wastewater systems without any effluent discharge data, which were 88, 64, 55, and 51%, respectively. The new Wastewater System Effluent Regulations, announced in July 2012, were designed to address this gap of exceeding effluent discharge limits and obligatory requirements of keeping discharge data records (Minister of Justice, 2016).

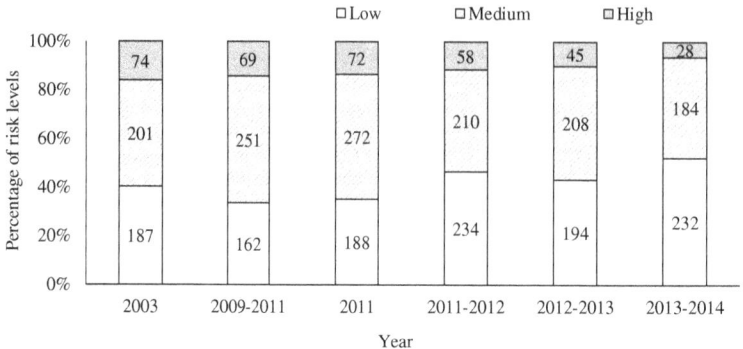

Figure 4. Trend of overall risk levels of WWTS in CFNC over the last 12 years since the first national assessment in 2003.

Notes: The value inside the bar represents the corresponding number of wastewater systems and the height of each bar represents the corresponding relative percentage at that risk level. Data adapted from ECCC (2016), INAC (2003) and Neegan Burnside (2011i). The number of lowrisk systems in 2014–2015 is 216, and medium- and high-risk systems data are not publicly available.

5. Evaluation of the risk levels of wastewater treatment systems in Canadian First Nations communities

According to the INAC Risk Level Evaluation Guidelines, the risk levels of WWTS in CFNC have been classified into three categories based on a scale of 1 to 10; low risk (1.0–4.0): systems having no problem or negligible minimal deficiencies, medium risk (4.1–7.0): systems requiring some repairs, and high risk (7.1–10): systems having potential health and safety issues. The overall risk of the system resides on five elemental weighing risks of effluent discharge (20%), system design (25%), operations (25%), reporting (10%), and operators (20%) of the respective wastewater treatment facility (INAC, 2011).

Recently, Environment and Climate Change Canada (ECCC, 2016) published a report of latest risk level scenario showing the improvement of WWTS in CFNC and indicated that high- and medium-risk systems have reduced from 14 and 51% of 2011 to 6 and 41% in 2013–2014, respectively.

The operational capacities of the First Nations WWTS have been depicted in the last assessment by Neegan Burnside (2011d, 2011i) that identified about 20% wastewater systems (113 systems) operating at or above the design capacity. Likewise, almost 40% wastewater systems of Saskatchewan (33 systems) and Alberta (27 systems) are operating in that concerned capacity range.

In 2003, nationally 16% (74 out of 462 systems) of the assessed systems were categorized as high risk, which needed immediate action to minimize potential health and safety concerns. The overall high-risk systems were slightly reduced to 14% (72 of 532 systems) in 2011. Moreover, the latest published data from the Environment and Climate Change Canada (ECCC, 2016) indicate a reduction in the high-risk systems to 6% of wastewater systems (28 out of 444 systems) in 2013–2014 (Figure 4). Ontario and Atlantic regions were more susceptible as a greater percentage of high-risk wastewater systems of about 36 and 25% were found in 2003 and 2011, respectively (Neegan Burnside, 2011i).

Although the percentage of high-risk WWTS in CFNC is in downward trend over the assessment period, the medium-risk level systems, which require certain improvements or system repairs, were increased from 44% (201 of 462 systems in 2003) to 51% (272 of 532 systems in 2011) (Figure 4). During the periods of 2011–2014, the medium-risk system was over 40%, which has slightly increased to 41% (184 of 444 systems) in 2013–2014. Alarmingly, more than 60% of the wastewater systems in the province of Alberta, Manitoba, and Quebec (60, 62, and 67%, respectively) have been categorized under the medium-risk level according to the National Assessment in 2011.

Despite the upward trend (increase from 40 to 52% between 2003 and 2013–2014) of the percentage of the low-risk systems in the national figure, the lowest value of 34% was found in 2009–2011 (Figure 4) and there were only 14 and 15% low-risk systems in Ontario and Quebec regions in 2011, respectively (Neegan Burnside, 2011i).

INAC developed four zones dividing the geographic areas of the First Nations communities based on the location and remoteness of the wastewater systems: Zone 1: First Nations located within 50 km of the service center with year-round road access, Zone 2: located between 50 km and 350 km of the service center with year-round road access, Zone 3: located over 350 km of the service center with year-round road access, and Zone 4: have no year-round road access to the service center.

The high-risk level is more prominent to the systems located in remote area and overall WWTS risks have increased with remoteness from 12% of the most readily accessible systems (Zone 1) to 29% of the least accessible systems (Zone 4). The reason could be the high transportation costs and unavailability of the resources. Number of people living in Zone 2 is much higher than any other zones where the high- and medium-risk systems are significantly greater than the low-risk systems. Although the medium-risk systems are almost evenly distributed among the zones, the number of medium-risks systems is almost half of the total systems (Table 2) in all zones. Similarly, 51% operators of WWTS had proper certification in the Zone 1, whereas only 27% operators of treatment systems were certified in Zone 4 (Neegan Burnside, 2011i).

6. Comparison of invested estimated cost for improving wastewater treatment systems

Although the main focus of this study was concentrated on First Nations wastewater services, a comparison study has been conducted between the actual investment and the estimated cost requirements. All the financial investment is pooled together for treating water and wastewater systems. There is lack of financial data solely for the WWTS (separated from water treatment system) in CFNC. Therefore, the related data of both water and wastewater systems have been considered from the previous assessments for the evaluation of invested amount and estimated costs, which include cost to upgrade existing systems to meet INAC's protocols, regulations, and federal and provincial standards and guidelines.

6.1. Investments by the Government of Canada

During the tenure of 1995 to 2003, the federal government provided $1.9 billion for the maintenance of water and wastewater services of the CFNC. Following the confirmation of significant risk to the quality and safety of drinking water of three-quarters of the First Nations systems (Auditor General of Canada, 2005), the Government of Canada considered First Nations water systems as a priority and approved $600 million in 2003 over the five-year period until 2008. This money was reserved for the improvement and repair of the water and wastewater systems under the First Nations Water Management Strategy (FNWMS), in addition to the regular annual departmental allocation of about

Table 2. First Nations' population, number of WWTS and percentage of systems at corresponding risk levels based on the four geographic zones

Zone	Population	No. of systems	High (%)	Medium (%)	Low (%)
Zone 1	95,212	177	12	44	44
Zone 2	127,587	238	9	56	35
Zone 3	13,558	24	4	58	38
Zone 4	68,522	93	29	57	14
Total	304,879	532	14	51	35

$200 million until 2016. In 2006, INAC secured $60 million from the previous budget and made it available over two years to achieve the goal of Plan of Action for the Drinking Water for the First Nations Community (INAC, 2007). In 2008, the FNWMS was reshaped to First Nations Water and Wastewater Action Plan (FNWWAP) and implemented the proposed action plan of INAC (2007).

Since 2008, the Government of Canada has approved $330 million over two years of four consecutive terms (about $165 million per year) until 2016 for the on-reserve First Nations water and wastewater systems' installation, operation, and maintenance through the FNWWAP. To complete the First Nations ongoing water and wastewater infrastructure, Canada's Economic Action Plan (CEAP) approved additional $165 million in 2009 budget for two years (INAC, 2015). In the budget of 2016, the Government of Canada approved $141.7 million for monitoring on-reserve water and $1.8 billion for construction, repair, and improvement of the on-reserve water and wastewater infrastructure over the next five years (2016–2021) (Department of Finance, 2016).

Overall, during the 10-year period from 2010–2011 to 2020–2021, the government of Canada has committed to providing about $4.2 billion (through the regular annual departmental allocation, FNWWAP, CEAP, and budget for 2016) for the Canadian First Nations' water and wastewater services.

6.2. Estimated costs by the national assessment

The national assessment report of INAC (2003) stated that there is a requirement of an estimated capital cost of $475–560 million to rectify the existing deficiencies, $185 million to address the back-logs, and $500 million to make water and wastewater services available to new houses Moreover, an additional $500 million is required to enrich the operational and maintenance practices, wastewater monitoring, proper training and education, developing standards, raising awareness, and to make emergency plans available in place. The estimated total cost was $ 1.7–1.8 billion over the five-year period (2003–2008) for the remedial action proposed by INAC (2003), which was almost equivalent to the total invested amount (about $1.7 billion) from the Government of Canada during the fiscal year of 2003–2008.

In the assessment report of Neegan Burnside (2011i), an estimate of $1.2 billion was claimed for the improvement of the existing systems, $4.7 billion for the recommended new servicing, and $419 million for the recommended operational and maintenance cost, which comprised a total of $6.3 billion over the 10-year period (2011–2021).

7. Recommendations: Challenges and opportunities

There are no publicly available data since the National Assessment (2011) report compiled by INAC, except for some partial information of the First Nations' wastewater risk levels data presented in the INAC departmental progress report. Therefore, to describe the First Nations' technological use, challenges to meet effluent discharge limits, zone-wise risk-level data, percentage of people receiving wastewater services, and many other vital issues have been presented in this review study based on the six-year-old data, and the recommendations are made based on the available information.

There are many operational and environmental concerns (for instance houses having no wastewater services, increase in septage surface discharge and effluent discharge guidelines, etc.) that have surfaced from the improper functioning of the WWTS in the CFNC, which were not well illustrated in the First Nations Water and Wastewater Action Plan (FNWWAP) (INAC, 2015); for example, occurrence of ground water contamination from leakage/drainage of liquids from lagoon, or the septic tank, or leaching bed drainage (Yates, 1985). These matters demonstrate the lack of available literature for the WWTS in CFNC, which are required for better understanding and to fix the flaws in the existing WWTS to minimize the health and environmental hazards to the First Nations people. Additionally, the regional roll-up reports of Neegan Burnside (2011a–2011i) also indicated that over-all 45% systems did not have effluent discharge data and a few of the provinces did not have a re-cord of any effluent discharge data at all. Therefore, it is highly recommended to monitor the discharged effluent limit to meet the INAC protocols and other federal and provincial regulations, and to instigate the researchers from the governmental organizations, corporate institutions, and university professors to overcome the difficulties faced by the CFNC.

The assessment report of Neegan Burnside (2011i) clearly specified that among the First Nations homes that were surveyed, 51% wastewater systems were identified at a medium overall risk, and most importantly 14% wastewater systems were classified at a high overall risk. Therefore, the ob-jectives of the national assessment of INAC (2003) were not met at the end of the five-year period in 2008, despite the allocation of $1.7 billion, which expresses the concern of either the incomplete-ness of the national assessment or the improper utilization (financial mismanagement) of the allo-cated funds towards the First Nations water and wastewater systems. Moreover, the amount of $4.2 billion declared in the Budget of 2016 (Department of Finance, 2016) to meet the FNWWAP for the 10-year period (2011–2021), is still lacking $2.1 billion from the estimated cost of $6.3 billion (Neegan Burnside, 2011i). For this reason, continuous tracking of the financial investment released along with the actual progress to meet the objective of the INAC protocol is very crucial and therefore, it is ad-visable to increase the budget as well as to concentrate more on proper utilization of the allocated budget to achieve the initial commitment of the Government of Canada. Furthermore, regular as-sessment of the risk level of the systems is recommended for the thorough evaluation of the First Nations wastewater systems.

Wastewater stabilization ponds are the most commonly used systems in CFNC. However, several incidents have indicated that nutrients' (Nitrogen and Phosphorus) concentration in the discharge effluent of sewer lagoons exceeded the threshold limit of effluent discharge set by Environment Canada. Hence, implementation of wastewater management strategies like constructed wetlands (where possible) or natural wetlands are advisable, which prevents releasing the sewer lagoon efflu-ent directly to the fresh source water or lake to minimize the nutrient loads. Moreover, to keep the nutrient discharge (unionized ammonia, nitrogen, phosphorous) within the threshold in the effluent discharge of the wastewater stabilization ponds in CFNC, it is advised to perform the feasibility study of using limestone rock filters, active slag filters, or some form of chemical treatment (aluminum or iron salts).

8. Conclusion

The risks level of First Nations water and wastewater systems greatly affect Canada's performance on key environmental sustainability issues and are considered as a part of the Canadian Environmental Sustainability Indicators (CESI) program. The risk-level indicator helps measure the progress toward the goals of the Federal Sustainable Development Strategy (2013–2016). From this study, it is concluded that since the commencement of national assessment on First Nations water and wastewater systems, an improvement is noticeable in First Nations wastewater systems. However, there are many First Nations houses that do not have minimum wastewater treatment services and many WWTS neither meet effluent discharge limits nor have effluent discharge records, which has become a serious health and environmental concern. High-risk wastewater systems have considerably reduced but the medium-risk systems have risen in numbers. The environmental con-cerns associated with these systems diminish the positive sign for Canada's performance on key environmental sustainability issues. Therefore, more funding would be required on a provincial and a federal level to meet the government's goal to continuously improve the wastewater treatment facilities of First Nations communities. Although the study is focused on the WWTS in CFNC, the find-ings and recommendations could also be applicable for the remote small communities outside of Canada where new wastewater regulations are implemented for the small and remote communi-ties, changes in law that affect the communities' effluent discharge limits or bringing new programs and policies for the improvement of the small communities WWTS.

Competing interests
The authors declare no competing interest.

Funding
This work was supported by Natural Sciences and
Engineering Research Council of Canada (NSERC) [CREATE
H2O].

Author details
Mofizul Islam[1]
E-mail: islammm6@myumanitoba.ca
Qiuyan Yuan[1]
E-mail: qiuyan.yuan@umanitoba.ca
[1] Department of Civil Engineering, University of Manitoba,
  Winnipeg, MB R3T 5V6 Canada.

References
AANDC. (2012a). *Water and wastewater infrastructure in First
    Nations communities: A national perspective.* Ottawa, ON:
    Author.
AANDC. (2012b). Water and wastewater report. Author.
    Retrieved from https://www.aadnc-aandc.gc.ca/DAM/
    DAM-INTER-HQ-ENR/STAGING/texte-text/enr_wtr_nawws
    _2010_2012a_1352820960474_1358000571860_eng.pdf
AANDC. (2015). Aboriginal affairs and northern development
    Canada and Canadian polar commission, 2014–15
    departmental performance report. Retrieved from https://
    www.aadnc-aandc.gc.ca/DAM/DAM-INTER-HQ-AI/
    STAGING/texte-text/dpr-14-15_1453385264963_eng.pdf
Assembly of First Nations. (2008). Canada-wide strategy for
    the management of municipal wastewater effluent and
    environment Canada's proposed regulatory framework
    for wastewater impacts for First Nations communities.
    Retrieved from http://www.afn.ca/uploads/files/regional-
    impact-analysis.pdf
Auditor General of Canada. (2005). *Report of the commissioner
    of the environment and sustainable development—
    drinking water in First Nations communities.* Minister of
    Public Works and Government Services Canada. Retrieved
    from http://www.oag-bvg.gc.ca/internet/docs/
    c20050905ce.pdf
Canadian Encyclopedia. (2015). *Indigenous people: Social
    conditions.* Author. Retrieved from http://www.
    thecanadianencyclopedia.ca/en/article/
    native-people-social-conditions/
Carlson, J. C., Anderson, J. C., Low, J. E., Cardinal, P., MacKenzie,
    S. D., Beattie, S. A., ... Hanson, M. L. (2013). Presence and
    hazards of nutrients and emerging organic
    micropollutants from sewage lagoon discharges into
    Dead Horse Creek, Manitoba, Canada. *Science of the Total
    Environment, 445,* 64–78. doi:10.1016/j.
    scitotenv.2012.11.100
CBC News. (2007). *Kashechewan: Water crisis in Northern
    Ontario.* Kashechewan, ON: Author. Retrieved from http://
    www.cbc.ca/news2/background/aboriginals/
    kashechewan.html
Department of Finance. (2016). *Budget 2016: Growing the
    middle class* (M. P. M. of F. William Francis Morneau, P.C.,
    Ed.). House of Commons. Retrieved from http://www.
    budget.gc.ca/2016/docs/plan/budget2016-en.pdf
ECCC. (2016). *Canadian environmental sustainability indicators:
    First Nations water and wastewater system risk.*
    Environment and Climate Change Canada. Retrieved from
    https://www.ec.gc.ca/indicateurs-indicators/EA902CF7-
    9D5E-4D92-9490-1E8031B5890A/

FirstNationsWaterFacilities_EN.pdf
FCM. (2011). *First Nations—municipal community infrastructure
    partnership program, service agreement toolkit* (2nd ed.).
    Ottawa, ON: Federation of Canadian Municipalities.
    Retrieved from https://www.fcm.ca/Documents/tools/
    CIPP/CIPP_Toolkit_EN.pdf
God's Lake First Nations. (2013). God's Lake First Nations—
    Sewage treatment plant and associated works. Retrieved
    December 28, 2016, from http://www.merx.com/English/
    SUPPLIER_Menu.Asp?WCE=Show&TAB=1&PORTAL=MERX&S
    tate=7&id=260643&src=osr&FED_ONLY=0&ACTION=&rowc
    ount=&lastpage=&MoreResults=&PUBSORT=2&CLOSESORT=
    0&IS_SME=N&hcode=JkDJIaKsVVPXIzBZmWMjMg%3D%3D
Government of Canada. (2016). Statement by Minister Carolyn
    Bennett—Government of canada steadfast in
    commitment to end long-term drinking water advisories
    on reserve. Retrieved December 27, 2016, from http://
    news.gc.ca/web/article-en.do?nid=1124109
Health Canada (2016). *Drinking water and wastewater–First
    Nations and Inuit Health Canada.* Author. Retrieved from
    http://www.hc-sc.gc.ca/fniah-spnia/promotion/public-
    publique/water-eau-eng.php#s3a
Heinke, G. W., Smith, A. N. D. W., & Finch, D. G. R. (1991).
    Guidelines for the planning and design of wastewater
    lagoon systems in cold climates. *Canadian Journal of Civil
    Engineering, 18,* 556–567. Retrieved from http://www.
    nrcresearchpress.com/doi/pdfplus/10.1139/l91-068
    https://doi.org/10.1139/l91-068
INAC. (2003). *National assessment of water and wastewater
    systems in First Nations communities.* Indian and
    Northern Affairs Canada. 35. Retrieved from http://www.
    aadnc-aandc.gc.ca/DAM/DAM-INTER-HQ/STAGING/texte-
    text/watw_1100100016374_eng.pdf
INAC. (2007). *Plan of action for drinking water in First Nations
    communities—progress report.* Retrieved from https://
    www.aadnc-aandc.gc.ca/DAM/DAM-INTER-HQ/STAGING/
    texte-text/pad07_1100100034969_eng.pdf
INAC. (2010a). *Protocol for centralised wastewater systems in
    First Nations communities. Standards for design,
    construction, operation, maintenance, and monitoring of
    centralised wastewater systems.* Indian and Northern
    Affairs Canada. Retrieved from http://www.ainc-inac.gc.
    ca/enr/wtr/index-eng.asp
INAC. (2010b). *Protocol for decentralised water and wastewater
    systems in First Nations communities (Decentralised
    Systems Protocol).* Indian and Northern Affairs Canada.
    Retrieved from http://www.aadnc-aandc.gc.ca/DAM/
    DAM-INTER-HQ/STAGING/texte-text/
    dsp_1100100034992_eng.pdf
INAC. (2011). *Fact Sheet—risk assessment of water and
    wastewater systems in First Nations communities.*
    Retrieved January 3, 2017, from http://www.aadnc-aandc.
    gc.ca/eng/1313687144247/1313687434335
INAC. (2014). *First Nations people in Canada.* Retrieved
    December 28, 2016, from https://www.aadnc-aandc.
    gc.ca/eng/1303134042666/1303134337338
INAC. (2015). *First Nations water and wastewater action plan.*
    Retrieved January 2, 2017, from https://www.aadnc-
    aandc.gc.ca/eng/1420658157414/1420658186376
INAC. (2016). *Roles and responsibilities.* Retrieved December
    27, 2016, from https://www.aadnc-aandc.gc.ca/
    eng/1314034319353/1314034564208
Mallett, L. (2010). A sustainable wastewater treatment
    screening tool for small remote and First Nation
    communities. *Masters Abstracts International, 49*(3), 115.
    Retrieved from http://search.proquest.com.uml.idm.oclc.
    org/espm/docview/912919141/6F3136FD3C3E465CPQ/1?
    accountid=14569
Minister of Justice. (2016). *Wastewater systems effluent
    regulations.* Author. Retrieved from http://laws-lois.justice.
    gc.ca/PDF/SOR-2012-139.pdf

Neegan Burnside. (2011a, January/April). *National assessment of First Nations water and wastewater systems: Alberta regional roll-up report final*. Indian and Northern Affairs Canada. Retrieved from http://www.aadnc-aandc.gc.ca/DAM/DAM-INTER-HQ/STAGING/texte-text/enr_wtr_nawws_rurnat_rurnat_1313761126676_eng.pdf

Neegan Burnside. (2011b, April). *National assessment of First Nations water and wastewater systems: Atlantic regional roll-up report final*. Indian and Northern Affairs Canada. Retrieved from http://www.aadnc-aandc.gc.ca/DAM/DAM-INTER-HQ/STAGING/texte-text/enr_wtr_nawws_rurnat_rurnat_1313761126676_eng.pdf

Neegan Burnside. (2011c, January). *National assessment of First Nations water and wastewater systems: British Columbia regional roll-up report final*. Indian and Northern Affairs Canada Prepared. Retrieved from https://www.aadnc-aandc.gc.ca/DAM/DAM-INTER-HQ/STAGING/texte-text/enr_wtr_nawws_rurbc_rurbc_1315622900946_eng.pdf

Neegan Burnside. (2011d, January). *National assessment of First Nations water and wastewater systems: Manitoba regional roll-up report*. Indigenous and Northern Affairs Canada. Retrieved from https://www.aadnc-aandc.gc.ca/eng/1314634863253/1314634934122#chp3_3_1

Neegan Burnside. (2011e, January). *National assessment of First Nations water and wastewater systems: Ontario regional roll-up report final*. Indigenous and Northern Affairs Canada. Retrieved from https://www.aadnc-aandc.gc.ca/DAM/DAM-INTER-HQ/STAGING/texte-text/enr_wtr_nawws_ruront_ruront_1314635179042_eng.pdf

Neegan Burnside. (2011f, January). *National assessment of First Nations water and wastewater systems: Quebec regional roll-up report final*. Indigenous and Northern Affairs Canada. Retrieved from https://www.aadnc-aandc.gc.ca/eng/1314634863253/1314634934122#chp3_3_1

Neegan Burnside. (2011g, January/April). *National assessment of first nations water and wastewater systems: Saskatchewan regional roll-up report final*. Indian and Northern Affairs Canada. Retrieved from http://www.aadnc-aandc.gc.ca/DAM/DAM-INTER-HQ/STAGING/texte-text/enr_wtr_nawws_rurnat_rurnat_1313761126676_eng.pdf

Neegan Burnside. (2011h, April). *National assessment of First Nations water and wastewater systems: Yukon regional roll-up report—final*. Indian and Northern Affairs Canada. Retrieved from http://www.aadnc-aandc.gc.ca/DAM/DAM-INTER-HQ/STAGING/texte-text/enr_wtr_nawws_rurnat_rurnat_1313761126676_eng.pdf

Neegan Burnside. (2011i, April). *National assessment of First Nations water and wastewater systems—national roll-up report*. Indigenous and Northern Affairs Canada. Retrieved from http://www.aadnc-aandc.gc.ca/DAM/DAM-INTER-HQ/STAGING/texte-text/enr_wtr_nawws_rurnat_rurnat_1313761126676_eng.pdf

Ontario Clean Water Agency. (2001). *Assessment study of water and wastewater systems and associated water management practices in Ontario First Nations communities*. Retrieved from http://www.cbc.ca/slowboil/pdfs/on/Kasabonika.pdf

Pictou Landing First Nation. (2012). *Sewer treatment plant*. Retrieved December 28, 2016, from http://www.plfn.ca/departments/capital-om-main-page/sewer-treatment-plant/

Pratt, C., Shilton, A., Pratt, S., Haverkamp, R. G., & Bolan, N. S. (2007). Phosphorus removal mechanisms in active slag filters treating waste stabilization pond effluent. *Environmental Science & Technology, 41*(9), 3296–3301. doi:10.1021/es062496b

Rusten, B., Eikebrokk, B., Ulgenes, Y., & Lygren, E. (2006). Design and operations of the Kaldnes moving bed biofilm reactors. *Aquacultural Engineering, 34*(3), 322–331. doi:10.1016/j.aquaeng.2005.04.002

Senate of Canada. (2007). *Safe drinking water for First Nations—Final report of the standing senate committee on aboriginal peoples*. Retrieved December 28, 2016, from http://www.parl.gc.ca/Content/SEN/Committee/391/abor/rep/rep08jun07-e.htm

Simeone, T. (2010). *Safe drinking water in first nations communities*. Ottawa: Library of Parliament (2010), (Publication No. 08-43-E). Retrieved from http://www.lop.parl.gc.ca/content/lop/ResearchPublications/prb0843-e.pdf

Spellman, F., & Drinan, J. (2014). *Wastewater stabilization ponds*. Retrieved from https://books.google.ca/books?hl=en&lr=&id=VmzOBQAAQBAJ&oi=fnd&pg=PP1&dq=wastewater+stabilization+ponds&ots=DEX2wqFjIg&sig=ZbCUq-ITTf4-Y4Tie-4tenVB tU https://doi.org/10.1201/b16787

Strang, T. J., & Wareham, D. G. (2006). Phosphorus removal in a waste-stabilization pond containing limestone rock filters. *Journal of Environmental Engineering and Science, 5*(6), 447–457. doi:10.1139/s06-017

Tomaszek, J. & Koszelnik, P. (Eds.). (2015). *Progress in environmental engineering*. Boca Raton, FL: CRC Press. https://doi.org/10.1201/b18261

US EPA. (2010). *Nutrient control design manual* (T. G. Moore, Ed.). Ohio: Author.

White, J. P., Murphy, L., & Spence, N. (2012). Water and Indigenous peoples: Canada's paradox. *International Indigenous Policy Journal, 3*(3). https://doi.org/10.18584/iipj.2012.3.3.3

Yates, M. V. (1985). Septic tank density and ground-water contamination. *Ground Water, 23*(5), 586–591. doi:10.1111/j.1745-6584.1985.tb01506.x

13

Potential lead hazards in pre-1978 childcare facilities in Southern Nevada

Melissa J. Marshall[1], Jessica M. Weislogel[1] and Shawn L. Gerstenberger[1]*

*Corresponding author: Shawn

L. Gerstenberger, Department of Environmental and Occupational Health, School of Community Health Sciences, University of Nevada, Las Vegas, 4505 S. Maryland Parkway, Box 453064, Las Vegas, NV 89154-3064, USA

E-mail: shawn.gerstenberger@unlv.edu

Reviewing editor: Conor Buggy, University College Dublin, Ireland

Abstract: Communities continue to find lead in buildings and manufactured goods, placing children at risk for negative health effects. This study sought to determine the presence or absence of traditional and non-traditional lead hazards in the total population of pre-1978 licensed childcare facilities in Clark County, Nevada ($N = 94$) through lead risk assessments. Analysis suggests that the pre-1978 structures in Clark County Nevada, USA do not follow national trends pertaining to the prevalence of lead-based paint, dust, and soil. Of the 94 facilities assessed: 30 (31.9%) contained lead-based paint, 41 (43.6%) contained leaded tile, 9 (9.5%) had dust exceeding EPA clearance standards, and 7 (7.4%) facilities had playground equipment test positive for lead. These results confirm the need for continued monitoring of traditional and non-traditional sources of lead to prevent unnecessary lead exposure in childcare facilities.

Subjects: Environment & Health; Environmental Health & Safety; Children and Youth; Community Health; Environmental health

Keywords: lead hazards; childcare facilities; lead risk assessment

1. Introduction

Human activities are the primary cause of adverse environmental and human health impacts caused by lead (Bellinger, 2004; Kessler, 1995). Environmental levels of lead remain on the rise due to

ABOUT THE AUTHORS

The authors' research focus is on childhood lead poisoning prevention and the impact of the built environment on human health. Gerstenberger, PhD, founded the Nevada Healthy Homes Partnership and has multiple collaborative grants with key community partners such as the Southern Nevada Health District, City of Henderson, City of Las Vegas, and the Nevada State Health Division. The authors have worked to identify and reduce lead and healthy homes hazards throughout Southern Nevada supported by grants from agencies such as the Centers for Disease Control and Prevention, US Department of Housing and Urban Development, and the Environmental Protection Agency.

PUBLIC INTEREST STATEMENT

Communities continue to find lead in buildings and manufactured goods, placing children at risk for negative health effects. This study sought to determine the presence or absence of traditional and non-traditional lead hazards in the total population of pre-1978 licensed childcare facilities in Clark County, Nevada ($N = 94$) through lead risk assessments. Analysis suggests that the pre-1978 structures in Clark County Nevada, USA do not follow national trends pertaining to the prevalence of lead-based paint, dust, and soil. Of the 94 facilities assessed: 30 (31.9%) contained lead-based paint, 41 (43.6%) contained leaded tile, 9 (9.5%) had dust exceeding EPA clearance standards, and 7 (7.4%) facilities had playground equipment test positive for lead. These results confirm the need for continued monitoring of traditional and non-traditional sources of lead to prevent unnecessary lead exposure in childcare facilities.

continued use in applications ranging from common household items to industrial applications. Lead can be found not only in traditional sources (paint, dust, soil, and water), but also in non-traditional sources like ceramic tile, plastic products, vinyl blinds, metal, jewelry, home remedies, and artificial turf (Gorospe & Gerstenberger, 2008; Jacobs et al., 2002; Maas, Patch, Morgan, & Pandolfo, 2005).

However, there are no beneficial uses for lead in the human body and it is especially toxic to children under the age of six years old (Lanphear, Dietrich, Auinger, & Cox, 2000). Reduced cognitive functioning, poor academic achievement, and behavioral issues are just a few of the well documented adverse effects of childhood lead poisoning (Needleman, 1980). The Centers for Disease Control and Prevention (CDC) estimates that there are at least 500,000 children under the age of six with blood lead levels above 5 µg/dL, the blood lead reference level for public health action (Lead, 2017). Therefore, it is extremely important to identify and limit childhood exposure to lead.

In 2006, the Southern Nevada Health District (SNHD) received a Childhood Lead Poisoning Prevention Program (CLPPP) grant from the CDC. A key aspect of the program was primary prevention aimed at determining what lead hazards existed in the community and how to abate them before children could become lead poisoned. The SNHD in partnership with the University of Nevada, Las Vegas (UNLV) implemented a lead hazard risk assessment program for all pre-1978, health-permitted childcare facilities located in Clark County, Nevada. This project focused on locating and identifying potential and actual lead hazards in childcare facilities. Limited research of lead in childcare facilities existed in the literature at the time of this project in 2008.

Weismann, Dusdieker, Cherryholmes, Hausler, & Dungy (1995) evaluated the amount of lead levels in six university affiliated day cares in 1995. Five of the centers were built before 1940 and the last was built in 1959. Their results showed that all had elevated lead-based paint levels; two had elevated window sill lead dust levels, and one had elevated window well lead dust levels (Weismann et al., 1995). Elevated water lead levels were also found in one facility, as were elevated soil levels of lead (Weismann et al., 1995). However, none of the facilities had elevated floor lead dust levels (Weismann et al., 1995). The Weisman et al. findings proved that lead could be found in day care facilities after the ban of residential lead-based paint in 1978.

Viverette et al. (1996) also conducted a study of four day care centers in New Orleans in 1996. They assessed public and private centers in both the inner city and outer city locations. Surprisingly, the inner city private center had more severe lead hazards than the in public center (Viverette et al., 1996). This was due to the differences in conditions and ground cover of the two facilities. The public day care had lower lead contamination because it had no bare soil and newer high quality play equipment, despite both being located in high vehicle traffic areas. Viverette et al. (1996) concluded that lead dust contamination had less to do with location of the day care center and was related more to the condition of the equipment and soil (Viverette et al., 1996).

The Department of Defense also conducted a study in 2002 of playground equipment to determine if any lead hazard risks were present (Belfit, Nix, & Graham, 2002). Belfit et al. found that 37% of US Department of Defense playground equipment exceeded the lead dust on residential floors standard, 40 µg/ft^2, set by the US Environmental Protection Agency (US EPA) (Belfit et al., 2002). Findings by both Viverette et al. and Belfit et al. demonstrated a need to test playground equipment at childcare facilities to ensure the safety of children's outdoor environments regardless of location.

Additional studies have been completed after the SNHD and UNLV project in 2008. Greenway and Gerstenberger (2010) documented that leaded toys existed at seven of the ten day care facilities they tested in Clark County, Nevada in 2010. Leaded day care toys (Sanders, Stolz, & Chacon-Baker,

2013) were also discovered in 12% of the overall toys tested in seven Southern New England facilities in 2013. Lead in drinking water above US EPA standard of 20 ppb (ppb) was also reported in 28% of first draw water samples in schools (Barn & Kosatsky, 2011) in Ontario, Canada in 2011 and 3.6% of first draw water samples from five South Central Kansas preschools and primary schools (Massey & Steele, 2012) in 2012. All of these studies confirm the continued need to monitor potential lead sources in childcare facilities.

Given the fact that dust, water, soil and playground equipment have been shown to contain to elevated lead levels in playgrounds and child care facilities, an investigation of all pre-1978 licensed childcare facilities was initiated. The purpose of this study was identify any possible sources of lead in facilities that posed a hazard to children, and suggest appropriate actions that need to be taken to attenuate or eliminate these lead hazards.

2. Methods

The SNHD is the regulating agency in charge of ensuring a safe, clean environment for children who are cared for in a group setting such as a Family Care Home (FCH), Group Care Home (GCH), and Commercial Childcare Center (CCC) in Clark County, NV. Childcare facilities are classified based on the maximum number of children that may attend each facility. A FCH may serve up to six children at one time and typically operates out of the caregivers private residence. Similarly, a GCH is permitted to care for up to twelve children at one time and; GCHs also typically operate out of the caregivers private residence. For the purpose of this study FCHs and GCHs were grouped into one category as they both typically operate out of private residences. Childcare Centers are permitted to care for twelve or more children and generally operate in a commercial center, large church or school facility. However, a few CCCs were found to operate out of converted private residences. Childcare facilities on whole are prohibited from operating out of a condominium or an apartment.

The SNHD issues permits to all FCHs, GCHs and CCCs that hold a business license in Clark County. At the time of the study, SNHD issued and regulated permits for over 600 childcare facilities, built both pre- and post-1978.

2.1. Selection of childcare facilities

Childcare facilities constructed in or before 1978 were selected to be tested for lead hazards. The SNHD maintains the addresses and owner information of the permitted childcare facilities in an internal database. The age of construction of each facility was determined by cross checking the SNHD address on file with the year of construction recorded on the Clark County Assessor's website (http://www.clarkcountynv.gov/assessor/Pages/RecordSearch.aspx). There were 94 permitted childcare facilities that were constructed in or pre-1978.

It is possible that new permits were added or deleted from the SNHD master list of childcare facilities during the course of the study. So all active pre-1978 childcare facilities on file at the SNHD as of January 31, 2009 were included. It was possible that unpermitted childcare facilities in Clark County were being run out of private residences operating as family care homes or group care homes without proper health permits or business licenses. These unpermitted facilities were not included in this study as they could not be accounted for.

All childcare facilities permitted by the SNHD received a letter from an inspector informing them of the SNHD Childcare Facility Lead Risk Assessment Program. Each center was informed that mandatory lead testing was to be performed at all childcare facilities constructed in or pre-1978. The Childcare Lead Risk Assessment Notice Letter also contained information on a voluntary toy risk assessment program offered by fellow graduate students at the University of Nevada, Las Vegas. SNHD inspectors focused on delivering the letters to all pre-1978 childcare facilities in their assigned districts. Districts were determined by zip codes to ensure accurate tracking of notice deliveries. The SNHD required letters to be delivered at least two weeks prior to a phone call that was placed to schedule an appointment for an on-site visit.

An excel spreadsheet was created for all of the scheduled lead risk assessments. The spreadsheet contained information on the childcare facility name, the lead case number (SNHD-CC-XXX), address, the SNHD permit number, the Clark County parcel number, the year of construction, the delivery date of the Childcare Lead Risk Assessment Notice Letter, the date of lead risk assessment, and if lead was present, where it was located.

2.2. Lead risk assessments

Upon arrival at a facility, the lead risk assessment team explained the lead risk assessment procedure to the facility operator and asked if there were any questions. An operator was either the facility's manager facility/homeowner or person in charge at the time of the inspection. The operators of the FCHs and GCHs were given the option to have their entire facility (house) screened for lead. If that offer was declined, it was noted in the final report and only the childcare areas (kitchen, bathrooms, outdoor play areas, and rooms that were utilized by the children) were screened, as those were the only areas permitted by the SNHD. Once all necessary paperwork was completed, the lead risk assessment began.

EPA-certified Lead Risk Assessors from the SNHD and UNLV completed inspections. Lead Risk Assessors used one of two Niton XLp300A series X-ray Fluorescence (XRF) analyzers to complete lead risk assessments in accordance with EPA and US Department of Housing and Urban Development (HUD) guidelines (U.S. Department of Housing & Urban Development, 1995). A minimum of three calibration readings were taken and recorded prior to the lead risk assessment, at the end, and every four hour interval the assessment lasted. Calibrations were taken using the National Institute of Standards and Technology (NIST) Standard Reference Material (SRM) nominal 1.0 mg/cm^2 paint film. The XRFs were used to test interior and exterior painted surfaces and structures for lead-based paint concentrations. A lead concentration of greater than or equal to 1.0 mg/cm^2 is identified as the threshold for a lead-based paint hazard by the EPA/HUD guidelines (U.S. Department of Housing & Urban Development, 1995). All XRF readings were recorded and photographs taken of lead positive surfaces or structures. A diagram of each facility was also created at the beginning of each risk assessment; rooms labeled "No Access" denoted non-childcare areas of FCH and GCH private dwellings.

2.3. Soil samples

Composite soil samples were collected in accordance with EPA/HUD guidelines using either the straight drip line or X formation technique for play areas, dependent on the location of bare soil (U.S. Department of Housing & Urban Development, 1995). Soil samples were taken by an EPA-certified Lead Risk Assessor using a soil corer to remove the top half inch of bare soil and collecting it in non-sterilized polyethylene 50 mL sample tubes. All soil samples were recorded and sample tubes labeled with location taken, date, and case number. Soil samples were collected from all facilities unless no bare soil was present at a facility; sandbox soil samples were also taken if present. Samples were to sent to a National Lead Laboratory Accreditation Program (NLLAP)—accredited laboratory for analysis via Graphite Furnace Atomic Absorption Spectroscopy (GFAAS) or Inductively Coupled Plasma Mass Spectrometer (ICP/MS). Samples did not require refrigeration and were stored in plastic bags labeled by case number at SNHD until several sets of samples were collected, and sent to the NLLAP-accredited laboratory. The longest hold time was approximately two weeks between sample collection and mailing.

2.4. Water samples

The facility operator or employee collected water samples from their respective childcare facilities. An EPA quick reference guide to the Lead and Copper Rule for schools and childcare facilities was provided to each facility along with directions for water sampling (U.S. Environmental Protection Agency, 1992). The SNHD inspector explained water collection directions and emphasized water must be undisturbed in the pipes for 6–8 h before the sample could be taken. SNHD also provided a one liter plastic sample bottle labeled with a case number, and a collection form to record the time and date of last water use, time and date of water collection, and signature of facility employee collecting the sample to each facility. The SNHD inspector retrieved the collected water sample from

the facilities after they were completed. All water samples were sent to an NLLAP-accredited laboratory for analysis using GFAAS. Samples did not require refrigeration and the one liter plastic bottles
labeled by case number were stored at SNHD until several sets of samples were collected, and sent
to the NLLAP-accredited laboratory. The longest hold time was approximately two weeks between
sample collection and mailing.

2.5. Dust wipe samples

HUD/EPA and American Society for Testing and Materials (ASTM) dust technician standards (American
Society for Testing & Materials International, 2003) for collecting dust samples version ASTM E1728.
EPA-certified Lead Risk Assessors collected dust wipes using Ghost Wipes (Environmental Express,
Mt. Pleasant, SC), non-sterilized polyethylene 50 mL sample tubes, and pre-formed plastic templates
measuring 0.5 ft^2 for floors and 0.25 ft^2 for window sills (BTS Laboratories, Richmond, VA). Dust wipe
samples were recorded on a sampling form indicating date, case number, location of sample, and
dimensions of the sample area. All sample tubes were labeled in the same manner and stored in a
collection bag labeled by case number and date. A chain of custody form was completed with this
information for each sample before sent for analysis.

A minimum of two dust wipe samples were collected from the entryway and play area of each
facility. These samples could be taken from floors, window sills, or other areas children frequented
throughout the childcare facility. Additional dust samples were collected from areas testing positive
for lead via XRF at the discretion of the Lead Risk Assessor. All dust wipe samples were sent to a
NLLAP-accredited laboratory for analysis using GFAAS or ICP/MS. Samples did not require refrigeration and were stored in the collection bags labeled by case number at SNHD until several sets of
samples were collected, and sent to the NLLAP-accredited laboratory. The longest hold time was
approximately two weeks between sample collection and mailing.

2.6. Playground equipment samples

Large pieces of anchored playground equipment were tested with the XRF. When a piece of equipment revealed a lead concentration above 1.0 mg/cm^2 standard for paint as determined by the XRF,
a dust sample was also collected of that area. The dust samples were sent to a NLLAP-accredited
laboratory for analysis using GFAAS or ICP/MS. Samples did not require refrigeration and were stored
in the collection bags labeled by case number at SNHD until several sets of samples were collected,
and sent to the NLLAP-accredited laboratory. The longest hold time was approximately two weeks
between sample collection and mailing.

2.7. Lead standards

The EPA and HUD have standards (EPA, 2000) for lead in dust, soil, and water (Table 1). This study
utilized the standards for floors and window sills because window troughs are difficult to test, as the
window units in Las Vegas, Nevada vary from the traditional windows with large troughs used in the
Midwest and on the East Coast and instead are narrow and traditionally made of vinyl or metal, not

Table 1. Lead clearance standards set by the United States Environmental Protection Agency (EPA) and Department of Housing and Urban Development (HUD)	
Sample type	**Standards**
XRF analysis	• 1 mg/cm^2
Dust samples	• Interior floors (carpeted and uncarpeted) = 40 g/ft^2 • Interior window sills = 250 μg/ft^2 • Window troughs = 400 ug/ft^2
Soil samples	• Bare play area = 400 ppm • Bare soil in non-play areas = 1,200 ppm • Abatement required = 5,000 ppm
Water samples	• 15 ug/L or 15 ppb or 0.015 mg/L

Source: U.S. Department of Housing and Urban Development (1995).

wood. Playground equipment was held to the standard of 40 µg/ft² for floors to be considered positive for this study.

2.8. Risk assessment reports

A final risk assessment report was provided to each childcare facility participating in this study. The report explained the condition of the components tested and if they were an imminent lead risk hazard. It also provided a detailed description of the lead hazards found on the premises, and noted the required course of action to remediate, contain, or abate the lead hazards that were present.

2.9. Data analysis

SPSS version 17.0 for Windows® was used to perform the statistical analyses for this project. Those analyses included the Mann–Whitney U Test, Pearson's Correlation, Linear Regression, a one way ANOVA, descriptive analyses, and an Independent T-test between two groups. All statistical tests ran had non significant results. Therefore, for the purpose of this study, the results will be reported using descriptive statistics.

2.10. Quality control measures

Quality control measures were used for both the collection of XRF data and in the collection of dust samples. A blank dust sample was submitted with each batch of samples collected from a facility. Blanks were submitted to capture the background lead dust concentrations in the air and ensure that samples were not contaminated from improper dust wipes or contaminated gloves. The laboratory utilized for this project followed in house quality assurance and control procedures approved by NLLAP.

3. Results

Lead risk assessments were performed on all 94 childcare facilities in Clark County, NV, built before 1978 holding active permits with the SNHD at the time of the study. There were 41 Family Care Homes, 4 Group Care Homes, and 49 Commercial Childcare Facilities assessed. Figure 1 illustrates the profile of the childcare facilities assessed by frequency of the year of construction. A total of 11,465 XRF readings, 91 soil samples, 93 water samples, and 424 dust samples were collected. XRF readings collected from non-structural items (i.e. garden hoses, garden pots) were excluded, as the focus of this study was structural components not non-structural items; these types of items were not consistently tested.

3.1. Descriptive statistics

A simple linear regression failed to find a correlation between the age of the dwelling or facility and the number of lead hazards; the number of samples available that exceeded clearance standards limited the analysis. Therefore, this section will present only the descriptive statistics.

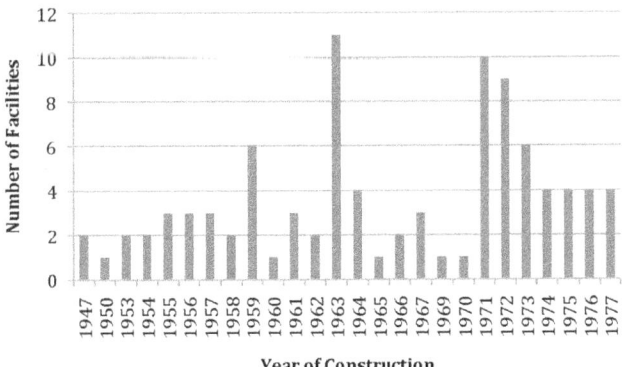

Figure 1. The number of Pre- 1978 permitted childcare facilities in Clark County, NV by year of construction.

3.2. Lead in paint and tile

Of the 94 facilities assessed, 30 (31.9%) contained lead-based paint. A positive reading for lead-based paint on the XRF is defined as a result greater than 1.0 mg/cm^2. A total of 9,076 total paint readings were taken; including quality control repeat readings. Of those, there were 194 (2.1%) lead-positive readings, excluding quality control readings. Values were determined to be paint readings by reviewing the substrate, component, color and notes that were recorded on the Excel spreadsheets and field forms. Lead in tile was also recorded during assessments. A total of 41 facilities (44%) contained lead-positive tile using the positive XRF lead-based paint definition. An independent T-test also found no statistical significance between the mean hazards for lead-based paint and leaded tile.

3.3. Lead in soil

A total of 91 soil samples were collected from the 64 facilities that had bare soil on the property. The range in the number of soil samples collected from the facilities was zero to eight samples; based on risk assessor discretion and the number of playgrounds or bare soil areas present. None of the samples collected exceeded the EPA's clearance standard of 400 ppm lead in soil for child play areas. The mean lead concentration for the 91 soil samples collected was 35.89 ppm, and the standard deviation (SD) was 40.77 ppm. The lead soil concentrations ranged from below the level of detection (LOD) less than 7–160 ppm. Soil concentrations of lead below the LOD were adjusted to 3.5 ppm to account for variations in laboratory detection limits. A total of 12 (12.7%) childcare facilities had soil lead concentrations equal or greater than 40 ppm, but none exceeded the clearance standard (400 ppm) for child play areas (Figure 2).

3.4. Lead in water

A total of 93 water samples were collected for the study. One facility did not have a water sample collected as it was remodeled in early 2000. None of the drinking water samples collected exceeded the EPA's Action Limit of 15 ppb lead in water. Five facilities (5.3%) had a detectable concentration of lead in the water ranging from 6–11 ppb (Figure 3). The remaining facilities had water concentrations below LOD of less than 5 ppb. Water samples below LOD were adjusted to 2.5 ppb to account for variations in laboratory detection limits.

3.5. Lead in dust

A total of 424 dust samples were collected from the childcare facilities, 323 (76%) of these samples were collected from interior floors and 101 (24%) samples were collected from window sills. The number of samples collected per facility ranged from two (study design minimum) to eighteen samples. The maximum number of samples collected was not established as each facility differed in square footage and number of lead hazards present; creating a need to collect a varied number of samples per facility.

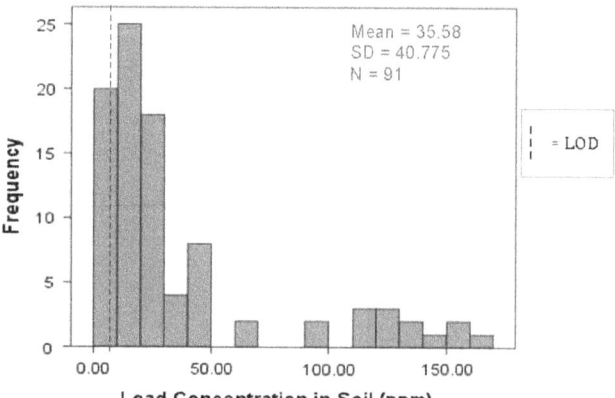

Figure 2. Soil lead concentration levels in parts per million (ppm) from Pre-1978 Clark County, NV permitted childcare facilities with bare soil.

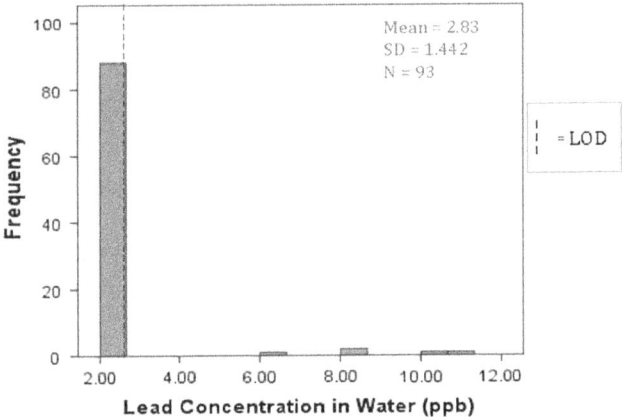

Figure 3. Lead concentration in water in parts per billion (ppb) from Pre-1978 Clark County, NV permitted childcare facilities.

The dust samples with lead concentrations below the LOD (floors 20 $\mu g/ft^2$, windowsills 40 $\mu g/ft^2$) were adjusted to half of the limit of detection, 10 $\mu g/ft^2$ for the floors and 20 $\mu g/ft^2$ for the window sills to account for variations in laboratory detection limits. Eleven (3.4%) floor dust samples were over the EPA clearance standard of 40 $\mu g/ft^2$ for interior floors, and 2 (1.9%) window sill samples exceeded the EPA clearance standard of 250 $\mu g/ft^2$ for window sills. The range of lead dust concentrations for interior floor samples was below LOD to 1200 $\mu g/ft^2$ with a mean of 15.33 $\mu g/ft^2$, SD = 66.76 $\mu g/ft^2$ (Figure 4). The range of lead dust concentration for window sills was below LOD to 740 $\mu g/ft^2$ with a mean of 42.48 $\mu g/ft^2$, SD = 96.32 $\mu g/ft^2$ (Figure 5).

3.6. Lead in playground equipment

A total of 50 childcare facilities had playground equipment that was tested for lead content using XRF techniques. Of the 50 facilities, a total of 98 individual pieces of playground equipment were tested resulting in 394 XRF readings. A total of 7 (7%) pieces of playground equipment were found to contain lead concentrations greater than 1 mg/cm^2 using the XRF. The substrates of these structures were categorized as metal or plastics. Positive readings were on 30.4% of painted metal and on 69.6% of plastics playground equipment. The range of positive lead concentrations was from 1.0 to 7.7 mg/cm^2, with a mean of 2.02 mg/cm^2. Pieces of playground equipment found to contain elevated lead levels had a dust wipe sample taken to see if lead was bioavailable via dust from the equipment. Dust samples collected from playground equipment were included in the floor dust analysis.

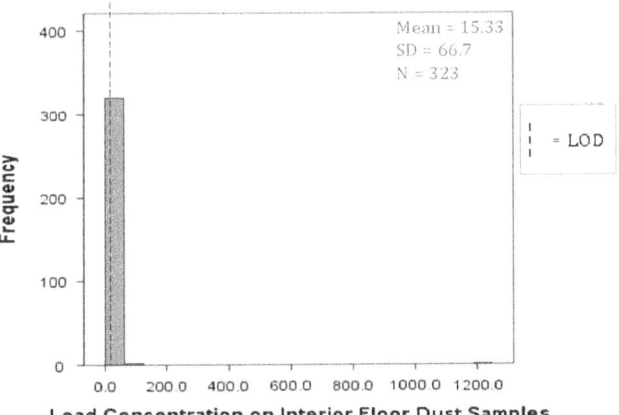

Figure 4. Lead concentration on interior floor dust samples in micrograms per foot squared ($\mu g/ft^2$) from Pre-1978 Clark County, NV permitted childcare facilities.

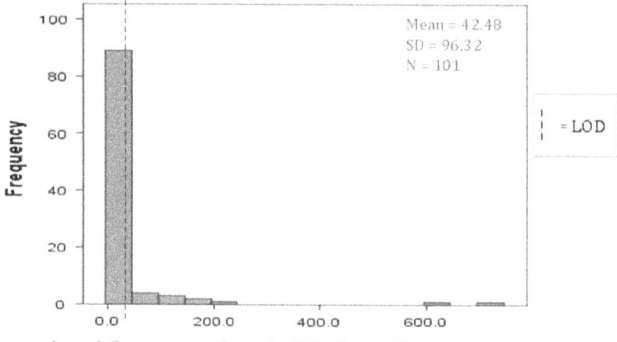

Figure 5. Lead concentration in window sill dust samples in micrograms per foot squared (µg/ft²) from Pre-1978 Clark County, NV permitted childcare facilities.

4. Discussion

There are many factors that contribute to the presence of lead hazards in a childcare facility. Studies have shown a direct correlation between the age of the dwelling or facility, and the number of lead hazards (Jacobs et al., 2002; Levin et al., 2008). This study was statistically unable to duplicate those results. However, descriptive statistics revealed that of the 94 facilities assessed for this study, 30 (31.9%) contained lead-based paint, 41 (43.6%) contained leaded tile, 9 (9.5%) had dust exceeding EPA clearance standards, and 7 (7.4%) facilities had positive playground equipment using the paint standard of greater than 1.0 mg/cm². None of the facilities contained lead concentrations in soil over clearance standards or had lead concentrations exceeding the Action Limit set for lead in water (U.S. Department of Housing & Urban Development, 1995; U.S. Environmental Protection Agency, 1992).

Most lead hazards occur due to the natural aging process of paints and substrates. When paints and substrates are not properly maintained, it can result in deteriorating paint that chips and flakes creating elevated lead concentrations in surrounding dust and soils (Jacobs et al., 2002). Fortunately, lead concentrations in soil do not appear to be the major route of exposure in the childcare facilities assessed in this study. Most of the soil samples collected correlated with the national background of lead levels (Agency for Toxic Substances & Disease Registry, 2007) of less than 10 to 30 ppm. This may be attributed to limited lead industries within urban areas of Clark County, NV and the newer housing stock built post-1978. However, a recent lead hazard control and healthy homes program in Henderson, NV reported 58 of 75 (77.3%) pre-1978 homes had lead-based paint identified during a lead inspection and risk assessment (Rufin, 2015). This finding is important for the FCHs and GCHs that are operated out of private residences. Despite having a smaller pre-1978 housing stock than other East Coast communities, the lead program in Henderson, NV still found a high percentage of lead-based paint that could potentially create lead hazards for children.

Most of the lead-based paint indentified in the study was found in good condition. This typically reduces the amount of lead dust produced. However, lead dust can result from alternative sources than paint. Environmental, take home, or non-traditional sources of lead can produce dust (Gorospe & Gerstenberger, 2008; Jacobs et al., 2002; Levin et al., 2008). One facility assessed did not have any lead-based paint present, but had an interior floor dust wipe with a lead concentration of 50 µg/ft², which is above the 40 µg/ft² clearance standard (U.S. Department of Housing & Urban Development, 1995). This sample was collected at the front entrance under a rubber welcome mat. Lead is used in a variety of consumer goods, including vinyl and polyvinyl chloride (Levin et al., 2008). Therefore, it is possible that the welcome mat itself was creating the lead dust above the clearance standard. Another facility had a positive interior floor sample with a lead concentration of 1,200 µg/ft². This sample was collected from a painted threshold that was found in deteriorated condition. SNHD mandated that this component be abated from the facility since it was a lead hazard. A set of mini-blinds located in the window sill also had lead dust above the clearance level (U.S. Department of Housing & Urban Development, 1995) of 250 µg/ft². The mini-blind dust wipe contained 600 µg/ft² and was removed by mandate of SNHD as well.

Playground equipment was also tested for lead due to increased awareness of lead in children's toys (Belfit et al., 2002; Greenway & Gerstenberger, 2010; Viverette et al., 1996; Weismann et al., 1995). It has been demonstrated that large playground structures contain lead in the paints and plastics to aid in color maintenance and integrity in extreme outdoor environments (Belfit et al., 2002; Greenway & Gerstenberger, 2010; Sanders et al., 2013; Viverette et al., 1996). Playground equipment assessed in this study contained both plastic and metal components. Seven of 94 (7.4%) large playground structures were found to have positive lead levels. Of the positive XRF readings, 30.4% were from painted metal, and 69.6% were from plastic components. Although only a few structures contained lead, dust wipes taken from a set of stairs on a slide contained 740 µg/ft^2 of lead dust. This result warrants further investigation and monitoring of playground equipment to prevent children from being exposed to lead dust that is bioavailable.

Clark County, NV is known for its extreme heat and intense summers. These factors could contribute to the breakdown of substrates or painted materials causing the release of leaded dust when lead is present in play equipment. The potential for this has already been considered for lead in artificial turf (Van Ulirsch et al., 2010). With the wear and tear, the outdoor playground equipment may produce leaded dust that could expose children. Furthermore, if proper hand washing techniques are not utilized after play, children could be further exposing themselves. Since Health Departments have jurisdiction over school and childcare facilities, they may want to implement annual screening of playground equipment for post-1978 facilities and upon addition of new equipment for leaded structures. All playground structures should be assessed using a Plastics XRF Analyzer using the lead in plastic standard of 600 ppm, and dust wiped to determine the bioavailability of lead to prevent future exposures (Ban of lead-containing paint & certain consumer products bearing lead-containing paint, 2001).

Another commonly found non-traditional source of lead in the childcare facilities was tile. The analysis determined that 43.6% of facilities had leaded tile, while only 31.9% of facilities had lead-based paint. While leaded tile was in a higher percentage of facilities than leaded paint, the US Consumer Product Safety Commission currently does not consider tile or ceramic glaze to fall under restrictions on the use of lead-based paint and products containing lead-based paint (Ban of lead-containing paint & certain consumer products bearing lead-containing paint, 2001). Jacobs et al. (2002) have already expressed their concern for leaded tile calling for further study to determine if it could be a source of lead exposure for children. A full investigation of the potential for tile to contribute to a lead dust hazard should be completed, especially during remediation and replacement efforts.

An additional unique non-traditional lead hazard was also discovered during the course of the study. Several childcare facility murals were painted using artists' paints. This would not be so interesting if artist's paints were included in the lead-based paint ban, however they were not (Ban of lead-containing paint & certain consumer products bearing lead-containing paint, 2001). These murals were a popular decoration in childcare facilities assessed in this study, and many were created using commercial paints. However, some of the positive lead-based paints identified during risk assessments were on murals created by artists' paints. SNHD recommended that childcare facilities ensure that the paint they use is of commercial grade that adheres to the standard (Ban of lead-containing paint & certain consumer products bearing lead-containing paint, 2001) set for lead in paints (600 ppm). Careful planning and education of employees can ensure that these paints are not used in facilities where children are present.

Lead was also a popular substrate used in plumbing before a 1,968 ban (Bryant, 2004). However, this was not a complete ban; lead was still permitted in certain plumbing materials up to specific allowable concentrations and did not address service lines containing lead and copper previously installed (Bryant, 2004). This is important because some childcare facilities, or even residential dwellings, may have leaded solder or lead service lines present post-1968 that contain lead. Without proper treatment applications by local water authorities, there is the risk of lead leaching into the water (Pieper, Tang, & Edwards, 2017). In Clark County, NV, the Southern Nevada Water Authority applies the treatment technique required to control for lead leaching and all of the water samples taken from childcare facilities were below the Action Limit of 15 ppb lead in water (U.S. Environmental Protection Agency, 1992). Only 5 of the 93 water samples collected during the study had lead concentrations above the laboratory's detection limit of 5 ppb. These samples ranged from 6 to 11 ppb of lead in water. However, the water crisis in Flint, Michigan is a prime example of how lead can leach into the water service lines when appropriate controls are negated (Pieper et al., 2017). Therefore, local authorities should not disregard the necessary chemical controls and potential lead hazard that may be located in water service lines.

4.1. Study limitations

The sample size (N = 94) for this study maybe considered small. However, this was the entire population of pre-1978 childcare facilities permitted as of January, 2009 in Clark County, NV. Since that time, new facilities operating out of pre-1978 structures have opened for business and have been assessed by the SNHD, but these facilities were not included here. This study was not able to account for childcare facilities operating without a required SNHD permit.

Each FCH or GCH that was assessed was given the option to have the entire facility/residential dwelling screened or only the child occupied areas. As per the SNHD health permit, only the areas that the children access are regulated by SHND. Therefore, the inspectors did not have the authority to assess the entire facility/dwelling without the owner's permission. Commercial childcare centers did not have this option because their structure was not a private residence. A few FCH and GCH operators chose to have their entire facility/dwelling screened, but most denied the offer. This is considered a limitation because additional lead hazards may have gone unscreened in alternative areas of the facility/dwelling. Although the child occupied areas are approved by SNHD, it is possible that children still enter these unapproved areas. A majority of the facilities tested were also constructed post-1950 when lead concentrations in paint were voluntarily being reduced by the industry (Mushak & Crocetti, 1990). These are all factors that could influence the amount of hazards identified per facility.

A total of six EPA Certified Lead Risk Assessors contributed to completing this study. Multiple inspectors completed the risk assessments, although they were all trained consistently the possibility of inter-rater reliability issues arise. Outdoor play equipment was one aspect of the assessment that varied by risk assessor. All risk assessors did not complete a follow-up dust wipe after identifying a positive XRF reading. This limits the ability to determine if the lead in the playground equipment is creating an actual lead dust hazard to the children.

It is also possible that additional pieces of playground equipment were positive based on the 600 ppm standard for lead in consumer products because the study team did not have access to an XRF Plastics Analyzer that could produce results in ppm (Ban of lead-containing paint & certain consumer products bearing lead-containing paint, 2001). Therefore, it is possible that the XRF readings below 1.0 mg/cm^2 could be above the 600 ppm standard since there is no conversion formula. It is recommended that future studies utilize a Plastics XRF Analyzer for testing lead concentration of playground equipment, followed by dust wipe sampling to determine the bioavailability of lead.

5. Conclusion

Overall, lead concentrations in soil, dust, water, and paint were very low and infrequent in the pre-1978 childcare facilities tested in Clark County, Nevada. Although this severely limits the discussion of the lead sources, more importantly it is excellent news for the residents of Clark County. Lead concentrations exceeding existing standards were certainly found in childcare facilities, but often were in isolated areas that could quickly and easily be removed or stabilized. Annual monitoring of traditional and non-traditional lead sources should be completed to ensure these facilities provide a safe environment for children. This allows us to take preventative measures instead of using children as bioindicators to find the lead sources.

Funding
This work was supported by Centers for Disease Control and Prevention [grant number 5H64EH000145-04].

Author details
Melissa J. Marshall[1]
E-mail: melissa.breunig@unlv.edu
Jessica M. Weislogel[1]
E-mail: newberryjess@gmail.com
Shawn L. Gerstenberger[1]
E-mail: shawn.gerstenberger@unlv.edu

[1] Department of Environmental and Occupational Health, School of Community Health Sciences, University of Nevada, Las Vegas, 4505 S. Maryland Parkway, Box 453064, Las Vegas, NV 89154-3064, USA.

References
Agency for Toxic Substances and Disease Registry. (2007). *Toxicological profile for lead* (CAS 7439-92-1). Atlanta, GA: U.S. Department of Health and Human Services, Public Health Service.

American Society for Testing and Materials International. (2003). *Standard practice for collection of settled dust samples using wipe sampling methods for subsequent lead determination* (ASTM E1728–03). West Conshohocken, PA: American Society for Testing and Materials.

Ban of lead-containing paint and certain consumer products bearing lead-containing paint. (2001). *US consumer product safety commission.* 6 CFR §1303.

Barn, P., & Kosatsky, T. (2011). Lead in school drinking water: Canada can and should address this important ongoing exposure source. *Canadian Journal of Public Health, 102*(2), 118–121.

Belfit, V. F., Nix, B. J., & Graham, S. C. (2002). Evaluation of high levels of lead in select plastic playground equipment. *Federal Facilities Environmental Journal, 13*(3), 115–122. https://doi.org/10.1002/ffej.10050

Bellinger, D. C. (2004). Lead. *Journal of Pediatrics, 13*(4), 1016–1022.

Bryant, S. D. (2004). Lead-contaminated drinking waters in the public schools of Philadelphia. *Journal of Toxicology: Clinical Toxicology, 42*(3), 287–294. https://doi.org/10.1081/CLT-120037429

EPA announces tough new standards for lead [news release]. (2000, December 6). Washington, DC: US Environmental Protection Agency. Retrieved April 18, 2017 from https://archive.epa.gov/epapages/newsroom_archive/newsreleases/51013a59f35fecb5852569c1005febf9.html

Gorospe, E. C., & Gerstenberger, S. L. (2008). Atypical sources of childhood lead poisoning in the United States: A systematic review from 1966–2006. *Clinical Toxicology, 46*(8), 728–737. https://doi.org/10.1080/15563650701481862

Greenway, J. A., & Gerstenberger, S. L. (2010). An evaluation of lead contamination in plastic toys collected from day care centers in the Las Vegas Valley, Nevada. *Bulletin of Environmental Contamination and Toxicology, 85*, 363–366. https://doi.org/10.1007/s00128-010-0100-3

Jacobs, D. E., Clickner, R. P., Zhou, J. Y., Viet, S. M., Marker, D. A., Rogers, J. W., ... Friedman, W. (2002). The prevalence of lead-based hazards in U.S. housing. *Environmental Health Perspectives, 110*(10), A599–A606. https://doi.org/10.1289/ehp.021100599

Kessler, E. (1995). Lead in the environment. *Ambio, 24*(1), 1.

Lanphear, B. P., Dietrich, K., Auinger, P., & Cox, C. (2000). Cognitive deficits associated with blood lead concentrations. *Public Health Reports, 115*(6), 521–529. https://doi.org/10.1093/phr/115.6.521

Lead. (2017, February 9). *Centers for Disease Control and Prevention website.* Retrieved April 5, 2017 from https://www.cdc.gov/nceh/lead/

Levin, R. B., Brown, M. J., Kashtock, M. E., Jacobs, D. E., Whelan, E. A., Rodman, J., ... Sinks, T. (2008). Lead exposures in U.S. children, 2008: Implications for prevention. *Environmental Health Perspectives, 116*(10), 1285–1293. https://doi.org/10.1289/ehp.11241

Maas, R. P., Patch, S. C., Morgan, D. M., & Pandolfo, T. J. (2005). Reducing lead exposure from drinking water: Recent history and current status. *Public Health Reports, 120*, 316–321. https://doi.org/10.1177/003335490512000317

Massey, A. R., & Steele, J. E. (2012). Lead in drinking water: Sampling primary schools and in South Central Kansas. *Journal of Environmental Health, 74*(7), 16–20.

Mushak, P., & Crocetti, A. (1990). Methods for reducing lead exposure in young children and other risk groups: An integrated summary of a report to the US Congress on childhood lead poisoning. *Environmental Health Perspectives, 89*, 125–135. https://doi.org/10.1289/ehp.9089125

Needleman, H. L. (1980). Lead exposure and human health: Recent data on an ancient problem. *Technol Review, 82*(5), 39–45.

Pieper, K. J., Tang, M., & Edwards, M. A. (2017). Flint water crisis caused by interrupted corrosion control: Investigating "ground zero" home. *Environmental Science & Technology, 51*(4), 2007–2014. https://doi.org/10.1021/acs.est.6b04034

Rufin, K. G. A. (2015). *Lead hazard control in Henderson, Nevada: Indentifying critical areas and the associated costs* [master's thesis]. Las Vegas, NV: University of Nevada.

Sanders, M., Stolz, J., & Chacon-Baker, A. (2013). Testing for lead in toys at day care centers. *Work, 44*, S29–S38.

U.S. Department of Housing and Urban Development. (1995). *HUD guidelines for the evaluation and control of lead-based paint hazards in housing.* Washington, DC: US Dept Housing and Urban Development.

U.S. Environmental Protection Agency. (1992). *Lead and copper monitoring guidance for water serving 501 to 3,300 persons* (EPA 812/B-92-005). Washington, DC: Office of Water.

Van Ulirsch, G., Gleason, K., Gerstenberger, S. L., Moffett, D. B., Pulliam, G., Ahmed, T., & Fagliano, J. (2010). Evaluating and regulating lead in synthetic turf. *Environmental Health Perspectives, 118*, 345–1349.

Viverette, L., Mielke, H. W., Brisco, M., Dixon, A., Schaefer, J., & Pierre, K. (1996). Environmental health in minority and other underserved populations: Benign methods for identifying lead hazards at day care centres of New Orleans. *Environmental Geochemistry and Health, 18*, 41–45. https://doi.org/10.1007/BF01757218

Weismann, D. N., Dusdieker, L. B., Cherryholmes, K. L., Hausler, W. J., & Dungy, C. I. (1995). Elevated environmental lead levels in a day care setting. *Archives of Pediatrics & Adolescent Medicine, 149*(8), 878–881. https://doi.org/10.1001/archpedi.1995.02170210052009

PERMISSIONS

The contributors of this book come from diverse backgrounds, making this book a truly international effort. This book will bring forth new frontiers with its revolutionizing research information and detailed analysis of the nascent developments around the world.

We would like to thank all the contributing authors for lending their expertise to make the book truly unique. They have played a crucial role in the development of this book. Without their invaluable contributions this book wouldn't have been possible. They have made vital efforts to compile up to date information on the varied aspects of this subject to make this book a valuable addition to the collection of many professionals and students.

This book was conceptualized with the vision of imparting up-to-date information and advanced data in this field. To ensure the same, a matchless editorial board was set up. Every individual on the board went through rigorous rounds of assessment to prove their worth. After which they invested a large part of their time researching and compiling the most relevant data for our readers.

The editorial board has been involved in producing this book since its inception. They have spent rigorous hours researching and exploring the diverse topics which have resulted in the successful publishing of this book. They have passed on their knowledge of decades through this book. To expedite this challenging task, the publisher supported the team at every step. A small team of assistant editors was also appointed to further simplify the editing procedure and attain best results for the readers.

Apart from the editorial board, the designing team has also invested a significant amount of their time in understanding the subject and creating the most relevant covers. They scrutinized every image to scout for the most suitable representation of the subject and create an appropriate cover for the book.

The publishing team has been an ardent support to the editorial, designing and production team. Their endless efforts to recruit the best for this project, has resulted in the accomplishment of this book. They are a veteran in the field of academics and their pool of knowledge is as vast as their experience in printing. Their expertise and guidance has proved useful at every step. Their uncompromising quality standards have made this book an exceptional effort. Their encouragement from time to time has been an inspiration for everyone.

The publisher and the editorial board hope that this book will prove to be a valuable piece of knowledge for researchers, students, practitioners and scholars across the globe.

LIST OF CONTRIBUTORS

Suhaib S. Salih and Tushar K. Ghosh
Chemical Engineering, University of Missouri-Columbia, 510 high street, Columbia, MI, 65201, USA
Department of Chemical Engineering, University of Tikrit, Iraq Nuclear Science and Engineering Institute, University of Missouri-Columbia, 416 S. Sixth Street, E2434 Lafferre Hall, Columbia, MI, 65211, USA

Andrea Olive
Political Science and Geography, University of Toronto Mississauga, 3359 Mississauga Rd, Mississauga, ON L5L 1C6, Canada

Karna Dahal and Jari Niemelä
Department of Environmental Sciences, University of Helsinki, Viikinkaari 2, Helsinki FI-00014, Finland

Sirkku Juhola
Department of Environmental Sciences, University of Helsinki, Viikinkaari 2, Helsinki FI-00014, Finland
Department of Built Environment, Aalto University, Espoo FI-00076, Finland

Izzet Ari and Ramazan Sari
Ministry of Development, Necatibey Cad., No: 110/A 06100, Yucetepe, Cankaya, Ankara, Turkey
Department of Business Administration, Middle East Technical Universities, Cankaya 06800, Ankara, Turkey
Department of Earth System Sciences, Middle East Technical Universities, Cankaya 06800, Ankara, Turkey

Asaminew Abiyu, Denghua Yan, Abel Girma, Xinshan Song and Hao Wang
College of Environmental Science and Engineering, Donghua University, Shanghai, China
State Key Laboratory of Simulation and Regulation of Water Cycle in River Basin, China Institute of Water Resources and Hydropower Research, Beijing, China

Benedicta Yayra Fosu-Mensah, Emmanuel Addae and Dzidzo Yirenya-Tawiah
Institute for Environment and Sanitation Studies (IESS), University of Ghana, Legon, Accra, Ghana

Emmanuel Mawuli Abalo, Seth Agyemang, Samuel Atio, Derrick Ofosu-Bosompem and Prince Peprah
Department of Geography and Rural Development, PMB, Kwame Nkrumah University of Science and Technology (KNUST), Kumasi, Ghana

Rita Ampomah-Sarpong
Department of Sociology and Social Work, PMB, Kwame Nkrumah University of Science and Technology (KNUST), Kumasi, Ghana

Ted R. Johnson
TRJ Environmental, Inc., 713 Shadylawn Rd, Chapel Hill NC 27514, USA

John E. Langstaff and Stephen Graham
U.S. Environmental Protection Agency, 109 TW Alexander Drive, Research Triangle Park, NC 27711, USA

Eric M. Fujita and David E. Campbell
Division of Atmospheric Sciences, Desert Research Institute, 2215 Raggio Parkway, Reno, NV 89512, USA

Barthelemy G. Honfoga
Faculty of Agronomic Sciences, School of Economics, Socio-Anthropology & Communication for Rural Development, University of Abomey-Calavi (UAC), 06 BP 1892 Cotonou, Akpakpa PK3, Abomey-Calavi, Benin

Agnes Oppong, David Azanu and Linda Aurelia Ofori
Department of Chemistry, Kwame Nkrumah University of Science and Technology, Kumasi, Ghana
Department of Laboratory Technology, Kumasi Technical University, Kumasi, Ghana
Department of Theoretical and Applied Biology, Kwame Nkrumah University of Science and Technology, Kumasi, Ghana

Marian Asantewah Nkansah, Godfred Darko, Francis Opoku, Thomas Bentum Essuman and Joshua Antwi-Boasiako
Department of Chemistry, Kwame Nkrumah University of Science and Technology, Kumasi, Ghana

Matt Dodd
School of Environment and Sustainability, Royal Roads University, Victoria, BC, Canada

Frank Nyame
Department of Earth Sciences, University of Ghana, Legon, Accra, Ghana

Mofizul Islam and Qiuyan Yuan
Department of Civil Engineering, University of Manitoba, Winnipeg, MB R3T 5V6 Canada

Melissa J. Marshall, Jessica M. Weislogel and Shawn L. Gerstenberger
Department of Environmental and Occupational Health, School of Community Health Sciences, University of Nevada, Las Vegas, 4505 S. Maryland Parkway, Las Vegas, NV 89154-3064, USA

Index

www.ingramcontent.com/pod-product-compliance
Lightning Source LLC
Chambersburg PA
CBHW080405190526
45161CB00003B/140